ENERGY POLICIES, POLITICS AND PRICES SERIES

# WORLDWIDE BIOMASS POTENTIAL: TECHNOLOGY CHARACTERIZATIONS

# ENERGY POLICIES, POLITICS AND PRICES SERIES

**Nuclear Power's Role in Generating Electricity**
*Perry G. Furham*
2009. ISBN: 978-1-60741-226-7

**OPEC, Oil Prices and LNG**
*Edward R. Pitt and Christopher N. Leung (Editors)*
2009. ISBN: 978-1-60692-897-4

**OPEC, Oil Prices and LNG**
*Edward R. Pitt and Christopher N. Leung (Editors)*
2009. ISBN: 978-1-60876-614-7 (online book)

**Dynamic Noncooperative Game Models for Deregulated Electricity Markets**
*Jose B. Cruz, Jr. and Xiaohuan Tan (Editors)*
2009. ISBN: 978-1-60741-078-2

**Energy Prices: Supply, Demand or Speculation?**
*John T. Perry (Editor)*
2009. ISBN: 978-1-60741-374-5

**Worldwide Biomass Potential: Technology Characterizations**
*R. L. Bain (Author)*
2010. ISBN: 978-1-60741-267-0

ENERGY POLICIES, POLITICS AND PRICES SERIES

# WORLDWIDE BIOMASS POTENTIAL: TECHNOLOGY CHARACTERIZATIONS

R.L. BAIN
AUTHOR

Nova Science Publishers, Inc.
New York

For permission to use material from this book please contact us:
Telephone 631-231-7269; Fax 631-231-8175
Web Site: http://www.novapublishers.com



**LIBRARY OF CONGRESS CATALOGING-IN-PUBLICATION DATA**

*Bain, R. L.*
*Worldwide biomass potential : technology characterizations / R.L. Bain.*
*p. cm.*
*Includes bibliographical references and index.*
*ISBN 978-1-60741-267-0 (hardcover)*
*1. Biomass energy. I. Title.*
*TP339.B34 2009*
*662'.88--dc22*
*2009028475*

*Published by Nova Science Publishers, Inc. ✢ New York*

# TABLE OF CONTENTS

# PREFACE

A joint EERE-PI project was completed to estimate the worldwide potential to produce and transport ethanol and other biofuels, with an emphasis on the 5 year and 10 year potential for biofuels supply to the United States. The range of countries and/or regions identified in the joint EERE-PI project includes Brazil, the Caribbean Basin, Central America, Mexico, Europe, China, India and Sub-Saharan Africa. This project, discussed in this book, included four specific tasks: 1) to identify the range of countries to be included in the study, 2) to assess the resource potential for production of ethanol from sugar and starch-based feedstocks and biodiesel, 3) to assess the resource potential for production of other biofuels and 4) to integrate results into the MARKAL energy policy model. Additionally, the NREL (National Renewable Energy Laboratory) examined in this book was primarily concerned with estimating the plant gate price (PGP) of liquid biofuels from selected biomass feedstocks for countries included in the study using representative existing and developing technologies. A methodology for comparing costs between countries was developed as well.

# WORLD BIOFUELS ASSESSMENT WORLDWIDE BIOMASS POTENTIAL: TECHNOLOGY CHARACTERIZATIONS

*R.L. Bain*

## ABSTRACT

A joint EERE-PI project was completed to estimate the worldwide potential to produce and transport ethanol and other biofuels, with an emphasis on the 5 year and 10 year potential for biofuels supply to the United States. The project included four specific tasks: 1) identify the range of countries to be included in the study, 2) assess the resource potential for production of ethanol from sugar and starch-based feedstocks, and biodiesel, 3) assess the resource potential for production of other biofuels, including lignocellulosic ethanol, pyrolysis oil, and renewable diesel, and 4) integrate results into the MARKAL energy policy model. The project team included DOE (Policy and International and the Office of the Biomass Program), Oak Ridge National Laboratory (feedstock supply curves), the National Renewable Energy Laboratory (conversion technology haracterizations), and Brookhaven National Laboratory (MARKAL analysis).

The NREL portion of this study was primarily concerned with estimating the plant gate price (PGP) of liquid biofuels (corn and wheat dry mill ethanol, cellulosic ethanol, biodiesel, renewable diesel, and pyrolytic fuel oil) from selected biomass feedstocks for countries included in the study using representative existing and developing technologies. A methodology for comparing costs between countries was developed. Plant sizes studies ranged from 25 MM GPY to 100 MM GPY. The results of the technology characterizations (capital costs, operating costs, plant gate prices) are presented in 2005 U.S. dollars and include estimates of comparative costs in each country.

## ACRONYMS

| | |
|---|---|
| AFUDC | Allowance for funds during construction |
| BC | Biochemical USA1Argentina5Brazil2China3 |
| BLS | Bureau of Labor Statistics |
| BNL | Brookhaven National Laboratory |
| bu | Bushel |
| CCM | Capital cost multiplier |

| | |
|---|---|
| CEPCI | Chemical Engineering plant cost index |
| COP | Cost of product |
| DCFROR | Discounted cash flow rate of return |
| DDG | Distillers dried grains |
| DFC | Direct fixed capital |
| DE | Domestic equipment |
| DOE | United States Department of Energy |
| FAME | Fatty acid methyl ester |
| GPM | Gross processing margin |
| GPY | Gallons per year |
| IE | Imported equipment |
| ISBL | Inside battery limits |
| kWh | Kilowatt hour |
| LC | Labor costs |
| LCM | Labor cost multiplier |
| LPM | Labor productivity multiplier |
| LRM | Labor rate multiplier |
| MACRS | Modified accelerated cost recovery system depreciation |
| MARKAL | MARKal ALlocation; a mathematical model of the energy system of one or several regions that provides a technology-rich basis for estimating energy dynamics of a multi-period horizon |
| MM | Million |
| MMT | Million metric tonnes |
| NREL | National Renewable Energy Laboratory |
| OBP | DOE Office of the Biomass Program |
| ORNL | Oak Ridge National Laboratory |
| OSBL | Outside battery limits |
| PCMI | Imported equipment multiplication factor |
| PCML | Local material multiplication factor |
| PC | Purchased equipment cost |
| PGP | Plant gate price |
| PI | DOE Policy and International Program |
| POX | Partial oxidation |
| SL | Straight line depreciation |
| TC | Thermochemical |
| TCC | Total capital cost |
| TIC | Total installed capital |
| TPI | Total plant investment |
| USD | United States dollars |
| VAT | Value added tax |

## TECHNOLOGY CHARACTERIZATIONS–EXECUTIVE SUMMARY

The overall objective of the joint EERE-PI project is to estimate the worldwide potential to produce and transport ethanol and other biofuels, with an emphasis on the 5 year and 10 year potential for biofuels supply to the United States. The project identifies four specific tasks:

- Task 1: Identify the range of countries to be included in the study
- Task 2: Assess the resource potential for production of ethanol from sugar and starch-based feedstocks, and biodiesel.
- Task 3: Assess the resource potential for production of other biofuels, including lignocellulosic ethanol, pyrolysis oil, and renewable diesel.
- Task 4: Integrate results into the MARKAL energy policy model

The project team includes DOE (Policy and International [PI] and the Office of the Biomass Program [OBP]), Oak Ridge National Laboratory (feedstock supply curves), the National Renewable Energy Laboratory (conversion technology characterizations), and Brookhaven National Laboratory (MARKAL analysis).

The NREL portion of this study is primarily concerned with estimating the plant gate price (PGP) of liquid biofuels from selected biomass feedstocks for countries included in the study using representative existing and developing technologies. The results of the technology characterizations are presented in 2005 U.S. dollars and include estimates of comparative capital and operating costs in each country.

The countries of interest selected by the project team for the study are:

- The United States
- Argentina
- Brazil
- Canada
- The Caribbean Basin
- China
- Colombia
- India
- Mexico

Supply curves in the years 2012, 2017, 2027 have been developed by Oak Ridge (ORNL 2007) for the following feedstocks: sugar cane, corn, soybeans, palm oil, bagasse, agricultural residues (corn stover and wheat straw), and woody/perennial materials. In some cases where data are not available, the ORNL feedstock prices represent single point values.

NREL has estimated PGPs for ethanol from corn dry mills, ethanol from sugar cane mills, ethanol from wheat mills, ethanol from biochemical conversion of lignocellulosic feedstocks, ethanol from thermochemical conversion of lignocellulosic feedstocks, residual fuel oil from pyrolysis of lignocellulosic feedstocks, biodiesel from soybeans, renewable diesel from soybeans, and biodiesel from palm oil.

While USA feedstock supply curve estimation is outside the scope of the present ORNL study, selected USA feedstock costs were used as a check on the reasonableness of the models developed. Although one complete PGP set of curves was developed (sugar cane ethanol in Brazil, Figure 1), the PGP comparisons presented in this section are based on a single feedstock cost (price at 50% of 2017 potential supply or single point value if a supply curve isn't available) for each country. Selected costs for sugar, starch and fatty acid feedstocks are given in Table 1, and costs of lignocellulosic feedstocks are given in Table 2.

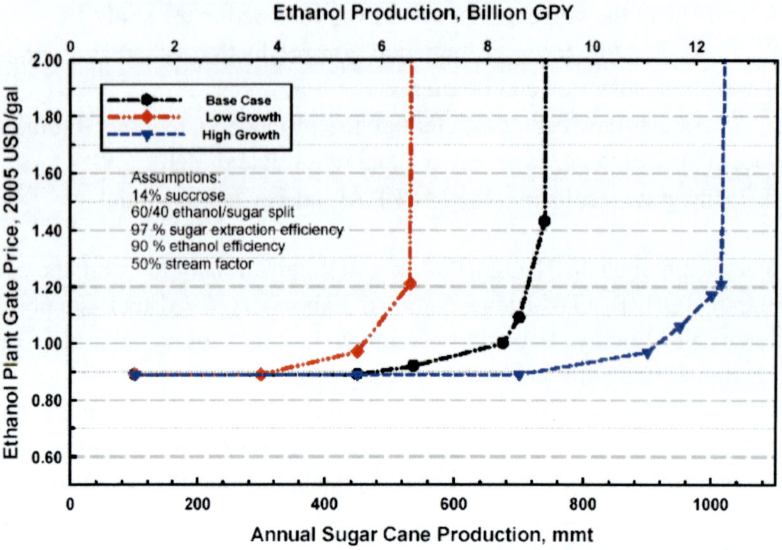


Figure 1. Plant gate prices for Brail 2017 sugar cane ethanol (based on 50 MMGPY plants): low, base, and high growth scenarios

**Table 1. Selected non-lignocellulosic feedstock prices (calculated from ORNL 2007)**

| | Sugar Cane | Corn | | Soy Beans | | Wheat | | Palm Oil |
|---|---|---|---|---|---|---|---|---|
| | $/ton | $/ton | $/bu | $/ton | $/bu | $/ton | $/bu | $/ton |
| **USA** | 29.00 | 107.0 | 3.00 | 187.6 | 5.60 | 113.3 | 3.40 | |
| **Argentina** | 34.30 | 127.5 | 3.57 | 143.6 | 4.29 | 92.5 | 2.78 | |
| **Brazil** | 10.45 | 135.0 | 3.78 | 175.0 | 5.22 | | | |
| **Canada** | | 86.0 | 2.41 | | | 102.0 | 3.06 | |
| **Caribbean Basin** | 15.00 | | | | | | | 275 |
| **Colombia** | 26.85 | | | | | | | 406 |
| **China** | 23.00 | 126.0 | 3.53 | 319.8 | 9.55 | 159.0 | 4.77 | |
| **India** | 15.00 | | | | | | | |
| **Mexico** | 29.00 | 130.0 | 3.64 | | | | | |

Product yields used in the study are given in Table 3.

The technology evaluations are based on a "factor" analysis–an approximate method that uses factors to estimate the capital costs of process systems based on purchased cost. Multiplier factors are based on historical published values, and represent plus forty percent/minus ten percent estimates. In this study factored cost estimates developed in one country, typically the United States, need to be converted to costs in other countries. Since location multipliers are different for capital and labor, estimates of equipment/labor splits for the costing factors are used, based on discussions given in Valle-Riestra (1983).

**Table 2. Selected lignocellulosic feedstock prices (calculated from ORNL 2007)**

| Country | Bagasse | Ag Residues (2005 USD/ST) | Woody /Perennial |
|---|---|---|---|
| USA | 35.0 | 35.0 | 35.0 |
| Argentina | 9.1 | 47.2 | 47.2 |
| Brazil | 7.6 | 30.8 | 47.2 |
| Canada | -- | 10.0 | 16.3-47.2 (31.8) |
| Caribbean Basin | 7.6 | 14.5 | 47.2 |
| China | 9.1 | 28.1 | 47.2 |
| Colombia | 14.5 | 14.5 | 47.2 |
| India | 7.3 | 7.3 | 47.2 |
| Mexico | 15.4 | 47.2 | 47.2 |

Plant gate prices in other countries are converted to selected country costs using published conceptual equipment and labor cost estimating factors. Capital costs are divided into domestic equipment and imported equipment. For imported equipment duty, freight, and Value Added Tax (VAT) are considered in determining a multiplication factor. A local material index is used for domestic equipment. Labor costs are based on USA costs, with labor rate multipliers and labor productivity multipliers. The factors, in combination with estimated equipment/labor percentages, are used to convert base case capital and operating costs to the specified country. Where factors are not available default values for imported materials (1), and depreciation schedule (10% straight line) are used.

Capital and operating cost estimates are based on published studies, performed over a period of time. For consistency all estimated are normalized to 2005 using the Chemical Engineering Plant Cost Index. The CEPCI is a standard cost index used to update cost estimates to a given year. It is made up of 41 Bureau of Labor Statistics (BLS) producer price indexes, 12 BLS labor cost indexes, and a calculated labor productivity multiplier. A summary of the capital costs are presented in Figure 4, expressed in terms of capital intensity. Capital intensity is the plant installed cost divided by the annual plant product capacity, assuming a 100 percent operating factor. As such, capital intensity incorporates plant size economies and process conversion efficiency. For the most part capital intensity is relatively constant for a given technology. The exceptions are Canada with higher than average capital intensity, and Mexico with lower than average capital intensity. The variations are due primarily to high construction labor costs in Canada and low construction labor costs in Mexico. The capital intensity of sugar cane ethanol is higher than corn ethanol because of the low capacity factor, 50%. Renewable diesel has low capital intensity because of the assumption of locating the process in a petroleum refinery where hydrogen can be purchased across the fence instead of investing in a hydrogen plant. Pyrolytic fuel oil capital intensity is low, but further upgrading will be needed to produce a transportation fuel. Cellulosic ethanol processes are more capital intensive than corn and sugar ethanol processes, reflecting the additional processing needed to convert lignocellulosic feeds.

**Table 3. Technology characterization product yields**

| Product - Feedstock | Yield | | | |
|---|---|---|---|---|
| | gal tonne | Btu/gal | Btu/gal gasoline | gal Gasoline Equiv tonne |
| Ethanol - Corn (a)[a] | 112.2 | 84,100 | 124,238 | 76.0 |
| Ethanol - Sugar Cane* | 13.6 | 84,100 | 124,238 | 9.2 |
| Ethanol (TC) - Bagasse** | 88.3 | 84,100 | 124,238 | 59.8 |
| Ethanol (TC) - Ag Residues** | 88.3 | 84,100 | 124,238 | 59.8 |
| Ethanol (TC) - Wood/Perennials** | 88.3 | 84,100 | 124,238 | 59.8 |
| Ethanol (BC) - Bagasse | 98.9 | 84,100 | 124,238 | 66.9 |
| Ethanol (BC) - Ag Residues | 98.9 | 84,100 | 124,238 | 66.9 |
| Ethanol (BC) - Wood/Perennials | 98.9 | 84,100 | 124,238 | 66.9 |
| Ethanol – Wheat[b] | 93.3 | 84,100 | 124,238 | 63.2 |
| Biodiesel - Soybeans | 49.8 | 129,500 | 124,238 | 51.9 |
| Renewable Diesel - Soybeans | 47.0 | 123,129 | 124,238 | 46.6 |
| Pyrolytic Fuel Oil - Bagasse | 151.0 | 83,600 | 124,238 | 101.6 |
| Pyrolytic Fuel Oil - Ag Residues | 151.0 | 83,600 | 124,238 | 101.6 |
| Pyrolytic Fuel Oil - Wood/Perennials | | 151.0 | 83,600 | 124,238 |
| Biodiesel - Palm Oil | | 270.0 | 129,500 | 124,238 |
| Renewable Diesel - Palm Oil | | 255.3 | 123,129 | 124,238 |

(a) DDG = 0.33ton/ton corn

(b) DDG = 0.37 ton/ton wheat

* at a 60/40 ethanol/sugar ratio

* * in addition, 15.4 gal/tonne higher alcohols (11.8 gal gasoline equivalents/tonne)

For different plant sizes, capital and operating costs are scaled using a 0.7 default scaling factor or individual scaling factor where known (0.50 – 0.85). Peters and Timmerhaus (2003) state "It is often necessary to estimate the cost of a piece of equipment when no cost data are available for the particular size of operational capacity involved. Good results can be obtained by using the logarithmic relationship known as the 'six-tenths-factor rule,' if the new piece of equipment is similar to one of another capacity for which cost data are available. According to this rule, if the cost of a given unit at one capacity is known, the cost of a similar unit with X times the capacity of the first is approximately $(X)^{0.6}$ times the cost of the initial unit." Valle-Riesta (1983) states "A logical consequence of the 'sixth-tenths-factor' rule for characterizing the relationship between equipment capacity and cost is that a similar relationship should hold for the direct fixed capital of specific plants ... In point of fact, the capacity exponent for plants, on the average, turns out to be closer to 0.7."

The feedstock costs and capital costs are used in a discounted cash flow analysis to estimate plant gate prices of products. The project and financial assumptions for the cash flow analyses are given in Table 5. The assumptions include 100% equity financing, a 10% return on equity, a 3-year permitting and construction period, a 20-year operating period, and 1.9% inflation (consistent with the DOE H2A model).

**Table 4. Process capital intensity, plant size — 100MM GPY**

| Product - Feed | USA | Argentina | Brazil | Canada | Caribbean | China | Colombia | India | Mexico |
|---|---|---|---|---|---|---|---|---|---|
| | Basin | | | | | | | | |
| | 2005 USD/annual gal | | | | | | | | |
| Ethanol - Corn | 0.49 | 0.52 | 0.50 | 0.63 | | 0.54 | | | 0.39 |
| Ethanol - Sugar Cane | 1.11 | 1.19 | 1.14 | | 1.25 | 1.22 | 1.11 | 1.19 | 0.89 |
| Ethanol (TC) - Cellulosic | 1.66 | 1.78 | 1.69 | 2.16 | 1.87 | 1.82 | 1.65 | 1.77 | 1.33 |
| Ethanol (BC) - Cellulosic | 1.66 | 1.78 | 1.69 | 2.16 | 1.87 | 1.82 | 1.65 | 1.77 | 1.33 |
| Ethanol - Wheat | 0.60 | 0.64 | | 0.78 | | 0.66 | | | |
| Biodiesel - Soybeans | 0.54 | 0.58 | 0.56 | | | 0.60 | 0.54 | | |
| Renewable Diesel - Soybeans | 0.11 | 0.12 | 0.11 | | | 0.12 | 0.11 | | |
| Pyrolytic Fuel Oil - Cellulosic | 0.32 | 0.37 | 0.35 | 0.45 | 0.39 | 0.38 | 0.34 | 0.37 | 0.28 |
| Biodiesel - Palm Oil | | | | | 0.61 | | 0.54 | | |

**Table 5. Project and financial assumptions**

| Factor | |
|---|---|
| Discount Rate | 10% |
| Equity / Debt | 100 / 0 % |
| Cost of Capital – for debt/equity sensitivity analysis | 7% |
| Debt Term – for debt/equity sensitivity analysis | 20 years |
| Inflation | 1.9% |
| Feed Escalation | 0% |
| Product Escalation | 0% |
| Construction Period | 3 years |
| Construction Investment Schedule | 8% / 60% / 32% |
| Stream Factor in Start-Up Year | 75% of later year stream factor |
| Stream Factor | 95% (50% for sugar mills) |
| Operating Life | 20 years |
| Annual Replacement Capital Allowance | 0% |
| Depreciation | Country Specific |
| USA | 7, 20 year double declining balance |
| Other Countries | Straight line, variable number of years |
| Project basis | Corporate, after tax |
| Year basis | 2005 |
| Effective Tax Rate – country specific | e.g., USA = 39% |

The feedstock costs (Tables 1 and 2), capital costs (derived from Table 3), process yields (Table 4), and financial factors (Table 5) are used to estimate plant gate prices, as shown in Table 6 for 100 MM GPY facilities. PGPs for corn dry mill ethanol vary from \$1.05/gal in Canada to \$1.53/gal in Brazil. Sugar cane ethanol ranges from \$0.82 in Colombia to \$1.66 in the United States, with Brazil estimated at \$1.05/gal. Bagasse appears to be an attractive opportunity for cellulosic ethanol, with PGPs ranging from \$0.54/gal in Brazil and Mexico to 0.65/gal in China. The USA bagasse ethanol number is high at \$1.17/gal; however, there is no supply curve and a conservative high bagasse price was assumed, \$35.00.ton. Biodiesel and renewable diesel prices are primarily a function of soybean prices (80-90% of total cost), and vary widely depending on the country, from \$0.92/gal biodiesel in Argentina (\$1.81/bu soybeans) to \$3.42/gal biodiesel in China (\$9.55/bu soybeans). Renewable diesel is incrementally higher than biodiesel, \$0.06 – 0.19/gal; however renewable diesel has a non-quantified advantage in cetane value (+30 to 50 cetane numbers) and in use in existing fuel blends and distribution systems. Pyrolytic fuel oil has attractive costs, \$0.26 - \$0.60/gal, but is not a transportation fuel; it's primarily role would be as a refinery feedstock. Upgrading costs and yields for pyrolysis-derived diesel and gasoline have not been estimated in this report.

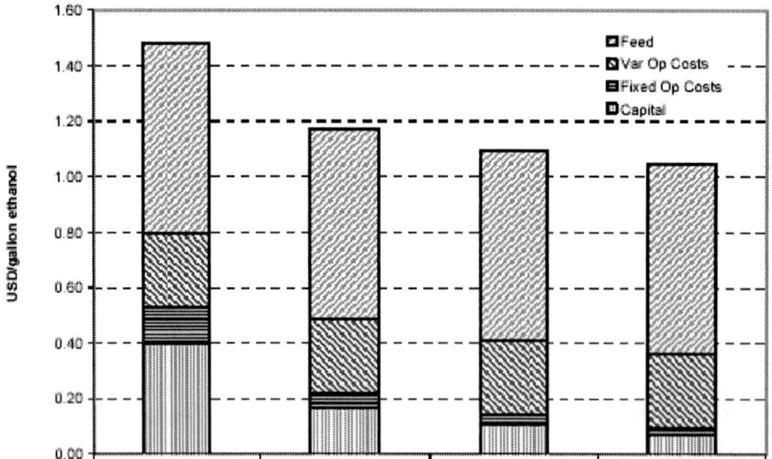


| | 4.6 | 25 | 50 | 100 |
|---|---|---|---|---|
| Feed | 0.68 | 0.68 | 0.68 | 0.68 |
| Var Op Costs | 0.27 | 0.27 | 0.27 | 0.27 |
| Fixed Op Costs | 0.13 | 0.05 | 0.03 | 0.02 |
| Capital | 0.40 | 0.17 | 0.11 | 0.07 |

Plant Size (MMGPY EtOH) with 60/40 Ethanol/Sucrose

Figure 2. Plant gate price of sugar cane ethanol in Brazil

Two study results are shown in Figure 2 and Figure 3 to illustrate the relative impact of capital and feedstock on plant gate prices. In Figure 2, for sugar cane ethanol production in Brazil, the capital intensity is low and feedstock plays a large role in price, especially as the plant size increases. Prices are assumed to be independent of plant size, while capital cost decreases according to the 0.7 rule. In this example feedstock represents about 50 % of price at 5 MM GPY (capital 22%), while at 50 MM GPY feedstock represents about 67% of price (capital 7.5%). Figure 3 is representative of the opposite extreme, a capital intensive process with low feedstock cost–thermochemical ethanol from bagasse in Brazil. In this case feedstock represents only 10% of price at a 25 MM GPY scale, increasing to 17% of price at 100 MM GPY; at the same time capital represents 55% of price at 25 MM GPY, decreasing to 48% at 100 MM GPY.

**Table 6. Plant gate prices, 2005$, 100 MM GPY Plant**

| Product - Feed | USA | Argentina | Brazil | Canada | Caribbean | China | Colombia | India | Mexico |
|---|---|---|---|---|---|---|---|---|---|
| | Basin | | | | | | | | |
| | 2005 USD/gal product | | | | | | | | |
| Ethanol - Corn | 1.29 | 1.49 | 1.54 | 1.06 | | 1.48 | | | 1.31 |
| Ethanol - Sugar Cane | 1.66 | | 1.05 | | 1.06 | 1.42 | 0.82 | 1.05 | 1.67 |
| Ethanol (TC) - Bagasse | 0.93 | 0.62 | 0.54 | | 0.62 | 0.65 | 0.63 | 0.57 | 0.54 |
| Ethanol (TC) - Ag Residues | 0.93 | 1.10 | 0.83 | 0.79 | 0.71 | 0.88 | 0.63 | 0.57 | 0.93 |
| Ethanol (TC) - Wood/Perennials | 0.93 | 1.10 | 1.04 | 1.06 | 1.12 | 1.13 | 1.04 | 1.05 | 0.93 |
| Ethanol (BC) - Bagasse | 1.17 | 0.90 | 0.83 | | 0.90 | 0.92 | 0.90 | 0.81 | 0.81 |
| Ethanol (BC) - Ag Residues | 1.17 | 1.32 | 1.09 | 1.06 | 0.98 | 1.14 | 0.90 | 0.81 | 1.17 |
| Ethanol (BC) - Wood/Perennials | 1.17 | 1.32 | 1.27 | 1.30 | 1.34 | 1.35 | 1.27 | 1.28 | 1.17 |
| Ethanol - Wheat | 1.39 | 1.20 | | 1.31 | | 1.89 | | | |
| Biodiesel - Soybeans | 2.11 | 0.92 | 2.00 | | | 3.42 | 0.98 | | |
| Renewable Diesel - Soybeans | 2.27 | 0.98 | 2.14 | | | 3.61 | 0.97 | | |
| Biodiesel - Palm Oil | | | | | 1.44 | | 1.91 | | |
| Pyrolytic Fuel Oil - Bagasse | 0.50 | 0.30 | 0.26 | | 0.30 | 0.32 | 0.32 | 0.26 | 0.29 |
| Pyrolytic Fuel Oil - Ag Residues | 0.50 | 0.58 | 0.43 | 0.38 | 0.35 | 0.46 | 0.32 | 0.26 | 0.52 |
| Pyrolytic Fuel Oil - Wood/Perennials | 0.50 | 0.58 | 0.55 | 0.54 | 0.59 | 0.60 | 0.56 | 0.55 | 0.52 |

Figures 4 - 6 present comparisons of the plant gate price of ethanol from three biomass sources–sugar cane, corn, and bagasse. For sugar cane, Figure 4, plant gate prices are primarily impacted by feedstock costs; the fraction of PGP representing feedstock ranges from 61% in the Caribbean to 78% in Mexico. There are small variations in capital and non-feedstock operating costs between countries. Comparable trends are seen for ethanol from corn dry mills, Figure 5. Again, capital and non-feedstock PGP contribution variations are small between countries. Feedstock has the largest contribution to PGP, and shows the most variation between countries. Feedstock cost percentages range from 46% in Canada to 66% in Brazil. Figure 6, shows a comparison for cellulosic ethanol from bagasse. The USA PGP assumes a high cost for bagasse, $35/ton, which is a value equal to that assumed in modeling for woody feedstocks. At this time there is no USA supply curve for bagasse. For the USA case feedstock represents 35% and capital 40%. For other countries, the bagasse cost estimates were estimated from the ORNL report, and represent between 11% (Caribbean Basin), and 23% (Colombia).

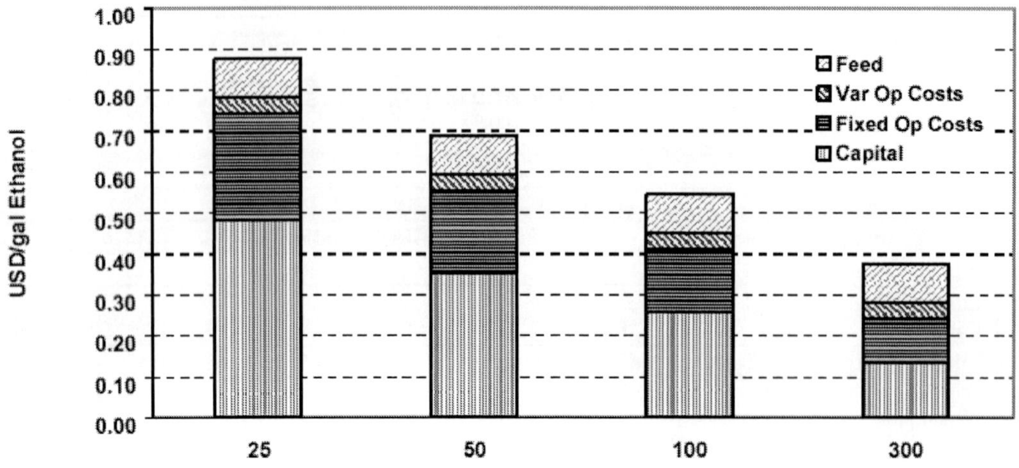


| | 25 | 50 | 100 | 300 |
|---|---|---|---|---|
| Feed | 0.09 | 0.09 | 0.09 | 0.09 |
| Var Op Costs | 0.04 | 0.04 | 0.04 | 0.04 |
| Fixed Op Costs | 0.26 | 0.20 | 0.15 | 0.11 |
| Capital | $0.48 | $0.35 | $0.26 | $0.14 |

Plant Size (Mil gal/yr)

Figure 3. Plant gate price for thermochemical cellulosic ethanol from bagasse in Brazil

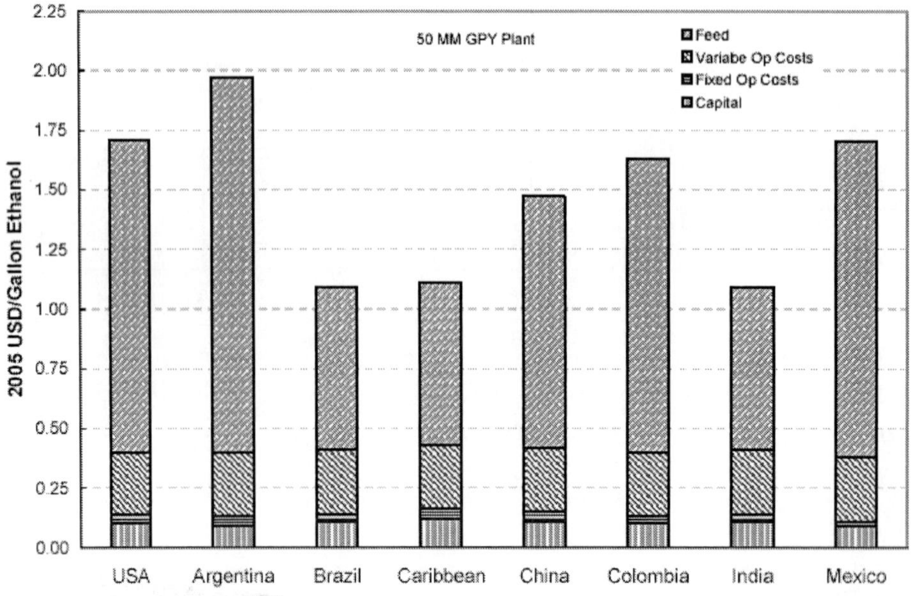


Figure 4. Comparison of sugar cane ethanol plant gate prices

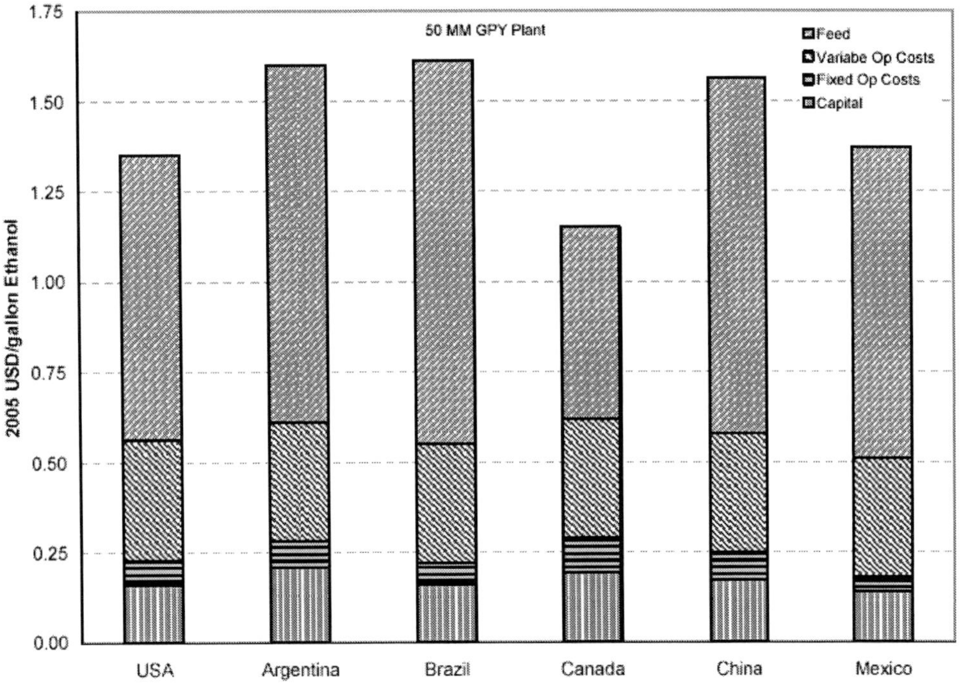


Figure 5. Comparison of corn dry mill ethanol plant gate prices

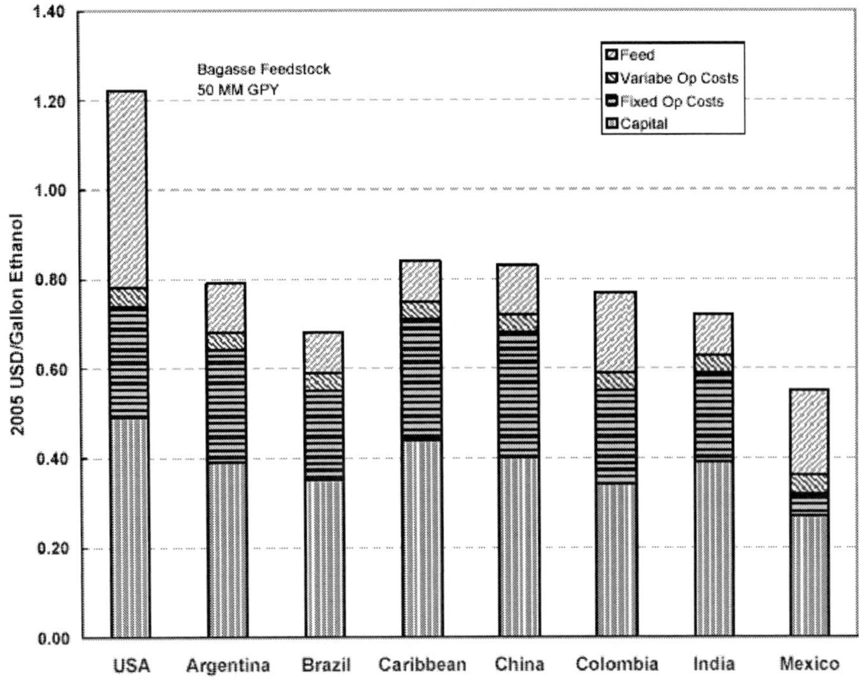


Figure 6. Comparison of cellulosic ethanol (TC) plant gate prices

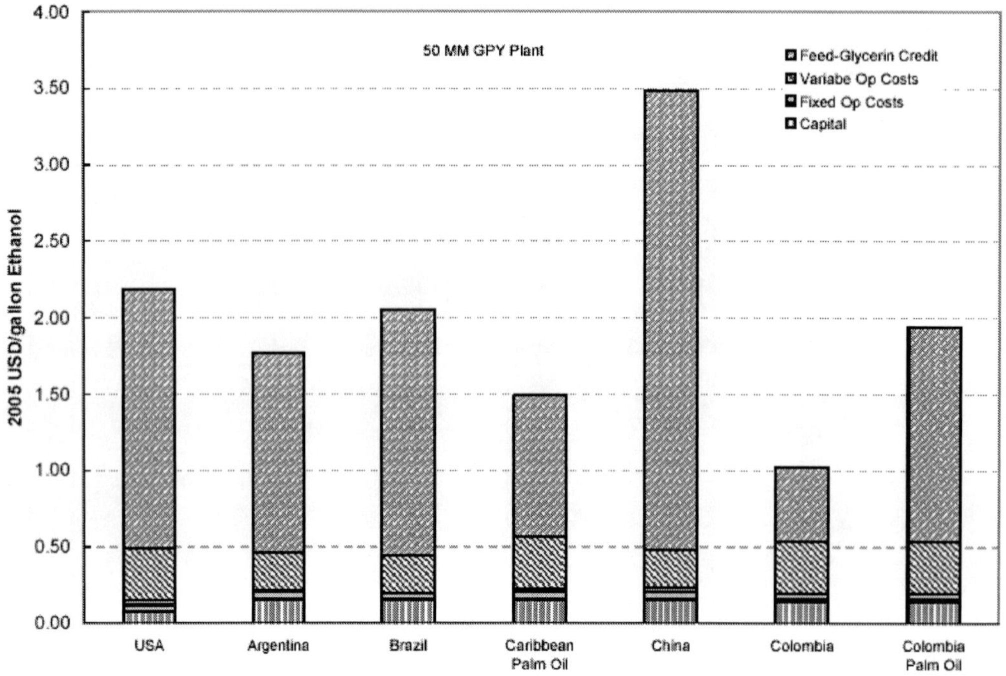


Figure 7. Comparison of biodiesel plant gate prices

Figure 7. shows a comparison of biodiesel PGPs for both biodiesel from soybeans and from palm oil. PGPs are primarily a function of feedstock cost.

# 1. Introduction

The overall objective of the joint EERE-PI project is to estimate the worldwide potential to produce and transport ethanol and other biofuels, with an emphasis on the 5 year and 10 year potential for biofuels supply to the United States. The project identifies four specific tasks:

- Task 1: Identify the range of countries to be included in the study
- Task 2: Assess the resource potential for production of ethanol from sugar and starch-based feedstocks, and biodiesel.
- Task 3: Assess the resource potential for production of other biofuels, including lignocellulosic ethanol and other biofuels, such as biomass gasification with gas-to-liquids technology.
- Task 4: Integrate results into the MARKAL energy policy model

The first task was completed by all participants in consultation with DOE. The next two tasks, which are largely data collection, analysis, and synthesis, have been integrated into the MARKAL model to improve its global representation of biofuels. ORNL was responsible for the completion of Task 2; NREL was responsible for Task 3; and BNL was responsible for the MARKAL model integration, Task 4.

The time frame for the study is designed to coincide with the President's "20 in 10" initiative and thus will be focused on two periods – what can be produced in 5 years and 10 years. The study also considers a longer-term period of 20 years in which barriers, constraints, and study assumptions (e.g., crop yields) can be somewhat more relaxed. The longer time period allows consideration of scenarios more comparable to the joint DOE-USDA billion-ton assessment. The NREL effort focused on the conversion technologies for ethanol from starch and sugar, lignocellulosic ethanol, biodiesel, and residual fuel oil via pyrolysis; conversion scale-up and infrastructure requirements, and estimated costs of biofuels in the context of conventional fuels and financial operating conditions.

The range of countries and/or regions identified in the joint EERE-PI project includes Brazil, the Caribbean Basin, Central America, Mexico, Europe, China, India, and Sub-Saharan Africa. These countries and regions are not entirely consistent with those used in the DOE MARKAL 15-region world energy model. The outcome of this task is an agreed upon list of countries and regions for subsequent assessment and analysis. The countries selected for this study are Argentina, Brazil, Canada, the Caribbean Basin, China, Colombia, India, and Mexico.

This task estimated the range of biofuels production plant gate prices (PGP), for example, ethanol and biodiesel, for each of the identified feedstock/conversion pathways. Estimated minimum biofuel selling prices for each period was compared to predicted prices of conventional fuels. The task took into account variation in feedstock costs and likely progress made in conversion technology either through reductions in capital and operating or changes in biorefinery conversion efficiencies. All appropriate costing assumptions, methods, and models have been documented. Costs are presented using U.S. costs as the base model, and converted to costs for the selected countries/regions using cost multipliers. Plant gates prices are presented on a post-tax basis to permit comparison with domestic production of biofuels.

# 2. Methodology

## 2.1. Approach

The technology evaluations are based on a 'factor' analysis (Valle-Riestra 1983). This is an approximate method that involves the use of factors for estimating the capital cost of process systems or process equipment based on purchased equipment costs (PC). Factor estimates are based on published values from a number of sources (Valle-Riestra 1983, Peters and Timmerhaus 1991), and primarily use studies of individual plant construction records. When applied to a new process the factors give approximate results, typically in the plus forty/minus ten percent range. They are used to give preliminary results before moving to the more costly estimates based on detailed engineering design and vendor quotes. The factors used are based on values developed at NREL for a number of detailed process design studies (Spath et al. 2005), and provide estimates of installation, instruments and controls, piping and insulation, electrical facilities, buildings, yard improvements, and auxiliary facilities. Table 7. Capital equipment costing factors gives the values of the factors used in this study.

**Table 7. Capital equipment costing factors**

| Total Installed Cost (TIC) | |
|---|---|
| | % of PC |
| Purchased Equipment Cost (PC) | 100 |
| Installation | 39 |
| Instrumentation and controls | 26 |
| Piping & Insulation | 31 |
| Electrical facilities | 10 |
| Buildings | 29 |
| Yard improvements | 12 |
| Auxiliary facilities (OSBL) | 0 |
| Total | 247 |

In this study factored cost estimates developed in one country, typically the United States, need to be converted to costs in other countries. Since location multipliers are different for capital and labor, estimates of equipment/labor splits for each of the costing factors are used, based on discussions given in Valle-Riestra (1983), as shown in Table 8.

Purchased equipment plus installation costs are added together to give total installed capital (TIC) cost. A number of indirect project costs must be added to determine total plant investment (TPI); including engineering costs, construction costs, contractor and legal costs, and project contingency costs. Not included as an indirect cost in this study are owner's costs, although many of the individual costs such as prepaid royalties and land are included. Costing factors for these indirect costs are given in Table 9. The project contingency used is 10% of PC (4% of TIC). This is typical of detailed cost estimates and is low for order of magnitude cost factor estimates. However, the project contingency is consistent with 'nth' plant analyses. The added uncertainty and risk associated with cost factor analysis is evaluated through sensitivity analysis.

**Table 8. Capital/labor ratios for costing factors**

| | | % Capital | % Labor | Capital Cost Multiplier (CCM) | Labor Cost Multiplier (LCM) |
|---|---|---|---|---|---|
| | % of PC | | | % of PC | % of PC |
| Purchased Equipment Cost (PC) | 100 | 100 | 0 | 100.0 | 0.0 |
| Installation | 39 | 25 | 75 | 9.8 | 29.3 |
| Instrumentation & Controls | 26 | 50 | 50 | 13.0 | 13.0 |
| Piping & Insulation | 31 | 50 | 50 | 15.5 | 15.5 |
| Electrical Facilities | 10 | 25 | 75 | 2.5 | 7.5 |
| Buildings | 29 | 50 | 50 | 14.5 | 14.5 |
| Yard improvements | 12 | 25 | 75 | 3.0 | 9.0 |
| Auxiliary facilities (OSBL) | 0 | 25 | 75 | 0.0 | 0.0 |

**Table 9. Capital cost indirect cost factors**

| Indirect Costs | % of TIC |
|---|---|
| Engineering | 13 |
| Construction | 10 |
| Legal & Contractors Fees | 7 |
| Project Contingency | 4 |
| Total | 34 |

**Table 10. Fixed costs**

| Fixed Cost Category | Value |
|---|---|
| Labor | Based on Phillips, et al. 2007 |
| Maintenance | 3% of TIC |
| General Overhead | 65% of Labor and Maintenance |
| Direct Overhead | 45% of Labor |
| Insurance | 0.5% of TIC |

Total plant cost (TPC) is determined by adding TPI and 'allowance for funds used during construction' (AFUDC), also referred to as interest during construction. Since this analysis assumes equity financing, no AFUDC is estimated. Total capital cost (TCC) is estimated by adding land and start-up costs to TPC.

Operating costs use a combination of cost factor analysis and estimated costs, e.g., feedstock cost and methanol cost for biodiesel. Labor costs are estimated based on costs developed for the thermochemical ethanol process (Phillips et al. 2007). Standard fixed costs and multipliers (Valle-Riestra 1983) are given in Table 10.

## 2.2. Financial Analysis

The technologies are financially compared using a discounted cash flow rate of return (DCFROR) analysis with a zero net present value (NPV) at a fixed return on equity of 10 percent. Project financial assumptions are given in Table 11.

All of the financial factors are standard. Since information on 'loss carry-forward' or 'depreciation carry-forward' is not available for all of the countries being studied, a corporate analysis is used. A project basis analysis assumes that depreciation deductions in any year are limited to taxable income. Excess depreciation can be carried forward to future years. The number of years of allowable carry-forward is specified by tax law. If a corporate basis is used the assumption is made that the business entity has other income to offset excess project depreciation, and all of the yearly depreciation can be used in estimating yearly cash flow.

**Table 11. Financial analysis assumptions**

| Factor | |
|---|---|
| Discount Rate | 10% |
| Equity / Debt | 100 / 0 % |
| Cost of Capital – for Debt/Equity | 7% |
| Debt Term – for Debt/Equity | 20 years |
| Inflation | 1.9% |
| Feedstock Escalation | 0% |
| Product Escalation | 0% |
| Construction Period | 3 years |
| Construction Investment Schedule | 8% / 60% / 32% |
| Stream Factor in Start-Up Year | 75% of later year stream factor |
| Stream Factor | 95% (50% for sugar mills) |
| Operating Life | 20 years |
| Annual Replacement Capital Allowance | 0% |
| Depreciation | Country specific |
| USA | 7, 20 year double declining balance |
| Other Countries | Straight line, variable number of years |
| Project Basis | Corporate, after tax |
| Year Basis | 2005 |
| Effective Tax Rate – Country Specific | e.g., USA = 39% |

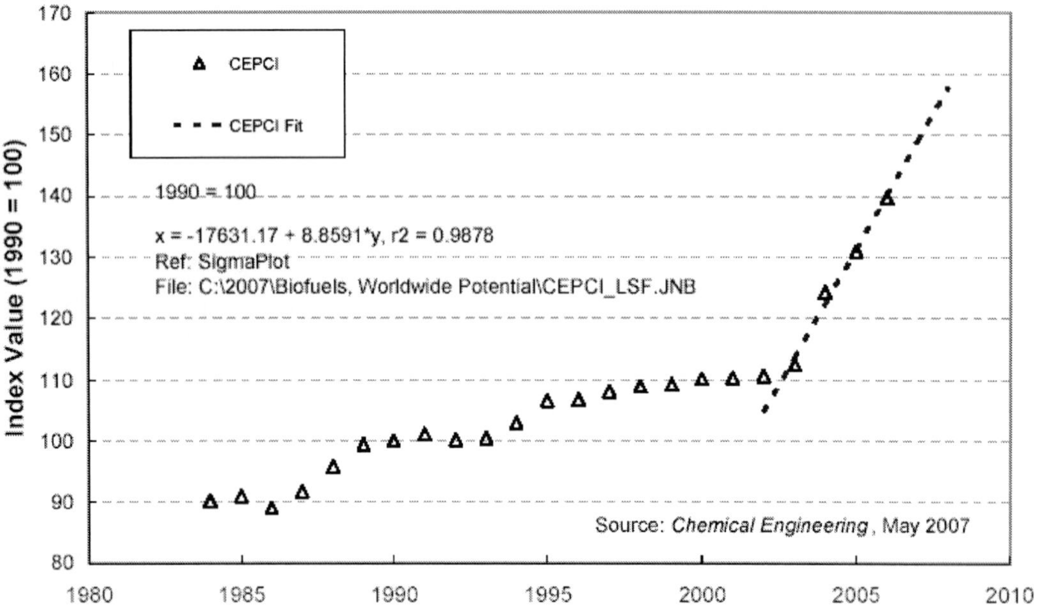


Figure 8. Chemical Engineering plant cost index

**Table 12. Plant size and yield information for cellulosic ethanol (TC), USA basis**

| Country | USA | Code = | 1 | | | | |
|---|---|---|---|---|---|---|---|
| **Cost component** | **Units** | **Cost Factor** | **Scale Factor** | **USA Cellulosic Ethanol TC)** | **USA Cellulosic Ethanol TC)** | **USA Cellulosic Ethanol TC)** | **USA Cellulosic Ethanol TC)** |
| Year $ | $ | | | 2005 | 2005 | 2005 | 2005 |
| Cost Index (2 = M&S, 3 = CE) | | | 3 | 130.93 | 130.93 | 130.93 | 130.93 |
| Plant Size | dry short ton/day | | | 1,800 | 900 | 3,601 | 10,801 |
| | dry tonne/day | | | 1,632 | 816 | 3,264 | 9,793 |
| Stream Factor | % | | | 95% | 95% | 95% | 95% |
| Feed | Dry short ton/yr | | | 6.242E+05 | 3.121E+05 | 1.248E+06 | 3.745E+06 |
| | GJ/short ton | | | 1.794E+01 | 1.794E+01 | 1.794E+01 | 1.794E+01 |
| | GJ/yr | | | 1.119E+07 | 5.598E+06 | 2.239E+07 | 6.717E+07 |
| Feed Cost | USD/dt | | | **$ 35.00** | **$ 35.00** | **$ 35.00** | **$ 35.00** |
| Yield (gal/Dry US Ton) | | | | | | | |
| Ethanol | gal/short ton | | | 80.1 | 80.1 | 80.1 | 80.1 |
| Mixed alcohols | gal/short ton | | | 94.1 | 94.1 | 94.1 | 94.1 |
| Process Efficiency - to ethanol | % HHV | | | 39.6% | 39.6% | 39.6% | 39.6% |
| Process efficiency - overall | % HHV | | | 47.4% | 47.4% | 47.4% | 47.4% |
| Ethanol | gal/yr | | | 5.000E+07 | 2.500E+07 | 1.000E+08 | 3.000E+08 |
| | Mil Gal/YR | | | **50.00** | **25.00** | **100.00** | **300.00** |
| | tpy | | | 1.657E+05 | 8.288E+04 | 3.315E+05 | 9.945E+05 |
| | GJ/yr | | | 4.436E+06 | 2.218E+06 | 8.873E+06 | 2.662E+07 |
| | gal/Stream day | | | 1.442E+05 | 7.210E+04 | 2.884E+05 | 8.652E+05 |
| | bbl/s stream day | | | 3.433E+03 | 1.717E+03 | 6.867E+03 | 2.060E+04 |
| Higher Alcohols | | | | | | | |
| | gal/ton | | | 14.0 | 14.0 | 14.0 | 14.0 |
| | gal/yr | | | 8.74E+06 | 4.37E+06 | 1.75E+07 | 5.24E+07 |

### Table 13. Capital costs, cellulosic ethanol (TC), USA basis

| Country | USA | Code = | 1 | | | | |
|---|---|---|---|---|---|---|---|
| Cost component | Units | Cost Factor | Scale Factor | USA Cellulosic Ethanol (TC) | USA Cellulosic Ethanol (TC) | USA Cellulosic Ethanol (TC) | USA Cellulosic Ethanol (TC) |
| **Capital Cost (million USD)** | | | | | | | |
| Feed Handling & Drying | | | 0.70 | 23.20 | 14.28 | 37.69 | 81.32 |
| Gasification | | | 0.80 | 12.90 | 7.41 | 22.46 | 54.09 |
| Tar Reforming & Quench | | | 0.70 | 38.40 | 23.64 | 62.38 | 134.60 |
| Acid Gas & Sulfur Removal | | | 0.85 | 14.50 | 8.05 | 26.14 | 66.50 |
| Alcohol synthesis - Compression | | | 0.65 | 16.00 | 10.20 | 25.11 | 51.28 |
| Alcohol Synthesis - Other | | | 0.65 | 4.60 | 2.93 | 7.22 | 14.74 |
| Alcohol Separation | | | 0.65 | 7.20 | 4.59 | 11.30 | 23.08 |
| Steam System & Power Generation | | | 0.70 | 16.80 | 10.34 | 27.29 | 58.89 |
| Cooling Water & Other Utilities | | | 0.70 | 3.60 | 2.22 | 5.85 | 12.62 |
| **Direct Fixed Capital (DFC), also called TIC** | | | | 137.20 | 83.65 | 225.44 | 497.12 |
| Engineering | DFC × MF | 0.12 | | 16.46 | 10.04 | 27.05 | 59.65 |
| Construction | DFC × MF | 0.13 | | 17.84 | 10.88 | 29.31 | 64.63 |
| Contractor & Legal | DFC × MF | 0.08 | | 10.98 | 6.69 | 18.04 | 39.77 |
| Process/Project Contingency | DFC × MF | 0.0412 | | 5.65 | 3.45 | 9.29 | 20.48 |
| **Total Plant Cost (TPC)** | | | | 188.13 | 114.71 | 309.13 | 681.65 |
| AFUDC | | | | | | | |
| **Total Plant Investment (TPI)** | | | | 188.13 | 114.71 | 309.13 | 681.65 |
| Land | | | 0.60 | 2.21 | 2.21 | 5.08 | 15.23 |
| Startup | | | | 0.00 | 0.00 | 0.00 | 0.00 |
| **Total Capital Cost (TCC)** | | | | 190.34 | 116.92 | 314.21 | 696.88 |
| **Contingency/TPI** | | | | | | | |
| Working Capital | DFC ×MF | 0.05 | | 9.41 | 5.74 | 15.46 | 34.08 |

**Table 14. Operating costs, cellulosic ethanol (TC), USA basis**

| Country | USA | Code = | 1 | | | | |
|---|---|---|---|---|---|---|---|
| **Cost component** | **Units** | **Cost Factor** | **Scale Factor** | **USA Cellulosic Ethanol (TC)** | **USA Cellulosic Ethanol (TC)** | **USA Cellulosic Ethanol (TC)** | **USA Cellulosic Ethanol (TC)** |
| **Variable Operating Costs (million USD/yr)** | | | | | | | |
| Feed | | | | 21.85 | 10.92 | 43.70 | 131.09 |
| Utilities | | | | 0.30 | 0.12 | 0.49 | 1.47 |
| Other | | | | 1.00 | 0.50 | 2.00 | 6.00 |
| Catalysts and Chemicals | | | | 0.70 | 0.35 | 1.40 | 4.20 |
| **Total** | | | | 23.85 | 11.90 | 47.59 | 142.75 |
| Fixed Operating Costs (million USD/yr) | | | | | | | |
| Labor (SF from V-R) | DFC × MF | 0.0165 | 0.25 | 2.26 | 1.90 | 2.69 | 3.54 |
| Maintenance (3% of A) | | | 0.03 | 4.12 | 2.51 | 6.76 | 14.91 |
| General Overhead (65% of labor + maint) | | | 0.65 | 4.15 | 2.87 | 6.15 | 12.00 |
| Direct Overhead (45% of Labor) | | | 0.45 | 1.02 | 0.86 | 1.21 | 1.59 |
| Insurance (0.5% of TIC) | | | 0.005 | 0.95 | 0.58 | 1.57 | 3.48 |
| **Total** | | | | 12.50 | 8.72 | 18.38 | 35.53 |

All costs are presented in 2005 U.S. dollars (USD). For cost estimates developed in other years the *Chemical Engineering* chemical plant cost index (CEPCI) is used to convert to 2005 dollars (Chemical Engineering 2007).

The CEPCI is a standard cost index that NREL uses to update cost estimates to a given year. It is made up of 41 Bureau of Labor Statistics (BLS) producer price indexes, 12 BLS labor cost indexes, and a calculated labor productivity multiplier. A plot of the CEPCI is given by Figure 8. The least squares curve fit is used to extrapolate a preliminary value for 2007 dollars.

## 2.3. Base Case Analysis

For each technology evaluated in this study a base case technology is used, and the costs converted to costs in the other countries using capital and operating cost factors, individual country feedstock costs, and individual country tax laws. To the extent possible, USA costs are used to develop the base case estimates. The only exception to this is in the case of sugar cane sugar/ethanol, where USA capital costs are not available. For sugar cane sugar/ethanol Brazil is used for the base case. The results for individual technologies are presented in later sections of this report.

Capital and operating costs are scaled using a 0.7 default scaling factor or individual scaling factor where known (0.50 – 0.85). Peters and Timmerhaus (2003) state "It is often necessary to estimate the cost of a piece of equipment when no cost data are available for the particular size of operational capacity involved. Good results can be obtained by using the logarithmic relationship known as the 'six-tenths-factor rule,' if the new piece of equipment is similar to one of another capacity for which cost data are available. According to this rule, if the cost of a given unit at one capacity is known, the cost of a similar unit with X times the capacity of the first is approximately (X)0.6 times the cost of the initial unit." Valle-Riesta (1983) states " A logical consequence of the 'sixth-tenths-factor' rule for characterizing the relationship between equipment capacity and cost is that a similar relationship should hold for the direct fixed capital of specific plants...In point of fact, the capacity exponent for plants, on the average, turns out to be closer to 0.7." The exception to this rule happens when plant capacity is increased by change in efficiency, not change in equipment size. In this case, capital cost in dollars remains constant, and capital cost in $/size decreases in proportion to efficiency increase.

In this section the cost sheets and cash flow results for cellulosic ethanol (thermochemical) are presented as an example. Plant size and yield information are given in Table 12. The capital costs (Phillips et al. 2007) are given in Table 13 as a function of plant size based on ethanol (primary product). Operating costs are given in Table 14. The information in Tables 12-14 is used in a cash flow calculation. The input cash flow information is given in Table 15.

Table 16. gives the depreciation schedules used in the study.

Table 17 presents the cash flow for the process. Other summary information can also be estimated from the DCFROR analysis. Figure 9 shows the 50 MM GPY project cumulative cash flow. The payback is about 7.5 years after start of operation, which is typical for 10% return. The breakdown of costs as a function of plant size is given in Figure 10.

**Table 15. DCFROR input, DCFROR analysis, cellulosic ethanol (TC), 50 MM GPY, USA basis**

| Case No: | Cellulosic Ethanol (TC):2000-2005:Jul2007 | |
|---|---|---|
| Project Description: | TC Ethanol - 1800 TPD -2005 | |
| Construction Period | 3 yrs | |
| | | Solve |
| Operating Life | 20 yrs | |
| EtOH Production | 4,436,016* GJ/yr | |
| EtOH COP | $ 12.36 /GJ | $ 1.10 /gal |
| Capital Cost | 190,338,640* | |
| Depreciable Capital | 188,128,640* | |
| 3 year | | |
| 5 year | | |
| 7 year | 160,156,160 | |
| 10 year | | |
| 15 year | | |
| 20 year | 27,972,480* | |
| 5 year SL | | |
| 9 year SL | | |
| 10 year SL | | |
| 15 year SL | | |
| 20 year SL | | |
| % capital year 1 | 8% | |
| % capital year 2 | 60% | |
| % capital year 3 | 32% | |
| Inflated Capital | 194,845,006 | |
| Inflated depreciable capi | 192,582,683 | |
| Replacement Capital | | |
| Variable Cost (I = 0) | | |
| year 1 | | |
| year 2 | | |
| year 3 | 0* | |
| year 4 | 17,884,811* | |
| year 5 to n | 23,846,415* | |
| Fixed Cost | 12,497,073* | |
| Working Cap | 9,406,432* | |
| Revenues | | |
| Ethanol | 54,834,054 | |
| Capacity Paymt | | |
| Coproduct | 10,049,351 Co-Prod Val | 1.15 *USD/gal |
| Total Revenue | 64,883,405 | |
| Yr 4 Prod, % | 75.0% | |
| Inflation rate, % | 1.90% | |
| Percent Debt | 0.00% | |
| Cost of Debt | 7.0% | |
| Term | 20 | |
| Taxes | | |
| Federal | | 35.0% |
| State | | 3.9% |

**Table 16. Depreciation schedules**

| MACRS with half-year rule | | | | | | Straight Line | | | | |
|---|---|---|---|---|---|---|---|---|---|---|
| 3-yr | 5-yr | 7-yr | 10-yr | 15-yr | 20-yr | 5-yr | 7-yr | 10-yr | 11-yr | 20-yr |
| 33.33 | 20.00 | 14.29 | 10.00 | 5.00 | 3.750 | 20.000 | 14.286 | 10.000 | 9.091 | 5.000 |
| 44.45 | 32.00 | 24.49 | 18.00 | 9.50 | 7.219 | 20.000 | 14.286 | 10.000 | 9.091 | 5.000 |
| 14.81 | 19.20 | 17.49 | 14.40 | 8.55 | 6.677 | 20.000 | 14.286 | 10.000 | 9.091 | 5.000 |
| 7.41 | 11.52 | 12.49 | 11.52 | 7.70 | 6.177 | 20.000 | 14.286 | 10.000 | 9.091 | 5.000 |
| | 11.52 | 8.93 | 9.22 | 6.93 | 5.713 | 20.000 | 14.286 | 10.000 | 9.091 | 5.000 |
| | 5.76 | 8.92 | 7.37 | 6.23 | 5.285 | | 14.286 | 10.000 | 9.091 | 5.000 |
| | | 8.93 | 6.55 | 5.90 | 4.888 | | 14.286 | 10.000 | 9.091 | 5.000 |
| | | 4.46 | 6.55 | 5.90 | 4.522 | | | 10.000 | 9.091 | 5.000 |
| | | | 6.56 | 5.91 | 4.462 | | | 10.000 | 9.091 | 5.000 |
| | | | 6.55 | 5.90 | 4.461 | | | 10.000 | 9.091 | 5.000 |
| | | | 3.28 | 5.91 | 4.462 | | | | 9.091 | 5.000 |
| | | | | 5.90 | 4.461 | | | | | 5.000 |
| | | | | 5.91 | 4.462 | | | | | 5.000 |
| | | | | 5.90 | 4.461 | | | | | 5.000 |
| | | | | 5.91 | 4.462 | | | | | 5.000 |
| | | | | 2.95 | 4.461 | | | | | |
| | | | | | 4.462 | | | | | |
| | | | | | 4.461 | | | | | |
| | | | | | 4.462 | | | | | |
| | | | | | 4.461 | | | | | |
| | | | | | 2.231 | | | | | |

## 2.4. Country Costs

Plant gate prices in other countries are converted to selected country costs using conceptual equipment and labor cost estimating factors based on a number of references (ASPEN Tech 2007, Humphries 2005a, Humphries 2005b, KMPG 2006, Richardson Engineering 1997). Capital costs are divided into domestic equipment (DE) and imported equipment (IE). For imported equipment duty, freight, and Value Added Tax (VAT) are considered in determining a multiplication factor (PCMI). A local material index (PCML) is used for domestic equipment.

$$\text{If } PCC_x = \text{purchased equipment cost in country X, then}$$
$$PCC_x = \text{USA PC} * (\%IE * PCMI + \%DE * PCML) \quad (1)$$

Labor costs ($LC_x$) are based on USA costs, with labor rate multipliers (LRM) and labor productivity multipliers (LPM).

$$LRM_x = \text{Country x labor rate/USA labor rate} \quad (2)$$
$$LC_x = LRM_x * LPM_x \quad (3)$$

These factors, in combination with the estimated equipment/labor percentages shown in Table 8 are used to convert base case capital and operating costs to the specified country. Country factors are given in Table 18. Where factors are not available default values for imported materials (1), and depreciation schedule (10% SL) are used.

**Table 17. Cash flow, cellulosic ethanol, 50 MM GPY, USA basis**

| Year | Income | Production Credit | Equity | Debt | Debt Service | Remaining Capital | Principal Paymt | Working Capital | Fixed Op cost | Replacement Capital |
|---|---|---|---|---|---|---|---|---|---|---|
| 1 | | | 15,227,091 | 0 | 0 | | | | | |
| 2 | | | 116,373,044 | 0 | 0 | | | | | |
| 3 | | | 63,244,871 | 0 | 0 | 0 | | | | |
| 4 | 51,489,355 | 0 | | | 0 | 0 | 0 | 9,952,850 | 13,223,026 | |
| 5 | 69,956,870 | 0 | | | 0 | 0 | 0 | | 13,474,264 | |
| 6 | 71,286,051 | 0 | | | 0 | 0 | 0 | | 13,730,275 | |
| 7 | 72,640,486 | 0 | | | 0 | 0 | 0 | | 13,991,150 | |
| 8 | 74,020,655 | 0 | | | 0 | 0 | 0 | | 14,256,982 | |
| 9 | 75,427,047 | 0 | | | 0 | 0 | 0 | | 14,527,865 | |
| 10 | 76,860,161 | 0 | | | 0 | 0 | 0 | | 14,803,894 | |
| 11 | 78,320,504 | 0 | | | 0 | 0 | 0 | | 15,085,168 | |
| 12 | 79,808,594 | 0 | | | 0 | 0 | 0 | | 15,371,786 | |
| 13 | 81,324,957 | 0 | | | 0 | 0 | 0 | | 15,663,850 | |
| 14 | 82,870,131 | 0 | | | 0 | 0 | 0 | | 15,961,463 | |
| 15 | 84,444,664 | 0 | | | 0 | 0 | 0 | | 16,264,731 | |
| 16 | 86,049,112 | 0 | | | 0 | 0 | 0 | | 16,573,761 | |
| 17 | 87,684,046 | 0 | | | 0 | 0 | 0 | | 16,888,663 | |
| 18 | 89,350,042 | 0 | | | 0 | 0 | 0 | | 17,209,547 | |
| 19 | 91,047,693 | 0 | | | 0 | 0 | 0 | | 17,536,529 | |
| 20 | 92,777,599 | 0 | | | 0 | 0 | 0 | | 17,869,723 | |
| 21 | 94,540,374 | 0 | | | 0 | 0 | 0 | | 18,209,247 | |
| 22 | 96,336,641 | 0 | | | 0 | 0 | 0 | | 18,555,223 | |
| 23 | 98,167,037 | 0 | | | 0 | 0 | 0 | (9,952,850) | 18,907,772 | |

| | | | | |
|---|---|---|---|---|
| | | | Project Internal Rate of Return | |
| | | | Rate of Return Estimate | |
| COP | 12.36 $/GJ | $ 1.10 /gal | | 15.00% |
| PROD CRE | 0.00 | | Internal Rate of Return | |
| TAX CREDI | 0.00 | | | 10.00% |
| | | | Desired Return | 10.00% |
| | | | NPV | 0 |
| | | $ 1.10 $/gal EtOH eq | | |

## Table 17. (Continued )

| Year | Variable Op cost | Pre-Dep Income | Deprec Schedule | Depreciation | Pre-tax Calculation | Tax | After-Tax Income | Cash Flow | Cum Cash Flow |
|---|---|---|---|---|---|---|---|---|---|
| 1 | | (15,227,091) | | | (15,227,091) | 0 | (15,227,091) | (15,227,091) | (15,227,091) |
| 2 | | (116,373,044) | | | (116,373,044) | 0 | (116,373,044) | (116,373,044) | (131,600,136) |
| 3 | | (63,244,871) | | | (63,244,871) | 0 | (63,244,871) | (63,244,871) | (194,845,006) |
| 4 | 18,923,737 | 9,389,741 | 0.1272 | 24,501,964 | (15,112,223) | 0 | (15,112,223) | 9,389,741 | (185,455,266) |
| 5 | 25,711,051 | 30,771,555 | 0.2192 | 42,217,993 | (11,446,438) | 0 | (11,446,438) | 30,771,555 | (154,683,711) |
| 6 | 26,199,561 | 31,356,214 | 0.1588 | 30,586,437 | 769,778 | 299,444 | 470,334 | 31,056,771 | (123,626,940) |
| 7 | 26,697,353 | 31,951,983 | 0.1155 | 22,245,866 | 9,706,117 | 3,775,679 | 5,930,437 | 28,176,303 | (95,450,637) |
| 8 | 27,204,603 | 32,559,070 | 0.0845 | 16,276,454 | 16,282,616 | 6,333,938 | 9,948,679 | 26,225,132 | (69,225,504) |
| 9 | 27,721,490 | 33,177,693 | 0.0838 | 16,137,502 | 17,040,190 | 6,628,634 | 10,411,556 | 26,549,059 | (42,676,446) |
| 10 | 28,248,198 | 33,808,069 | 0.0833 | 16,040,217 | 17,767,851 | 6,911,694 | 10,856,157 | 26,896,374 | (15,780,071) |
| 11 | 28,784,914 | 34,450,422 | 0.0447 | 8,606,941 | 25,843,481 | 10,053,114 | 15,790,367 | 24,397,308 | 8,617,237 |
| 12 | 29,331,828 | 35,104,980 | 0.0066 | 1,277,682 | 33,827,298 | 13,158,819 | 20,668,479 | 21,946,161 | 30,563,398 |
| 13 | 29,889,132 | 35,771,975 | 0.0066 | 1,277,396 | 34,494,579 | 13,418,391 | 21,076,188 | 22,353,583 | 52,916,981 |
| 14 | 30,457,026 | 36,451,642 | 0.0066 | 1,277,682 | 35,173,960 | 13,682,670 | 21,491,290 | 22,768,972 | 75,685,953 |
| 15 | 31,035,709 | 37,144,223 | 0.0066 | 1,277,396 | 35,866,827 | 13,952,196 | 21,914,632 | 23,192,027 | 98,877,981 |
| 16 | 31,625,388 | 37,849,964 | 0.0066 | 1,277,682 | 36,572,281 | 14,226,617 | 22,345,664 | 23,623,346 | 122,501,327 |
| 17 | 32,226,270 | 38,569,113 | 0.0066 | 1,277,396 | 37,291,717 | 14,506,478 | 22,785,239 | 24,062,635 | 146,563,962 |
| 18 | 32,838,569 | 39,301,926 | 0.0066 | 1,277,682 | 38,024,244 | 14,791,431 | 23,232,813 | 24,510,495 | 171,074,457 |
| 19 | 33,462,502 | 40,048,663 | 0.0066 | 1,277,396 | 38,771,267 | 15,082,023 | 23,689,244 | 24,966,640 | 196,041,097 |
| 20 | 34,098,290 | 40,809,587 | 0.0066 | 1,277,682 | 39,531,905 | 15,377,911 | 24,153,994 | 25,431,676 | 221,472,773 |
| 21 | 34,746,157 | 41,584,969 | 0.0066 | 1,277,396 | 40,307,573 | 15,679,646 | 24,627,927 | 25,905,323 | 247,378,096 |
| 22 | 35,406,334 | 42,375,084 | 0.0066 | 1,277,682 | 41,097,402 | 15,986,889 | 25,110,512 | 26,388,195 | 273,766,291 |
| 23 | 36,079,054 | 53,133,061 | 0.0066 | 1,277,396 | 51,855,665 | 20,171,854 | 31,683,811 | 32,961,207 | 306,727,498 |
| | | | | | | Project Internal Rate of Return | | | |
| | | | | | | Rate of Return Estimate | | | |
| COP | COP | 12.36 $/GJ | | | | | 15% | | |
| PROD CRE | PROD CREDIT | 0 | | | | Internal Rate of Return | | | |
| TAX CREDI | TAX CREDIT | 0 | | | | | 10% | | |
| | | | | | | Desired Return | 10% | | |
| | | | | | | NPV | 0 | | |
| | | | | 1.10 $/gal EtOH eq | | | | | |

## Table 18. International cost location factors

| Country/Region<br>Cost Variable | | USA | Brazil | China | India | Argentina | Colombia | Canada | Mexico | Caribbean Basin |
|---|---|---|---|---|---|---|---|---|---|---|
| **Country Code** | | **1** | **2** | **3** | **4** | **5** | **6** | **7** | **8** | **9** |
| % domestic equipment | | 100% | 80% | 80% | 50% | 50% | 50% | 75% | 50% | 25% |
| % imported equipment | | 0% | 20% | 20% | 50% | 50% | 50% | 25% | 50% | 75% |
| Exchange Rate in 2005 | Country = 'x" USD | 1 | 0.425 | 0.124 | 0.022 | 0.328 | 0.438 | 0.859 | 0.094 | 0.131 |
| Imported Material | Duty, % | 0 | 14 | 8 | 65 | 1 | 1 | 0 | 7 | 1 |
| | Freight, % | 0 | 10 | 10 | 20 | 1 | 1 | 5 | 5 | 1 |
| | VAT, % | 0 | 18 | 17 | 0 | 1 | 1 | 13.5 | 10 | 1 |
| Imported Material Index | | 1 | 1.42 | 1.35 | 1.85 | 1.25 | 1.25 | 1.07 | 1.22 | 1.25 |
| Local Material Index | | 1 | 1.34 | 1.03 | 1.06 | 1.15 | 1.14 | 1.1 | 1.02 | 1.14 |
| Labor | $/hr | 20.00 | 4.00 | 6.07 | 2.27 | 7.63 | 5.65 | 24.00 | 2.47 | 8.39 |
| USA | Productivity | 1.00 | 2.13 | 3.66 | 3.30 | 2.23 | 2.25 | 1.40 | 1.95 | 2.29 |
| | Location Cost Factor | 1 | 1 | 1 | 1 | 1 | 1 | 1 | 1 | 1 |
| Tax Rate | | 0.39 | 0.34 | 0.33 | 0.3366 | 0.35 | 0.35 | 0.3 | 0.35 | 0.3 |
| Depreciation | | 7,20 y DDB | 10% SL 4 % SL bldgs | 9% SL 4.5% SL bldgs | 15% SL 10% SL bldgs | 10% SL 2% SL bldgs | 10% SL 5% SL bldgs | 10 % SL | 5% SL 74% LS | 10 % SL |

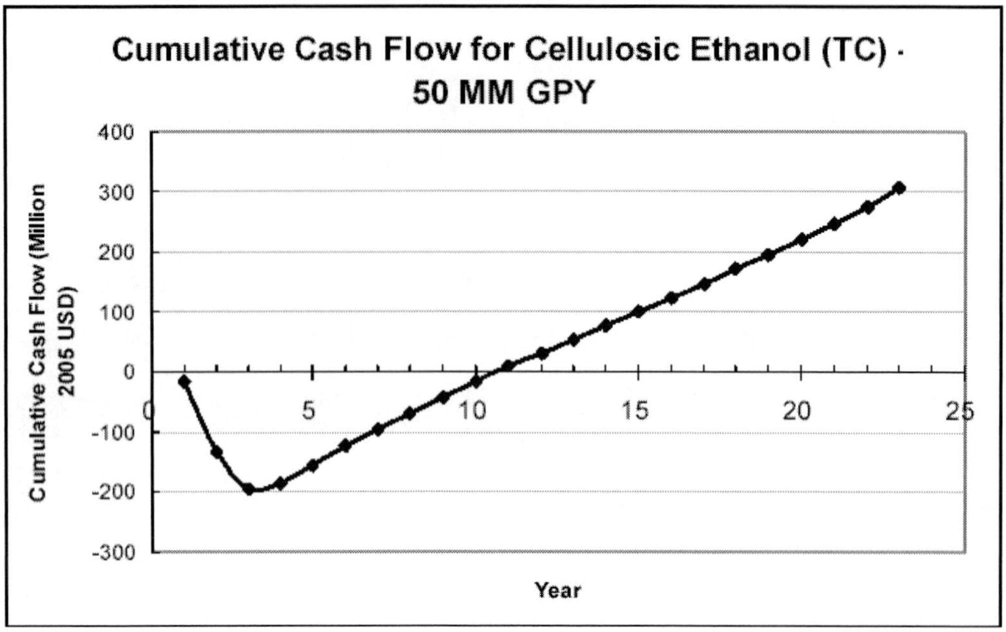


Figure 9. Cash flow, cellulosic ethanol (TC), 50 MM GPY

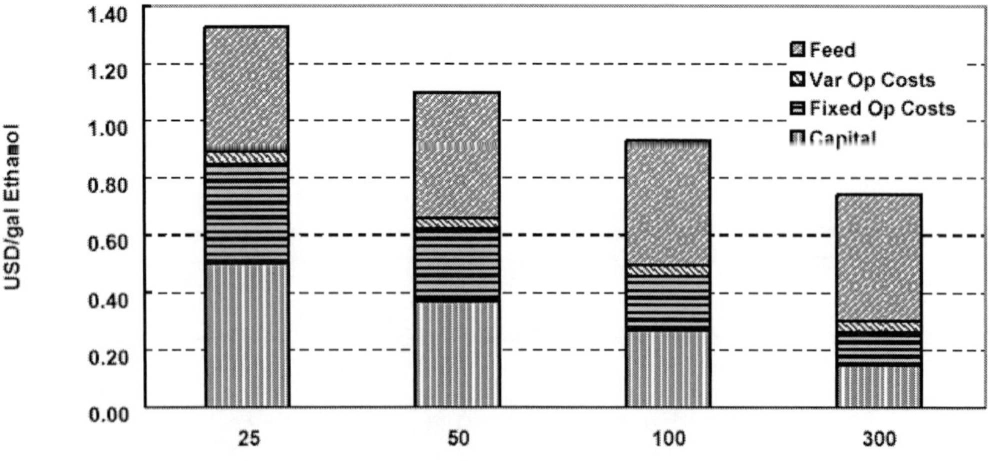


| | 25 | 50 | 100 | 300 |
|---|---|---|---|---|
| Feed | 0.44 | 0.44 | 0.44 | 0.44 |
| Var Op Costs | 0.04 | 0.04 | 0.04 | 0.04 |
| Fixed Op Costs | 0.35 | 0.25 | 0.18 | 0.12 |
| Capital | $0.50 | $0.37 | $0.27 | $0.15 |

Plant Size (Mil gal/yr)

Figure 10. Cellulosic ethanol (TC) cost contributions

The feedstock supply curves developed by Oak Ridge National Laboratory give costs versus supply for a base case, low growth, and high growth. The initial technology costing effort is intended to provide non-feedstock capital and operating cost information to the market modeling effort by Brookhaven National Laboratory. To gain insights on comparative technology costs one representative feedstock cost for each feedstock in each country was chosen–a pseudo-average baseline supply curve feedstock cost. These values are given in Table 19. The cellulosic feedstock supply costs from the ORNL study (ORNL 2007) were separated into bagasse, agricultural residues (primarily corn stover and wheat straw), and woody/perennial feeds. Costs of these materials by country are given in Table 20.

**Table 19. Representative feedstock costs for base modeling**

| | Sugar Cane $/ton | Corn $/ton | $/bu | Soy Beans | | Wheat | | Palm Oil $/ton |
|---|---|---|---|---|---|---|---|---|
| | | | | $/ton | $/bu | $/ton | $/bu | |
| **USA** | 29.00 | 107.0 | 3.00 | 187.6 | 5.60 | 113.3 | 3.40 | |
| **Argentina** | 34.30 | 127.5 | 3.57 | 143.6 | 4.29 | 92.5 | 2.78 | |
| **Brazil** | 10.45 | 135.0 | 3.78 | 175.0 | 5.22 | | | |
| **Canada** | | 86.0 | 2.41 | | | 102.0 | 3.06 | |
| **Caribbean Basin** | 15.00 | | | | | | | 275 |
| **Colombia** | 26.85 | | | | | | | 406 |
| **China** | 23.00 | 126.0 | 3.53 | 319.8 | 9.55 | 159.0 | 4.77 | |
| **India** | 15.00 | | | | | | | |
| **Mexico** | 29.00 | 130.0 | 3.64 | | | | | |

**Table 20. Feedstock prices for bagasse, agriculture residues, and wood/perennials**

| Country | Bagasse | Ag Residues (2005 USD/ST) | Woody /Perennial |
|---|---|---|---|
| USA | 35.0 | 35.0 | 35.0 |
| Argentina | 9.1 | 47.2 | 47.2 |
| Brazil | 7.6 | 30.8 | 47.2 |
| Canada | -- | 10.0 | 16.3-47.2 (31.8) |
| Caribbean Basin | 7.6 | 14.5 | 47.2 |
| China | 9.1 | 28.1 | 47.2 |
| Colombia | 14.5 | 14.5 | 47.2 |
| India | 7.3 | 7.3 | 47.2 |
| Mexico | 15.4 | 47.2 | 47.2 |

## 3. Ethanol – Corn Dry Mill

The discussion of the corn dry mill is taken verbatim from McAloon et al. (2000). Figure 11 depicts the dry mill process. The majority of the flowsheet information was provided by Delta-T Corporation, which designs, constructs, and operates corn ethanol plants (Delta-T Corporation 1997).

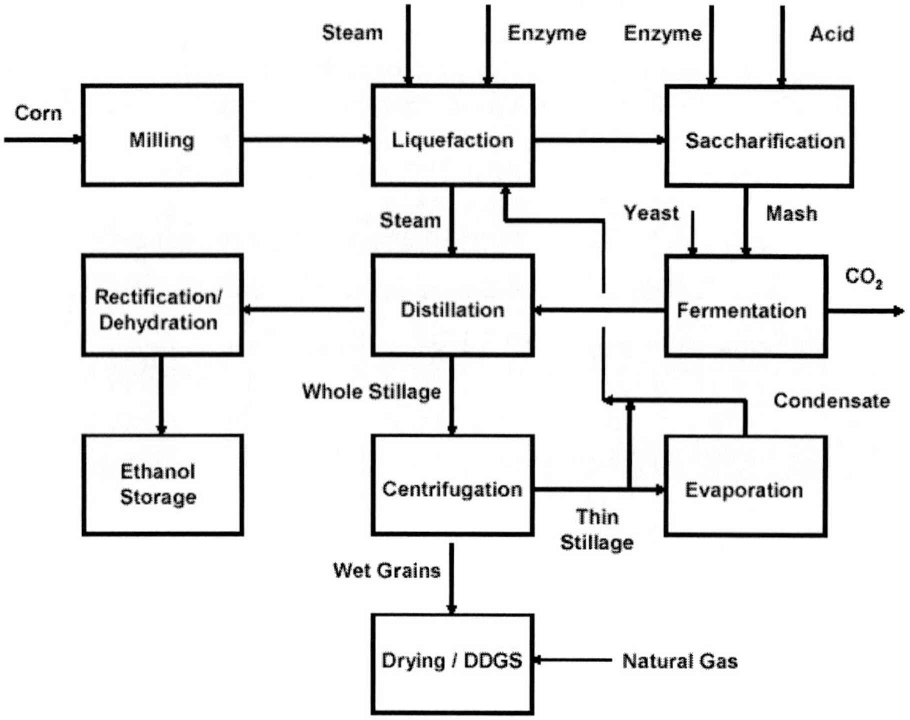


Figure 11. Corn dry mill process flow diagram

Corn is received and conveyed to two storage silos, having a combined capacity of 10 days. Stored corn is conveyed to grain-cleaning equipment where trash such as tramp metal and rocks (0.4%) is removed, and then to two hammer mills (two operating mills, plus one standby). The corn meal is metered to a continuous liquefaction tank, where it is mixed with hot evaporator condensate, and purchased alpha-amylase enzyme. The condensate is heated with steam to maintain 88°C (190°F) in the tank. Used caustic from the clean-in-place system and lime are also added to provide optimum pH (6) and calcium for the alpha-amylase. Urea is added to provide nitrogen to the yeast fermentation. After liquefaction, backset (recycled thin stillage fro the centrifuge) is added, amounting to 15 % by volume of the final mash. Then the mash is heated to 110°C (230°F), held for 20 minutes, and cooled to 60°C (140°F). Continuous saccharification takes place in a stirred tank where purchased glucoamylase is added with sulfuric acid for pH control (4.4). Residence time in the saccharification tank is 6 hours. The saccharified mash is cooled to 32°C (89°F) and fed to four continuous cascade fermentors where yeast is added. Total residence time in the fermenter is 46 hours.

Temperature is maintained below 34°C (93°F) by recirculation through two external heat exchangers, and pH is maintained above 3.5. Recirculating the off-gas through a compressor mixes the airlift fermentors. The concentration of ethanol in the whole beer leaving the fermentors is 9% by weight (12% by volume).

In liquefaction, the alpha-amylase attacks the starch polymer randomly, producing maltose (di-glucose) and higher oligomers. In saccharification, the gluco-amylase attacks the non-reducing end of maltose and higher oligomers, splitting off glucose. In addition to the alpha 1-4 linkages, there are alpha 1-6 branch points. These are attacked by pullulanase. This enzyme is probably found as a minor constituent of commercial enzymes, which are not pure enzyme preparations, but complex mixtures. The latest development in dry-mill ethanol enzymes is alpha amylase containing some protease that makes some of the corn protein available for yeast nutrition.

The whole beer is heated, degassed, and fed to the beer column. Steam and cooling water for heating and cooling of the mash, whole beer, and whole stillage are conserved by the use of heat recovery exchangers. Fermenter off-gas and vapors from degassing the whole beer are sent to a water scrubber where ethanol vapor is removed and recycled. The scrubbed $CO_2$ is released to the atmosphere. The whole stillage leaves the bottom of the beer column at less than 0.1% by weight ethanol. The overhead vapors pass to the bottom of the rectifier, where the concentration of the ethanol is increased from 45% to 91% by weight. The bottoms of the rectifier are pumped to the top of the stripper. The bottoms from the stripper (less than 0.1 % by weight ethanol) are recycled to the liquefaction tank along with the evaporator condensate. The concentrated vapor from the rectifier is superheated and passes through one of two dehydrating molecular sieve beds; one is used while the other is regenerated. Vapors from the regenerated bed are condensed and recycled to the rectifier. The superheated vapor passing through the molecular sieve bed contains more than 99% by weight ethanol. The product is condensed, cooled, stored, denatured with gasoline (5% by volume), and shipped. Ethanol storage is 12 days.

The whole stillage is partially evaporated in the first three stages of a six-effect vacuum evaporator. The partially evaporated whole stillage is separated in a decanter centrifuge (one operating plus one standby). The wet grains leave the centrifuge at 35% by weight total solids. The thin stillage from the centrifuge is partially recycled as backset, and the remainder concentrated in the three final stages of the evaporator syrup containing 55% by weight solids. To conserve steam and cooling water, the condensation of overhead vapors from the rectifier to provide reflux for distillation is accomplished in the evaporator. The syrup and wet grains are mixed and dried in a gas-fired rotary dryer. The distillers dry grains (DDG) leaving the dryer contain 9% moisture by weight. The process is designed to be essentially zero discharge. Makeup water is added only for the cooling tower and the $CO_2$ scrubber, and no wastewater is produced.

The base case for the corn dry-mill is the United States. Capital and operating costs (McAloon et al. 2000; Ibsen et al. 2005; Jechura 2005) are given in Table 21. A summary, based on 3 USD/bu corn is shown in Figure 12. Plant gate price decreases from 1.41 USD/gal for a 4.6 MM GPY plant to 1.25 USD/gallon for a 200 MM GPY. Corn cost is the largest component at 1.05 USD/gallon, and DDGS credit is 0.27 USD/gallon. Figure 13 shows the impact of corn price on plant gate price for 3 plant sizes - 25, 50, and 100 MM GPY. The slope of the curves is 0.35 (USD/gallon) / (USD/bu).

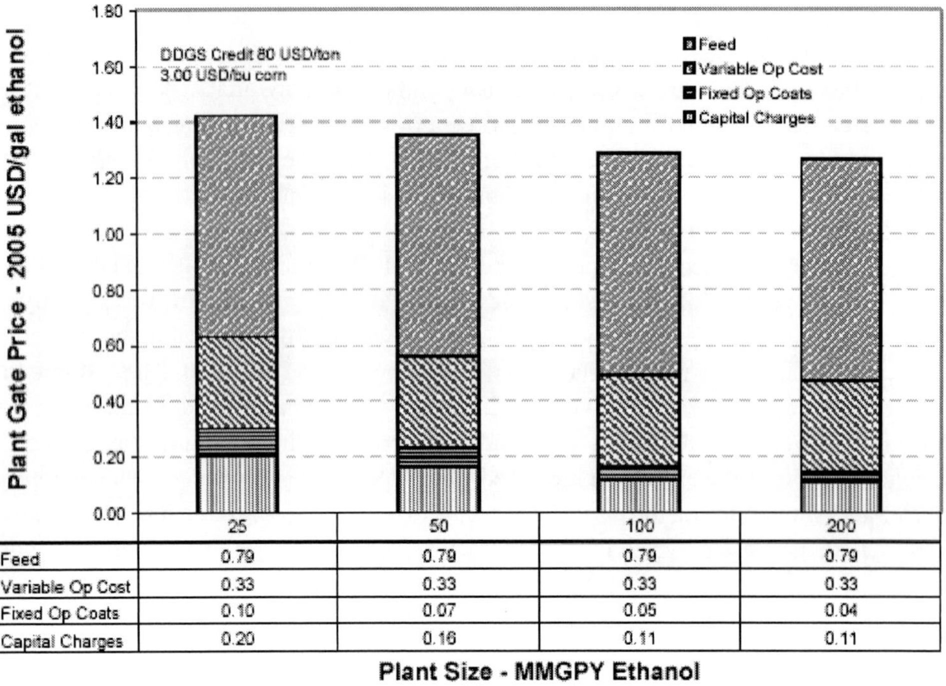


| | 25 | 50 | 100 | 200 |
|---|---|---|---|---|
| Feed | 0.79 | 0.79 | 0.79 | 0.79 |
| Variable Op Cost | 0.33 | 0.33 | 0.33 | 0.33 |
| Fixed Op Coats | 0.10 | 0.07 | 0.05 | 0.04 |
| Capital Charges | 0.20 | 0.16 | 0.11 | 0.11 |

Figure 12. USA base cost of ethanol, corn dry mill

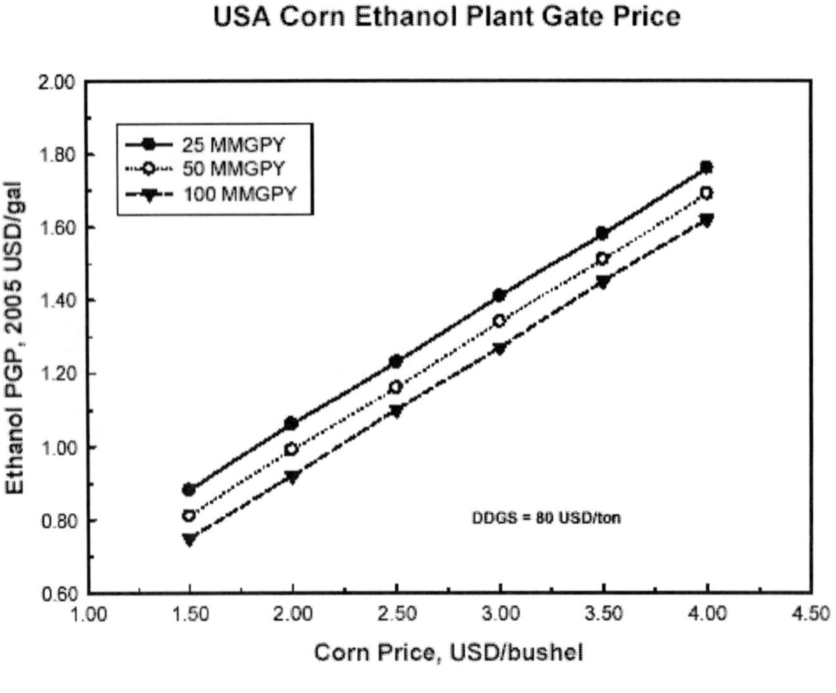


Figure 13. USA ethanol plant gate prices, impact of corn price and plant size

**Table 21. Corn dry mill capital and operating costs, USA base case**

| Country | USA | Code = | 1 | | | | |
|---|---|---|---|---|---|---|---|
| Cost component | Units | Cost Factor | Scale Factor | USA Dry Mill Ethanol | USA Dry Mill Ethanol | USA Dry Mill Ethanol | USA Dry Mill Ethanol |
| **Year $** | **$** | | | **2005** | **2005** | **2005** | **2005** |
| Cost Index (2 = M&S, 3 = CE) | | | 3 | 130.93 | 130.93 | 130.93 | 130.93 |
| Plant Size | dry short ton/day | | | 708 | 1,415 | 2,833 | 5,666 |
| | dry tonne/day | | | 642 | 1,284 | 2,570 | 5,140 |
| Stream Factor | % | | | 95% | 95% | 95% | 95% |
| Feed | Dry short ton/yr | | | 2.454E+05 | 4.908E+05 | 9.823E+05 | 1.965E+06 |
| | GJ/short ton | | | 1.688E+01 | 1.688E+01 | 1.688E+01 | 1.688E+01 |
| | GJ/yr | | | 4.142E+06 | 8.284E+06 | 1.658E+07 | 3.316E+07 |
| Feed Cost | USD/dt | | | $ 107.14 | $ 107.14 | $ 107.14 | $ 107.14 |
| Yield (gal/Dry US Ton) | | | | | | | |
| Ethanol | gal/short ton | | | 101.8 | 101.8 | 101.8 | 101.8 |
| DDGs | dry ton/ton | | | 0.33 | 0.33 | 0.33 | 0.33 |
| Process Efficiency - to ethanol | % HHV | | | 53.5% | 53.5% | 53.5% | 53.5% |
| Process efficiency - overall | % HHV | | | | | | |
| Ethanol | gal/yr | | | 2.498E+07 | 4.996E+07 | 1.000E+08 | 2.000E+08 |
| | Mil Gal/YR | | | 25.0 | 50.0 | 100.0 | 200.0 |
| | tpy | | | 8.281E+04 | 1.656E+05 | 3.315E+05 | 6.630E+05 |
| | GJ/yr | | | 2.216E+06 | 4.433E+06 | 8.872E+06 | 1.774E+07 |
| | gal/Stream day | | | 7.204E+04 | 1.441E+05 | 2.884E+05 | 5.768E+05 |
| | bbl/s stream day | | | 1.715E+03 | 3.431E+03 | 6.866E+03 | 1.373E+04 |
| DDGs | dry ton/yr | | | 8.098E+04 | 1.620E+05 | 3.242E+05 | 6.483E+05 |

**Table 21. (Continued)**

| Country | USA | Code = | 1 | | | | |
|---|---|---|---|---|---|---|---|
| **Cost component** | **Units** | **Cost Factor** | **Scale Factor** | **USA Dry Mill Ethanol** | **USA Dry Mill Ethanol** | **USA Dry Mill Ethanol** | **USA Dry Mill Ethanol** |
| **Year $** | **$** | | | **2005** | **2005** | **2005** | **2005** |
| Corn density | lb/bu | | | 56.0 | 56.0 | 56.0 | 56.0 |
| Corn | $/bu | | | 3.00 | 3.00 | 3.00 | 3.00 |
| Corn HHV | Btu/lb | | | 8.00E+03 | 8.00E+03 | 8.00E+03 | 8.00E+03 |
| | MMBtu/ton | | | 1.60E+01 | 1.60E+01 | 1.60E+01 | 1.60E+01 |
| **Capital Cost (million USD)** | | | | | | | |
| Feed Handling | | | 0.70 | 1.95 | 3.16 | 5.14 | 8.36 |
| Saccharification | | | 0.70 | 1.61 | 2.62 | 4.26 | 6.93 |
| Fermentation | | | 0.70 | 3.53 | 5.74 | 9.33 | 15.16 |
| Distillation | | | 0.70 | 3.91 | 6.36 | 10.34 | 16.79 |
| Solid/Syrup Separation | | | 0.70 | 8.63 | 14.02 | 22.79 | 37.03 |
| Storage/Load out | | | 0.70 | 1.09 | 1.77 | 2.87 | 4.67 |
| Wastewater Treatment | | | 0.70 | 0.52 | 0.85 | 1.37 | 2.23 |
| Air compressor | | | 0.70 | 0.10 | 0.16 | 0.26 | 0.43 |
| Steam Gen & Cooling Water | | | 0.70 | 1.37 | 2.22 | 3.61 | 5.87 |
| **Direct Fixed Capital (DFC), also called TIC** | | | | 22.72 | 36.91 | 59.99 | 97.46 |
| Engineering | DFC × MF | 0.12 | | 2.73 | 4.43 | 7.20 | 11.69 |
| Construction | DFC × MF | 0.13 | | 2.95 | 4.80 | 7.80 | 12.67 |
| Contractor & Legal | DFC × MF | 0.08 | | 1.82 | 2.95 | 4.80 | 7.80 |
| Process/Project Contingency | DFC × MF | 0.0412 | | 0.94 | 1.52 | 2.47 | 4.02 |
| **Total Plant Cost (TPC)** | | | | 31.15 | 50.61 | 82.26 | 133.63 |
| AFUDC | | | | | | | |

**Table 21. (Continued)**

| Country | USA | Code = | 1 | | | | |
|---|---|---|---|---|---|---|---|
| **Cost component** | **Units** | **Cost Factor** | **Scale Factor** | **USA Dry Mill Ethanol** | **USA Dry Mill Ethanol** | **USA Dry Mill Ethanol** | **USA Dry Mill Ethanol** |
| **Year $** | **$** | | | **2005** | **2005** | **2005** | **2005** |
| **Total Plant Investment (TPI)** | | | | 31.15 | 50.61 | 82.26 | 133.63 |
| Land | | | 0.60 | 2.21 | 3.35 | 5.08 | 10.16 |
| **Startup** | | | | **0** | **0.00** | **0.00** | **0.00** |
| **Total Capital Cost (TCC)** | | | | 33.36 | 53.96 | 87.34 | 143.79 |
| **Contingency/TPI** | | | | | | | |
| Working Capital | DFC x MF | 0.05 | | 1.56 | 2.53 | 4.11 | 6.68 |
| **Variable Operating Costs (million USD/yr)** | | | | | | | |
| Feed | | | | 26.29 | 52.58 | 105.25 | 210.50 |
| Nat Gas, GJ/y | 8.51E+05 | 6.33 | USD/GJ | 5.39 | 10.78 | 21.57 | 43.14 |
| Electricity, kWh/y | 2.01E+07 | 0.05 | $/kWh | 1.01 | 2.01 | 4.02 | 8.05 |
| Catalysts and Chemicals, Misc | | | | 1.79 | 3.58 | 7.17 | 14.33 |
| **Total** | | | | 34.48 | 68.95 | 138.01 | 276.02 |
| Fixed Operating Costs (million USD/yr) | | | | | | | |
| Labor (SF from V-R) | DFC × MF | 0.0255 | 0.25 | 0.58 | 0.69 | 0.82 | 0.97 |
| Maintenance (3% of A) | | | 0.03 | 0.68 | 1.11 | 1.80 | 2.92 |
| General Overhead (65% of labor + maint) | | | 0.65 | 0.82 | 1.17 | 1.70 | 2.53 |
| Direct Overhead (45% of Labor) | | | 0.45 | 0.26 | 0.31 | 0.37 | 0.44 |
| Insurance (0.5% of TIC) | | | 0.005 | 0.17 | 0.27 | 0.44 | 0.72 |
| **Total** | | | | 2.51 | 3.54 | 5.13 | 7.59 |

**Table 22. Corn dry mill ethanol: Argentina capital and operating costs**

| Country | Argentina | Code= | 5 | | | | | |
|---|---|---|---|---|---|---|---|---|
| **Cost component** | **Units** | **Cost Factor** | **Scale Factor** | **USA Dry Mill Ethanol** | **Argentina Dry Mill Ethanol** | **Argentina Dry Mill Ethanol** | **Argentina Dry Mill Ethanol** | **Argentina Dry Mill Ethanol** |
| **Year $** | **$** | | | **2005** | **2005** | **2005** | **2005** | **2005** |
| Cost Index (2 = M&S, 3 = CE) | | | 3 | 130.93 | 130.93 | 130.93 | 130.93 | 130.93 |
| Plant Size | dry short ton/day | | | 708 | 708 | 1,415 | 2,833 | 5,666 |
| | dry tonne/day | | | 642 | 642 | 1,284 | 2,570 | 5,140 |
| Stream Factor | % | | | 95% | 95% | 95% | 95% | 95% |
| Feed | Dry short ton/yr | | | 2.454E+05 | 2.454E+05 | 4.908E+05 | 9.823E+05 | 1.965E+06 |
| | GJ/short ton | | | 1.688E+01 | 1.688E+01 | 1.688E+01 | 1.688E+01 | 1.688E+01 |
| | GJ/yr | | | 4.142E+06 | 4.142E+06 | 8.284E+06 | 1.658E+07 | 3.316E+07 |
| Feed Cost | USD/dt | | | $ 125.00 | $ 127.50 | $ 127.50 | $ 127.50 | $ 127.50 |
| Yield (gal/Dry US Ton) | | | | | | | | |
| Ethanol | gal/short ton | | | 101.8 | 101.8 | 101.8 | 101.8 | 101.8 |
| DDGs | dry ton/ton | | | 0.33 | 0.33 | 0.33 | 0.33 | 0.33 |
| Process Efficiency - to ethanol | % HHV | | | 53.5% | 53.5% | 53.5% | 53.5% | 53.5% |
| Process efficiency - overall | % HHV | | | | | | | |
| Ethanol | gal/yr | | | 2.498E+07 | 2.498E+07 | 4.996E+07 | 1.000E+08 | 2.000E+08 |
| | Mil Gal/YR | | | **25.0** | **25.0** | **50.0** | **100.0** | **200.0** |
| | tpy | | | 8.281E+04 | 8.281E+04 | 1.656E+05 | 3.315E+05 | 6.630E+05 |
| | GJ/yr | | | 2.216E+06 | 2.216E+06 | 4.433E+06 | 8.872E+06 | 1.774E+07 |
| | gal/Stream day | | | 7.204E+04 | 7.204E+04 | 1.441E+05 | 2.884E+05 | 5.768E+05 |
| | bbl/s stream day | | | 1.715E+03 | 1.715E+03 | 3.431E+03 | 6.866E+03 | 1.373E+04 |
| DDGs | dry ton/yr | | | 8.098E+04 | 8.098E+04 | 1.620E+05 | 3.242E+05 | 6.483E+05 |

**Table 22. (Continued)**

| Country | Argentina | Code= | 5 | | | | | |
|---|---|---|---|---|---|---|---|---|
| **Cost component** | **Units** | **Cost Factor** | **Scale Factor** | **USA Dry Mill Ethanol** | **Argentina Dry Mill Ethanol** | **Argentina Dry Mill Ethanol** | **Argentina Dry Mill Ethanol** | **Argentina Dry Mill Ethanol** |
| **Year $** | **$** | | | **2005** | **2005** | **2005** | **2005** | **2005** |
| Corn density | lb/bu | | | 56.0 | 56.0 | 56.0 | 56.0 | 56.0 |
| Corn | $/bu | | | **3.50** | **3.57** | **3.57** | **3.57** | **3.57** |
| Corn HHV | Btu/lb | | | 8.00E+03 | 8.00E+03 | 8.00E+03 | 8.00E+03 | 8.00E+03 |
| | MMBtu/ton | | | 1.60E+01 | 1.60E+01 | 1.60E+01 | 1.60E+01 | 1.60E+01 |
| **Capital Cost (million USD)** | | | | | | | | |
| Feed Handling | | | 0.70 | 1.95 | 2.09 | 3.40 | 5.52 | 8.97 |
| Saccharification | | | 0.70 | 1.61 | 1.73 | 2.82 | 4.58 | 7.44 |
| Fermentation | | | 0.70 | 3.53 | 3.80 | 6.17 | 10.02 | 16.28 |
| Distillation | | | 0.70 | 3.91 | 4.20 | 6.83 | 11.10 | 18.03 |
| Solid/Syrup Separation | | | 0.70 | 8.63 | 9.27 | 15.06 | 24.47 | 39.76 |
| Storage/Load out | | | 0.70 | 1.09 | 1.17 | 1.90 | 3.09 | 5.01 |
| Wastewater Treatment | | | 0.70 | 0.52 | 0.56 | 0.91 | 1.48 | 2.40 |
| Air compressor | | | 0.70 | 0.10 | 0.11 | 0.17 | 0.28 | 0.46 |
| Steam Gen & Cooling Water | | | 0.70 | 1.37 | 1.47 | 2.39 | 3.88 | 6.30 |
| **Direct Fixed Capital (DFC), also called TIC** | | | | 22.72 | 24.40 | 39.63 | 64.42 | 104.65 |
| Engineering | DFC × MF | 0.12 | | 2.73 | 2.93 | 4.76 | 7.73 | 12.56 |
| Construction | DFC × MF | 0.13 | | 2.95 | 3.17 | 5.15 | 8.37 | 13.60 |
| Contractor & Legal | DFC ×MF | 0.08 | | 1.82 | 1.95 | 3.17 | 5.15 | 8.37 |
| Process/Project Contingency | DFC × MF | 0.0412 | | 0.94 | 1.01 | 1.63 | 2.65 | 4.31 |
| **Total Plant Cost (TPC)** | | | | 31.15 | 33.45 | 54.34 | 88.33 | 143.49 |
| AFUDC | | | | | | | | |
| **Total Plant Investment (TPI)** | | | | 31.15 | 33.45 | 54.34 | 88.33 | 143.49 |
| Land | | | 0.60 | 2.21 | 2.21 | 3.35 | 5.08 | 10.16 |

**Table 22. (Continued)**

| Country | Argentina | Code= | 5 | | | | | |
|---|---|---|---|---|---|---|---|---|
| **Cost component** | **Units** | **Cost Factor** | **Scale Factor** | **USA Dry Mill Ethanol** | **Argentina Dry Mill Ethanol** | **Argentina Dry Mill Ethanol** | **Argentina Dry Mill Ethanol** | **Argentina Dry Mill Ethanol** |
| **Year $** | **$** | | | **2005** | **2005** | **2005** | **2005** | **2005** |
| **Startup** | | | | **0** | **0** | **0.00** | **0.00** | **0.00** |
| **Total Capital Cost (TCC)** | | | | 33.36 | 35.66 | 57.69 | 93.41 | 153.65 |
| **Contingency/TPI** | | | | | | | | |
| Working Capital | DFC × MF | 0.05 | | 1.56 | 1.67 | 2.72 | 4.42 | 7.17 |
| **Variable Operating Costs (million USD/yr)** | | | | | | | | |
| Feed | | | | 30.67 | 31.29 | 62.57 | 125.24 | 250.49 |
| Nat Gas, GJ/y | 8.51E+05 | 6.33 | USD/GJ | 5.39 | 5.39 | 10.78 | 21.57 | 43.14 |
| Electricity, kWh/y | 2.01E+07 | 0.05 | $/kWh | 1.01 | 1.01 | 2.01 | 4.02 | 8.05 |
| Catalysts and Chemicals, Misc | | | | 1.79 | 1.79 | 3.58 | 7.17 | 14.33 |
| **Total** | | | | 38.86 | 39.47 | 78.94 | 158.01 | 316.01 |
| Fixed Operating Costs (million USD/yr) | | | | | | | | |
| Labor (SF from V-R) | DFC x MF | 0.0255 | 0.25 | 0.58 | 0.49 | 0.58 | 0.70 | 0.83 |
| Maintenance (3% of A) | | | 0.03 | 0.68 | 0.73 | 1.19 | 1.93 | 3.14 |
| General Overhead (65% of labor + maint) | | | 0.65 | 0.82 | 0.80 | 1.15 | 1.71 | 2.58 |
| Direct Overhead (45% of Labor) | | | 0.45 | 0.26 | 0.22 | 0.26 | 0.31 | 0.37 |
| Insurance (0.5% of TIC) | | | 0.005 | 0.17 | 0.18 | 0.29 | 0.47 | 0.77 |
| **Total** | | | | 2.51 | 2.42 | 3.48 | 5.12 | 7.68 |

## Table 23. Corn dry mill ethanol: Brazil capital and operating costs

| Country | Brazil | Code = | 2 | | | | | |
|---|---|---|---|---|---|---|---|---|
| **Cost component** | **Units** | **Cost Factor** | **Scale Factor** | **USA Dry Mill Ethanol** | **Brazil Dry Mill Ethanol** | **Brazil Dry Mill Ethanol** | **Brazil Dry Mill Ethanol** | **Brazil Dry Mill Ethanol** |
| **Year $** | **$** | | | **2005** | **2005** | **2005** | **2005** | **2005** |
| Cost Index (2 = M&S, 3 = CE) | | | 3 | 130.93 | 130.93 | 130.93 | 130.93 | 130.93 |
| Plant Size | dry short ton/day | | | 708 | 708 | 1,415 | 2,833 | 5,666 |
| | dry tonne/day | | | 642 | 642 | 1,284 | 2,570 | 5,140 |
| Stream Factor | % | | | 95% | 95% | 95% | 95% | 95% |
| Feed | Dry short ton/yr | | | 2.454E+05 | 2.454E+05 | 4.908E+05 | 9.823E+05 | 1.965E+06 |
| | GJ/short ton | | | 1.688E+01 | 1.688E+01 | 1.688E+01 | 1.688E+01 | 1.688E+01 |
| | GJ/yr | | | 4.142E+06 | 4.142E+06 | 8.284E+06 | 1.658E+07 | 3.316E+07 |
| Feed Cost | USD/dt | | | $ 125.00 | $ 134.29 | $ 134.29 | $ 134.29 | $ 134.29 |
| Yield (gal/Dry US Ton) | | | | | | | | |
| Ethanol | gal/short ton | | | 101.8 | 101.8 | 101.8 | 101.8 | 101.8 |
| DDGs | dry ton/ton | | | 0.33 | 0.33 | 0.33 | 0.33 | 0.33 |
| Process Efficiency - to ethanol | % HHV | | | 53.5% | 53.5% | 53.5% | 53.5% | 53.5% |
| Process efficiency - overall | % HHV | | | | | | | |
| Ethanol | gal/yr | | | 2.498E+07 | 2.498E+07 | 4.996E+07 | 1.000E+08 | 2.000E+08 |
| | Mil Gal/YR | | | **25.0** | **25.0** | **50.0** | **100.0** | **200.0** |
| | tpy | | | 8.281E+04 | 8.281E+04 | 1.656E+05 | 3.315E+05 | 6.630E+05 |
| | GJ/yr | | | 2.216E+06 | 2.216E+06 | 4.433E+06 | 8.872E+06 | 1.774E+07 |
| | gal/Stream day | | | 7.204E+04 | 7.204E+04 | 1.441E+05 | 2.884E+05 | 5.768E+05 |
| | bbl/s stream day | | | 1.715E+03 | 1.715E+03 | 3.431E+03 | 6.866E+03 | 1.373E+04 |
| DDGs | dry ton/yr | | | 8.098E+04 | 8.098E+04 | 1.620E+05 | 3.242E+05 | 6.483E+05 |

**Table 23. (Continued)**

| Country | Brazil | Code = | 2 | | | | | |
|---|---|---|---|---|---|---|---|---|
| **Cost component** | **Units** | **Cost Factor** | **Scale Factor** | **USA Dry Mill Ethanol** | **Brazil Dry Mill Ethanol** | **Brazil Dry Mill Ethanol** | **Brazil Dry Mill Ethanol** | **Brazil Dry Mill Ethanol** |
| **Year $** | **$** | | | **2005** | **2005** | **2005** | **2005** | **2005** |
| Corn density | lb/bu | | | 56.0 | 56.0 | 56.0 | 56.0 | 56.0 |
| Corn | $/bu | | | **3.50** | **3.76** | **3.76** | **3.76** | **3.76** |
| Corn HHV | Btu/lb | | | 8.00E+03 | 8.00E+03 | 8.00E+03 | 8.00E+03 | 8.00E+03 |
| | MMBtu/ton | | | 1.60E+01 | 1.60E+01 | 1.60E+01 | 1.60E+01 | 1.60E+01 |
| **Capital Cost (million USD)** | | | | | | | | |
| Feed Handling | | | 0.70 | 1.95 | 1.99 | 3.23 | 5.26 | 8.54 |
| Saccharification | | | 0.70 | 1.61 | 1.65 | 2.68 | 4.36 | 7.08 |
| Fermentation | | | 0.70 | 3.53 | 3.61 | 5.87 | 9.54 | 15.49 |
| Distillation | | | 0.70 | 3.91 | 4.00 | 6.50 | 10.56 | 17.16 |
| Solid/Syrup Separation | | | 0.70 | 8.63 | 8.82 | 14.33 | 23.29 | 37.84 |
| Storage/Load out | | | 0.70 | 1.09 | 1.11 | 1.81 | 2.94 | 4.77 |
| Wastewater Treatment | | | 0.70 | 0.52 | 0.53 | 0.86 | 1.40 | 2.28 |
| Air compressor | | | 0.70 | 0.10 | 0.10 | 0.16 | 0.27 | 0.43 |
| Steam Gen & Cooling Water | | | 0.70 | 1.37 | 1.40 | 2.27 | 3.69 | 6.00 |
| **Direct Fixed Capital (DFC), also called TIC** | | | | 22.72 | 23.22 | 37.72 | 61.31 | 99.59 |
| Engineering | DFC × MF | 0.12 | | 2.73 | 2.79 | 4.53 | 7.36 | 11.95 |
| Construction | DFC × MF | 0.13 | | 2.95 | 3.02 | 4.90 | 7.97 | 12.95 |
| Contractor & Legal | DFC × MF | 0.08 | | 1.82 | 1.86 | 3.02 | 4.90 | 7.97 |
| Process/Project Contingency | DFC × MF | 0.0412 | | 0.94 | 0.96 | 1.55 | 2.53 | 4.10 |
| **Total Plant Cost (TPC)** | | | | 31.15 | 31.84 | 51.72 | 84.06 | 136.56 |
| AFUDC | | | | | | | | |
| **Total Plant Investment (TPI)** | | | | 31.15 | 31.84 | 51.72 | 84.06 | 136.56 |
| Land | | | 0.60 | 2.21 | 2.21 | 3.35 | 5.08 | 10.16 |

**Table 23. (Continued)**

| Country | Brazil | Code = | 2 | | | | | |
|---|---|---|---|---|---|---|---|---|
| **Cost component** | Units | Cost Factor | Scale Factor | **USA Dry Mill Ethanol** | **Brazil Dry Mill Ethanol** | **Brazil Dry Mill Ethanol** | **Brazil Dry Mill Ethanol** | **Brazil Dry Mill Ethanol** |
| **Year $** | **$** | | | **2005** | **2005** | **2005** | **2005** | **2005** |
| **Startup** | | | | **0** | **0** | **0.00** | **0.00** | **0.00** |
| **Total Capital Cost (TCC)** | | | | 33.36 | 34.05 | 55.07 | 89.14 | 146.72 |
| Contingency/TPI | | | | | | | | |
| Working Capital | DFC × MF | 0.05 | | 1.56 | 1.59 | 2.59 | 4.20 | 6.83 |
| **Variable Operating Costs (million USD/yr)** | | | | | | | | |
| Feed | | | | 30.67 | 32.95 | 65.90 | 131.91 | 263.82 |
| Nat Gas, GJ/y | 8.51E+05 | 6.33 | USD/GJ | 5.39 | 5.39 | 10.78 | 21.57 | 43.14 |
| Electricity, kWh/y | 2.01E+07 | 0.05 | $/kWh | 1.01 | 1.01 | 2.01 | 4.02 | 8.05 |
| Catalysts and Chemicals, Misc | | | | 1.79 | 1.79 | 3.58 | 7.17 | 14.33 |
| **Total** | | | | 38.86 | 41.14 | 82.27 | 164.67 | 329.34 |
| Fixed Operating Costs (million USD/yr) | | | | | | | | |
| Labor (SF from V-R) | DFC × MF | 0.0255 | 0.25 | 0.58 | 0.25 | 0.29 | 0.35 | 0.42 |
| Maintenance (3% of A) | | | 0.03 | 0.68 | 0.70 | 1.13 | 1.84 | 2.99 |
| General Overhead (65% of labor + maint) | | | 0.65 | 0.82 | 0.61 | 0.93 | 1.42 | 2.21 |
| Direct Overhead (45% of Labor) | | | 0.45 | 0.26 | 0.11 | 0.13 | 0.16 | 0.19 |
| Insurance (0.5% of TIC) | | | 0.005 | 0.17 | 0.17 | 0.28 | 0.45 | 0.73 |
| **Total** | | | | 2.51 | 1.84 | 2.76 | 4.21 | 6.54 |

**Table 24. Corn dry mill ethanol: Canada capital and operating costs**

| Country | Canada | Code = | 7 | | | | | |
|---|---|---|---|---|---|---|---|---|
| **Cost component** | **Units** | **Cost Factor** | **Scale Factor** | **USA Dry Mill Ethanol** | **Canada Dry Mill Ethanol** | **Canada Dry Mill Ethanol** | **Canada Dry Mill Ethanol** | **Canada Dry Mill Ethanol** |
| **Year $** | **$** | | | **2005** | **2005** | **2005** | **2005** | **2005** |
| Cost Index (2 = M&S, 3 = CE) | | | 3 | 130.93 | 130.93 | 130.93 | 130.93 | 130.93 |
| Plant Size | dry short ton/day | | | 708 | 708 | 1,415 | 2,833 | 5,666 |
| | dry tonne/day | | | 642 | 642 | 1,284 | 2,570 | 5,140 |
| Stream Factor | % | | | 95% | 95% | 95% | 95% | 95% |
| Feed | Dry short ton/yr | | | 2.454E+05 | 2.454E+05 | 4.908E+05 | 9.823E+05 | 1.965E+06 |
| | GJ/short ton | | | 1.688E+01 | 1.688E+01 | 1.688E+01 | 1.688E+01 | 1.688E+01 |
| | GJ/yr | | | 4.142E+06 | 4.142E+06 | 8.284E+06 | 1.658E+07 | 3.316E+07 |
| Feed Cost | USD/dt | | | $ 125.00 | $ 80.00 | $ 80.00 | $ 80.00 | $ 80.00 |
| Yield (gal/Dry US Ton) | | | | | | | | |
| Ethanol | gal/short ton | | | 101.8 | 101.8 | 101.8 | 101.8 | 101.8 |
| DDGs | dry ton/ton | | | 0.33 | 0.33 | 0.33 | 0.33 | 0.33 |
| Process Efficiency - to ethanol | % HHV | | | 53.5% | 53.5% | 53.5% | 53.5% | 53.5% |
| Process efficiency - overall | % HHV | | | | | | | |
| Ethanol | gal/yr | | | 2.498E+07 | 2.498E+07 | 4.996E+07 | 1.000E+08 | 2.000E+08 |
| | Mil Gal/YR | | | **25.0** | **25.0** | **50.0** | **100.0** | **200.0** |
| | tpy | | | 8.281E+04 | 8.281E+04 | 1.656E+05 | 3.315E+05 | 6.630E+05 |
| | GJ/yr | | | 2.216E+06 | 2.216E+06 | 4.433E+06 | 8.872E+06 | 1.774E+07 |
| | gal/Stream day | | | 7.204E+04 | 7.204E+04 | 1.441E+05 | 2.884E+05 | 5.768E+05 |
| | bbl/s stream day | | | 1.715E+03 | 1.715E+03 | 3.431E+03 | 6.866E+03 | 1.373E+04 |

**Table 24. (Continued)**

| Country | Canada | Code = | 7 | | | | | |
|---|---|---|---|---|---|---|---|---|
| **Cost component** | **Units** | **Cost Factor** | **Scale Factor** | **USA Dry Mill Ethanol** | **Canada Dry Mill Ethanol** | **Canada Dry Mill Ethanol** | **Canada Dry Mill Ethanol** | **Canada Dry Mill Ethanol** |
| **Year $** | **$** | | | **2005** | **2005** | **2005** | **2005** | **2005** |
| DDGs | dry ton/yr | | | 8.098E+04 | 8.098E+04 | 1.620E+05 | 3.242E+05 | 6.483E+05 |
| Corn density | lb/bu | | | 56.0 | 56.0 | 56.0 | 56.0 | 56.0 |
| Corn | $/bu | | | 3.50 | 2.24 | 2.24 | 2.24 | 2.24 |
| Corn HHV | Btu/lb | | | 8.00E+03 | 8.00E+03 | 8.00E+03 | 8.00E+03 | 8.00E+03 |
| | MMBtu/ton | | | 1.60E+01 | 1.60E+01 | 1.60E+01 | 1.60E+01 | 1.60E+01 |
| **Capital Cost (million USD)** | | | | | | | | |
| Feed Handling | | | 0.70 | 1.95 | 2.54 | 4.12 | 6.70 | 10.89 |
| Saccharification | | | 0.70 | 1.61 | 2.10 | 3.42 | 5.56 | 9.02 |
| Fermentation | | | 0.70 | 3.53 | 4.61 | 7.48 | 12.16 | 19.75 |
| Distillation | | | 0.70 | 3.91 | 5.10 | 8.29 | 13.47 | 21.88 |
| Solid/Syrup Separation | | | 0.70 | 8.63 | 11.25 | 18.27 | 29.70 | 48.24 |
| Storage/Load out | | | 0.70 | 1.09 | 1.42 | 2.30 | 3.75 | 6.09 |
| Wastewater Treatment | | | 0.70 | 0.52 | 0.68 | 1.10 | 1.79 | 2.91 |
| Air compressor | | | 0.70 | 0.10 | 0.13 | 0.21 | 0.34 | 0.55 |
| Steam Gen & Cooling Water | | | 0.70 | 1.37 | 1.78 | 2.90 | 4.71 | 7.65 |
| **Direct Fixed Capital (DFC), also called TIC** | | | | 22.72 | 29.60 | 48.09 | 78.17 | 126.98 |
| Engineering | DFC × MF | 0.12 | | 2.73 | 3.55 | 5.77 | 9.38 | 15.24 |
| Construction | DFC × MF | 0.13 | | 2.95 | 3.85 | 6.25 | 10.16 | 16.51 |
| Contractor & Legal | DFC × MF | 0.08 | | 1.82 | 2.37 | 3.85 | 6.25 | 10.16 |
| Process/Project Contingency | DFC × MF | 0.0412 | | 0.94 | 1.22 | 1.98 | 3.22 | 5.23 |
| **Total Plant Cost (TPC)** | | | | 31.15 | 40.59 | 65.94 | 107.18 | 174.12 |
| AFUDC | | | | | | | | |
| **Total Plant Investment (TPI)** | | | | 31.15 | 40.59 | 65.94 | 107.18 | 174.12 |
| Land | | | 0.60 | 2.21 | 2.21 | 3.35 | 5.08 | 10.16 |

**Table 24. (Continued)**

| Country | Canada | Code = | 7 | | | | | |
|---|---|---|---|---|---|---|---|---|
| **Cost component** | **Units** | **Cost Factor** | **Scale Factor** | **USA Dry Mill Ethanol** | **Canada Dry Mill Ethanol** | **Canada Dry Mill Ethanol** | **Canada Dry Mill Ethanol** | **Canada Dry Mill Ethanol** |
| **Year $** | **$** | | | **2005** | **2005** | **2005** | **2005** | **2005** |
| **Startup** | | | | **0** | **0** | **0.00** | **0.00** | **0.00** |
| **Total Capital Cost (TCC)** | | | | 33.36 | 42.80 | 69.29 | 112.26 | 184.28 |
| **Contingency/TPI** | | | | | | | | |
| Working Capital | DFC × MF | 0.05 | | 1.56 | 2.03 | 3.30 | 5.36 | 8.71 |
| **Variable Operating Costs (million USD/yr)** | | | | | | | | |
| Feed | | | | 30.67 | 19.63 | 39.26 | 78.58 | 157.17 |
| Nat Gas, GJ/y | 8.51E+05 | 6.33 | USD/GJ | 5.39 | 5.39 | 10.78 | 21.57 | 43.14 |
| Electricity, kWh/y | 2.01E+07 | 0.05 | $/kWh | 1.01 | 1.01 | 2.01 | 4.02 | 8.05 |
| Catalysts and Chemicals, Misc | | | | 1.79 | 1.79 | 3.58 | 7.17 | 14.33 |
| **Total** | | | | 38.86 | 27.81 | 55.63 | 111.35 | 222.69 |
| Fixed Operating Costs (million USD/yr) | | | | | | | | |
| Labor (SF from V-R) | DFC x MF | 0.0255 | 0.25 | 0.58 | 0.97 | 1.16 | 1.38 | 1.64 |
| Maintenance (3% of A) | | | 0.03 | 0.68 | 0.89 | 1.44 | 2.35 | 3.81 |
| General Overhead (65% of labor + maint) | | | 0.65 | 0.82 | 1.21 | 1.69 | 2.42 | 3.54 |
| Direct Overhead (45% of Labor) | | | 0.45 | 0.26 | 0.44 | 0.52 | 0.62 | 0.74 |
| Insurance (0.5% of TIC) | | | 0.005 | 0.17 | 0.21 | 0.35 | 0.56 | 0.92 |
| **Total** | | | | 2.51 | 3.72 | 5.16 | 7.32 | 10.64 |

**Table 25. Corn dry mill ethanol: China capital and operating costs**

| Country | China | Code = | 3 | | | | | |
|---|---|---|---|---|---|---|---|---|
| **Cost component** | **Units** | **Cost Factor** | **Scale Factor** | **USA Dry Mill Ethanol** | **China Dry Mill Ethanol** | **China Dry Mill Ethanol** | **China Dry Mill Ethanol** | **China Dry Mill Ethanol** |
| **Year $** | **$** | | | **2005** | **2005** | **2005** | **2005** | **2005** |
| Cost Index (2 = M&S, 3 = CE) | | | 3 | 130.93 | 130.93 | 130.93 | 130.93 | 130.93 |
| Plant Size | dry short ton/day | | | 708 | 708 | 1,415 | 2,833 | 5,666 |
| | dry tonne/day | | | 642 | 642 | 1,284 | 2,570 | 5,140 |
| Stream Factor | % | | | 95% | 95% | 95% | 95% | 95% |
| Feed | Dry short ton/yr | | | 2.454E+05 | 2.454E+05 | 4.908E+05 | 9.823E+05 | 1.965E+06 |
| | GJ/short ton | | | 1.688E+01 | 1.688E+01 | 1.688E+01 | 1.688E+01 | 1.688E+01 |
| | GJ/yr | | | 4.142E+06 | 4.142E+06 | 8.284E+06 | 1.658E+07 | 3.316E+07 |
| Feed Cost | USD/dt | | | $ 125.00 | $ 126.43 | $ 126.43 | $ 126.43 | $ 126.43 |
| Yield (gal/Dry US Ton) | | | | | | | | |
| Ethanol | gal/short ton | | | 101.8 | 101.8 | 101.8 | 101.8 | 101.8 |
| DDGs | dry ton/ton | | | 0.33 | 0.33 | 0.33 | 0.33 | 0.33 |
| Process Efficiency - to ethanol | % HHV | | | 53.5% | 53.5% | 53.5% | 53.5% | 53.5% |
| Process efficiency - overall | % HHV | | | | | | | |
| Ethanol | gal/yr | | | 2.498E+07 | 2.498E+07 | 4.996E+07 | 1.000E+08 | 2.000E+08 |
| | Mil Gal/YR | | | **25.0** | **25.0** | **50.0** | **100.0** | **200.0** |
| | tpy | | | 8.281E+04 | 8.281E+04 | 1.656E+05 | 3.315E+05 | 6.630E+05 |
| | GJ/yr | | | 2.216E+06 | 2.216E+06 | 4.433E+06 | 8.872E+06 | 1.774E+07 |
| | gal/Stream day | | | 7.204E+04 | 7.204E+04 | 1.441E+05 | 2.884E+05 | 5.768E+05 |
| | bbl/s stream day | | | 1.715E+03 | 1.715E+03 | 3.431E+03 | 6.866E+03 | 1.373E+04 |

**Table 25. (Continued)**

| Country | China | Code = | 3 | | | | | |
|---|---|---|---|---|---|---|---|---|
| **Cost component** | **Units** | **Cost Factor** | **Scale Factor** | **USA Dry Mill Ethanol** | **China Dry Mill Ethanol** | **China Dry Mill Ethanol** | **China Dry Mill Ethanol** | **China Dry Mill Ethanol** |
| **Year $** | **$** | | | **2005** | **2005** | **2005** | **2005** | **2005** |
| DDGs | dry ton/yr | | | 8.098E+04 | 8.098E+04 | 1.620E+05 | 3.242E+05 | 6.483E+05 |
| Corn density | lb/bu | | | 56.0 | 56.0 | 56.0 | 56.0 | 56.0 |
| Corn | $/bu | | | 3.50 | 3.54 | 3.54 | 3.54 | 3.54 |
| Corn HHV | Btu/lb | | | 8.00E+03 | 8.00E+03 | 8.00E+03 | 8.00E+03 | 8.00E+03 |
| | MMBtu/ton | | | 1.60E+01 | 1.60E+01 | 1.60E+01 | 1.60E+01 | 1.60E+01 |
| **Capital Cost (million USD)** | | | | | | | | |
| Feed Handling | | | 0.70 | 1.95 | 2.14 | 3.48 | 5.66 | 9.19 |
| Saccharification | | | 0.70 | 1.61 | 1.78 | 2.89 | 4.69 | 7.62 |
| Fermentation | | | 0.70 | 3.53 | 3.89 | 6.32 | 10.27 | 16.68 |
| Distillation | | | 0.70 | 3.91 | 4.31 | 7.00 | 11.37 | 18.48 |
| Solid/Syrup Separation | | | 0.70 | 8.63 | 9.50 | 15.43 | 25.08 | 40.75 |
| Storage/Load out | | | 0.70 | 1.09 | 1.20 | 1.95 | 3.16 | 5.14 |
| Wastewater Treatment | | | 0.70 | 0.52 | 0.57 | 0.93 | 1.51 | 2.46 |
| Air compressor | | | 0.70 | 0.10 | 0.11 | 0.18 | 0.29 | 0.47 |
| Steam Gen & Cooling Water | | | 0.70 | 1.37 | 1.51 | 2.45 | 3.98 | 6.46 |
| **Direct Fixed Capital (DFC), also called TIC** | | | | 22.72 | 25.00 | 40.62 | 66.02 | 107.25 |
| Engineering | DFC × MF | 0.12 | | 2.73 | 3.00 | 4.87 | 7.92 | 12.87 |
| Construction | DFC × MF | 0.13 | | 2.95 | 3.25 | 5.28 | 8.58 | 13.94 |
| Contractor & Legal | DFC × MF | 0.08 | | 1.82 | 2.00 | 3.25 | 5.28 | 8.58 |
| Process/Project Contingency | DFC × MF | 0.0412 | | 0.94 | 1.03 | 1.67 | 2.72 | 4.42 |
| **Total Plant Cost (TPC)** | | | | 31.15 | 34.28 | 55.69 | 90.52 | 147.06 |
| AFUDC | | | | | | | | |
| **Total Plant Investment (TPI)** | | | | 31.15 | 34.28 | 55.69 | 90.52 | 147.06 |
| Land | | | 0.60 | 2.21 | 2.21 | 3.35 | 5.08 | 10.16 |

**Table 25. (Continued)**

| Country | China | Code = | 3 | | | | | |
|---|---|---|---|---|---|---|---|---|
| **Cost component** | **Units** | **Cost Factor** | **Scale Factor** | **USA Dry Mill Ethanol** | **China Dry Mill Ethanol** | **China Dry Mill Ethanol** | **China Dry Mill Ethanol** | **China Dry Mill Ethanol** |
| **Year $** | **$** | | | **2005** | **2005** | **2005** | **2005** | **2005** |
| **Startup** | | | | **0** | **0** | **0.00** | **0.00** | **0.00** |
| **Total Capital Cost (TCC)** | | | | 33.36 | 36.49 | 59.04 | 95.60 | 157.22 |
| **Contingency/TPI** | | | | | | | | |
| Working Capital | DFC × MF | 0.05 | | 1.56 | 1.71 | 2.78 | 4.53 | 7.35 |
| **Variable Operating Costs (million USD/yr)** | | | | | | | | |
| Feed | | | | 30.67 | 31.02 | 62.05 | 124.19 | 248.38 |
| Nat Gas, GJ/y | 8.51E+05 | 6.33 | USD/GJ | 5.39 | 5.39 | 10.78 | 21.57 | 43.14 |
| Electricity, kWh/y | 2.01E+07 | 0.05 | $/kWh | 1.01 | 1.01 | 2.01 | 4.02 | 8.05 |
| Catalysts and Chemicals, Misc | | | | 1.79 | 1.79 | 3.58 | 7.17 | 14.33 |
| **Total** | | | | 38.86 | 39.21 | 78.42 | 156.95 | 313.91 |
| Fixed Operating Costs (million USD/yr) | | | | | | | | |
| Labor (SF from V-R) | DFC × MF | 0.0255 | 0.25 | 0.58 | 0.64 | 0.77 | 0.91 | 1.08 |
| Maintenance (3% of A) | | | 0.03 | 0.68 | 0.75 | 1.22 | 1.98 | 3.22 |
| General Overhead (65% of labor + maint) | | | 0.65 | 0.82 | 0.91 | 1.29 | 1.88 | 2.80 |
| Direct Overhead (45% of Labor) | | | 0.45 | 0.26 | 0.29 | 0.34 | 0.41 | 0.49 |
| Insurance (0.5% of TIC) | | | 0.005 | 0.17 | 0.18 | 0.30 | 0.48 | 0.79 |
| **Total** | | | | 2.51 | 2.77 | 3.91 | 5.66 | 8.37 |

## Table 26. Corn dry-mill ethanol: Mexico capital and operating costs

| Country | Mexico | Code = | 8 | | | | | |
|---|---|---|---|---|---|---|---|---|
| **Cost component** | **Units** | **Cost Factor** | **Scale Factor** | **USA Dry Mill Ethanol** | **Mexico Dry Mill Ethanol** | **Mexico Dry Mill Ethanol** | **Mexico Dry Mill Ethanol** | **Mexico Dry Mill Ethanol** |
| **Year $** | **$** | | | **2005** | **2005** | **2005** | **2005** | **2005** |
| Cost Index (2 = M&S, 3 = CE) | | | 3 | 130.93 | 130.93 | 130.93 | 130.93 | 130.93 |
| Plant Size | dry short ton/day | | | 708 | 708 | 1,415 | 2,833 | 5,666 |
| | dry tonne/day | | | 642 | 642 | 1,284 | 2,570 | 5,140 |
| Stream Factor | % | | | 95% | 95% | 95% | 95% | 95% |
| Feed | Dry short ton/yr | | | 2.454E+05 | 2.454E+05 | 4.908E+05 | 9.823E+05 | 1.965E+06 |
| | GJ/short ton | | | 1.688E+01 | 1.688E+01 | 1.688E+01 | 1.688E+01 | 1.688E+01 |
| | GJ/yr | | | 4.142E+06 | 4.142E+06 | 8.284E+06 | 1.658E+07 | 3.316E+07 |
| Feed Cost | USD/dt | | | $ 125.00 | $ 113.57 | $ 113.57 | $ 113.57 | $ 113.57 |
| Yield (gal/Dry US Ton) | | | | | | | | |
| Ethanol | gal/short ton | | | 101.8 | 101.8 | 101.8 | 101.8 | 101.8 |
| DDGs | dry ton/ton | | | 0.45 | 0.45 | 0.45 | 0.45 | 0.45 |
| Process Efficiency - to ethanol | % HHV | | | 53.5% | 53.5% | 53.5% | 53.5% | 53.5% |
| Process efficiency - overall | % HHV | | | | | | | |
| Ethanol | gal/yr | | | 2.498E+07 | 2.498E+07 | 4.996E+07 | 1.000E+08 | 2.000E+08 |
| | Mil Gal/YR | | | **25.0** | **25.0** | **50.0** | **100.0** | **200.0** |
| | tpy | | | 8.281E+04 | 8.281E+04 | 1.656E+05 | 3.315E+05 | 6.630E+05 |
| | GJ/yr | | | 2.216E+06 | 2.216E+06 | 4.433E+06 | 8.872E+06 | 1.774E+07 |
| | gal/Stream day | | | 7.204E+04 | 7.204E+04 | 1.441E+05 | 2.884E+05 | 5.768E+05 |
| | bbl/s stream day | | | 1.715E+03 | 1.715E+03 | 3.431E+03 | 6.866E+03 | 1.373E+04 |
| DDGs | dry ton/yr | | | 8.559E+04 | 8.559E+04 | 1.712E+05 | 3.426E+05 | 6.852E+05 |
| Corn density | lb/bu | | | 56.0 | 56.0 | 56.0 | 56.0 | 56.0 |
| Corn | $/bu | | | **3.50** | **3.18** | **3.18** | **3.18** | **3.18** |
| Corn HHV | Btu/lb | | | 8.00E+03 | 8.00E+03 | 8.00E+03 | 8.00E+03 | 8.00E+03 |
| | MMBtu/ton | | | 1.60E+01 | 1.60E+01 | 1.60E+01 | 1.60E+01 | 1.60E+01 |
| **Capital Cost (million USD)** | | | | | | | | |
| Feed Handling | | | 0.70 | 1.95 | 1.57 | 2.55 | 4.14 | 6.72 |

Table 26. (Continued)

| Country | Mexico | Code = | 8 | | | | | |
|---|---|---|---|---|---|---|---|---|
| **Cost component** | **Units** | **Cost Factor** | **Scale Factor** | **USA Dry Mill Ethanol** | **Mexico Dry Mill Ethanol** | **Mexico Dry Mill Ethanol** | **Mexico Dry Mill Ethanol** | **Mexico Dry Mill Ethanol** |
| **Year $** | **$** | | | **2005** | **2005** | **2005** | **2005** | **2005** |
| Saccharification | | | 0.70 | 1.61 | 1.30 | 2.11 | 3.43 | 5.57 |
| Fermentation | | | 0.70 | 3.53 | 2.84 | 4.62 | 7.51 | 12.19 |
| Distillation | | | 0.70 | 3.91 | 3.15 | 5.12 | 8.31 | 13.51 |
| Solid/Syrup Separation | | | 0.70 | 8.63 | 6.94 | 11.28 | 18.33 | 29.78 |
| Storage/Load out | | | 0.70 | 1.09 | 0.88 | 1.42 | 2.31 | 3.76 |
| Wastewater Treatment | | | 0.70 | 0.52 | 0.42 | 0.68 | 1.11 | 1.80 |
| Air compressor | | | 0.70 | 0.10 | 0.08 | 0.13 | 0.21 | 0.34 |
| Steam Gen & Cooling Water | | | 0.70 | 1.37 | 1.10 | 1.79 | 2.91 | 4.72 |
| **Direct Fixed Capital (DFC), also called TIC** | | | | 22.72 | 18.28 | 29.69 | 48.25 | 78.39 |
| Engineering | DFC × MF | 0.12 | | 2.73 | 2.19 | 3.56 | 5.79 | 9.41 |
| Construction | DFC × MF | 0.13 | | 2.95 | 2.38 | 3.86 | 6.27 | 10.19 |
| Contractor & Legal | DFC × MF | 0.08 | | 1.82 | 1.46 | 2.38 | 3.86 | 6.27 |
| Process/Project Contingency | DFC × MF | 0.0412 | | 0.94 | 0.75 | 1.22 | 1.99 | 3.23 |
| **Total Plant Cost (TPC)** | | | | 31.15 | 25.06 | 40.71 | 66.17 | 107.49 |
| AFUDC | | | | | | | | |
| **Total Plant Investment (TPI)** | | | | 31.15 | 25.06 | 40.71 | 66.17 | 107.49 |
| Land | | | 0.60 | 2.21 | 2.21 | 3.35 | 5.08 | 10.16 |
| **Startup** | | | | | | | | |
| **Total Capital Cost (TCC)** | | | | 33.36 | 27.27 | 44.06 | 71.25 | 117.65 |
| **Contingency/TPI** | | | | | | | | |
| Working Capital | DFC × MF | 0.05 | | 1.56 | 1.25 | 2.04 | 3.31 | 5.37 |
| **Variable Operating Costs (million USD/yr)** | | | | | | | | |
| Feed | | | | 30.67 | 27.87 | 55.74 | 111.56 | 223.13 |
| Nat Gas, GJ/y | 8.51E+05 | 6.33 | USD/GJ | 5.39 | 5.39 | 10.78 | 21.57 | 43.14 |
| Electricity, kWh/y | 2.01E+07 | 0.05 | $/kWh | 1.01 | 1.01 | 2.01 | 4.02 | 8.05 |

**Table 26. (Continued)**

| Country | Mexico | Code = | 8 | | | | | |
|---|---|---|---|---|---|---|---|---|
| **Cost component** | **Units** | **Cost Factor** | **Scale Factor** | **USA Dry Mill Ethanol** | **Mexico Dry Mill Ethanol** | **Mexico Dry Mill Ethanol** | **Mexico Dry Mill Ethanol** | **Mexico Dry Mill Ethanol** |
| **Year $** | **$** | | | **2005** | **2005** | **2005** | **2005** | **2005** |
| Catalysts and Chemicals, Misc | | | | 1.79 | 1.79 | 3.58 | 7.17 | 14.33 |
| **Total** | | | | 38.86 | 36.05 | 72.11 | 144.32 | 288.65 |
| Fixed Operating Costs (million USD/yr) | | | | | | | | |
| Labor (SF from V-R) | DFC x MF | 0.0255 | 0.25 | 0.58 | 0.14 | 0.17 | 0.20 | 0.24 |
| Maintenance (3% of A) | | | 0.03 | 0.68 | 0.55 | 0.89 | 1.45 | 2.35 |
| General Overhead (65% of labor + maint) | | | 0.65 | 0.82 | 0.45 | 0.69 | 1.07 | 1.68 |
| Direct Overhead (45% of Labor) | | | 0.45 | 0.26 | 0.06 | 0.07 | 0.09 | 0.11 |
| Insurance (0.5% of TIC) | | | 0.005 | 0.17 | 0.14 | 0.22 | 0.36 | 0.59 |
| **Total** | | | | 2.51 | 1.33 | 2.04 | 3.16 | 4.96 |
| Capital Cost (TI) | USD/annual gal | | | 0.91 | 0.73 | 0.59 | 0.48 | 0.39 |
| | USD/daily bbl | | | 13246 | 10654 | 8654 | 7028 | 5708 |
| | USD/annual gal EtOH | | | 0.91 | 0.73 | 0.59 | 0.48 | 0.39 |
| | USD/daily bbl EtOH eq | | | 13,246 | 10,654 | 8,654 | 7,028 | 5,708 |
| | USD/daily bbl COE | | | 21,751 | 17,495 | 14,211 | 11,540 | 9,373 |
| Variable Operating Cost | USD/gal | | | 1.56 | 1.44 | 1.44 | 1.44 | 1.44 |
| Fixed Operating Cost | USD/gal | | | 0.10 | 0.05 | 0.04 | 0.03 | 0.02 |
| Feed | USD/gal | | | 1.23 | 1.12 | 1.12 | 1.12 | 1.12 |
| DDG Credit | | | 80 | 0.27 | 0.27 | 0.27 | 0.27 | 0.27 |

The capital and operating costs given in Tables 22 through 26 were used to generate plant gate costs for each country using representative corn prices. The corn price selected was taken from (ORNL 2007), assuming the corn price at 50% of 2017 maximum supply potential for low, base, and high growth scenarios. Corn production in billions of bushels (50% of maximum) is given by Table 27. For comparison, USA corn production in 2006 was 10.5 billion bushels (NCGA 2007).

**Table 27. Country corn production in 2017 (50% of maximum supply)**

| | Low | Base | High |
|---|---|---|---|
| | Billion bushels | | |
| Argentina | 0.303 | 0.484 | 0.917 |
| Brazil | 0.850 | 1.063 | 1.413 |
| Canada | 0.185 | 0.315 | 0.425 |
| China | 3.283 | 3.937 | 4.764 |
| Mexico | 0.425 | 0.528 | 0.655 |

Conversion Factor: 0.0394 bil bu/MMT

Table 28 and Table 29 give the corresponding corn prices at 50% production in USD/short ton, and USD/ bu, respectively.

**Table 28. 2017 Corn prices (50% of maximum supply) in USD/short ton**

| | Low | Base | High |
|---|---|---|---|
| | USD/short ton | | |
| Argentina | 127.50 | 127.50 | 127.50 |
| Brazil | 99.06 | 134.40 | 128.64 |
| Canada | 80.02 | 80.02 | 80.02 |
| China | 126.50 | 126.50 | 126.50 |
| Mexico | 117.70 | 113.60 | 113.60 |

Conversion Factor: 35.71 bu/short ton

**Table 29. 2017 Corn prices (50% of maximum supply) in USD/bushel**

| | Low | Base | High |
|---|---|---|---|
| | USD/bushel | | |
| Argentina | 3.57 | 3.57 | 3.57 |
| Brazil | 2.77 | 3.76 | 3.60 |
| Canada | 2.24 | 2.24 | 2.24 |
| China | 3.54 | 3.54 | 3.54 |
| Mexico | 3.30 | 3.18 | 3.18 |

Conversion Factor: 35.71 bu/short ton

These values were used to calculate plant gate prices. The summary results are given in Table 30 (25 MM GPY) and Table 31. Figure 14 presents the PGP results graphically.

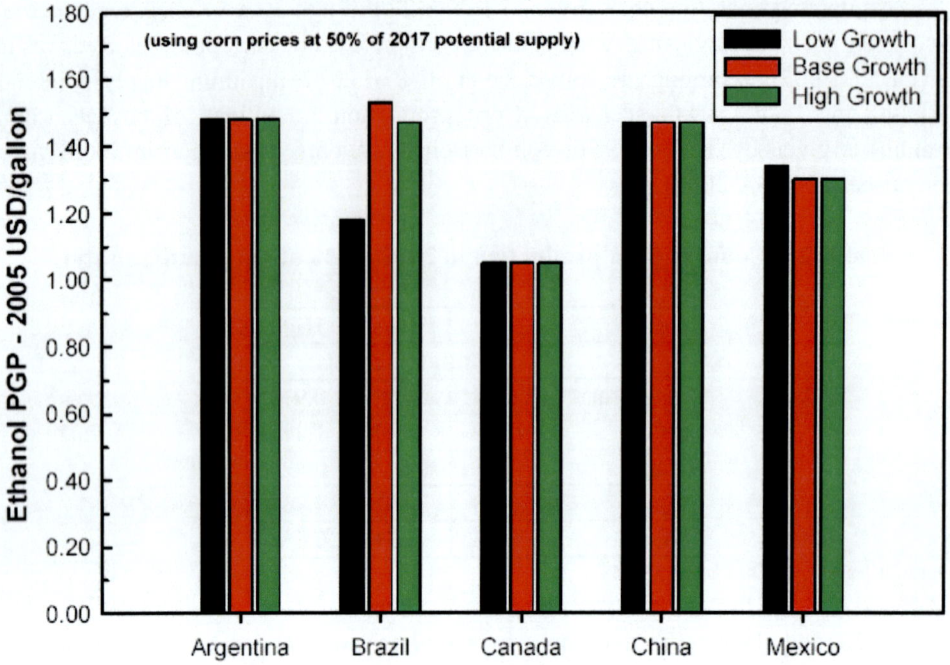


Figure 14. Country 2017 corn dry mill ethanol plant gate prices at 50% of corn supply

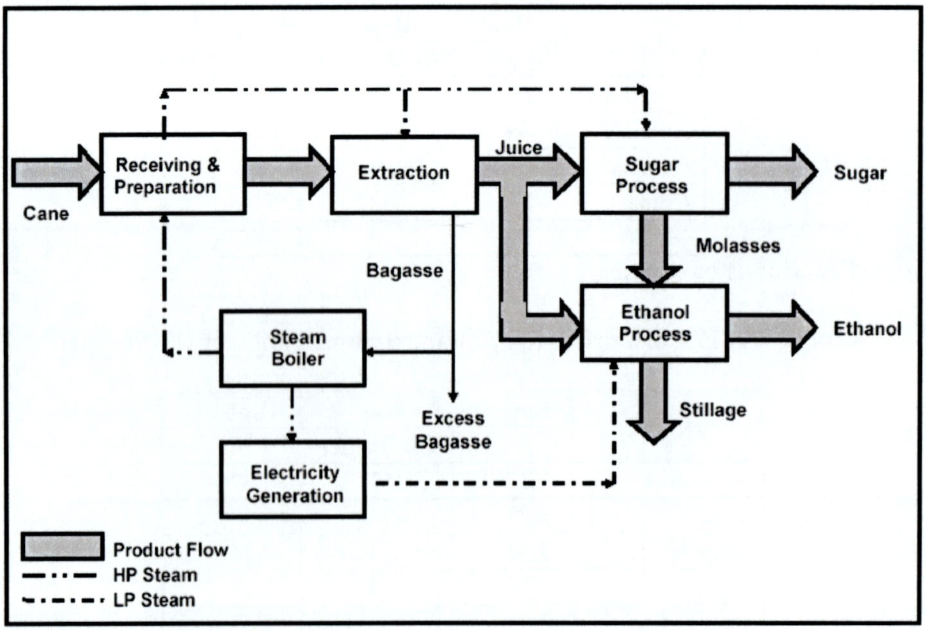


Figure 15. Brazilian sugar mill model (Oliverio and Ribeiro 2006)

# 4. Ethanol – Sugar Cane Mill

A review of the literature and personal communications (Turn 2007) showed no published information for estimates of capital and operating costs of an integrated sugar-ethanol mill in the United States. Therefore, Brazil was chosen as the base case model. A process flow diagram for a prototypical Brazilian mill is shown in Figure 15. The mill operator has the ability to change the amount of sucrose going to sugar and the amount going to ethanol depending upon market demands and the relative prices of sugar and ethanol. A review of the literature on the Brazilian sugar industry and personal communication (Walter 2007) showed no published detailed capital cost estimates for integrated sugar-ethanol mills. Two references (Bohlmann 2006, van den Wall Bake 2006) gave overall capital cost values, but a review of their literature sources showed the data were taken from a non-published report. Therefore, the total installed cost (TIC) data given by Bohlmann were used to generate a capital cost curve, Figure 16, for a Brazilian base case.

The curve fit of the data was used to estimate TIC costs for a base mill and the procedure outlined in the methodology section was then followed to estimate capital and operating costs for the suite of countries. The prototypical mill produces both sugar and ethanol. The financial analysis can be performed two ways. Either a known plant gate price for one of the products can be used and the second calculated, or both can be calculated. In the second case a protocol is needed for the estimate. In this study the energy contents of sugar and ethanol are used to estimate yearly total product energy content; the product price in USD/GJ is estimated using a NPV=0 calculation; and then separate product prices are estimated based on energy content.

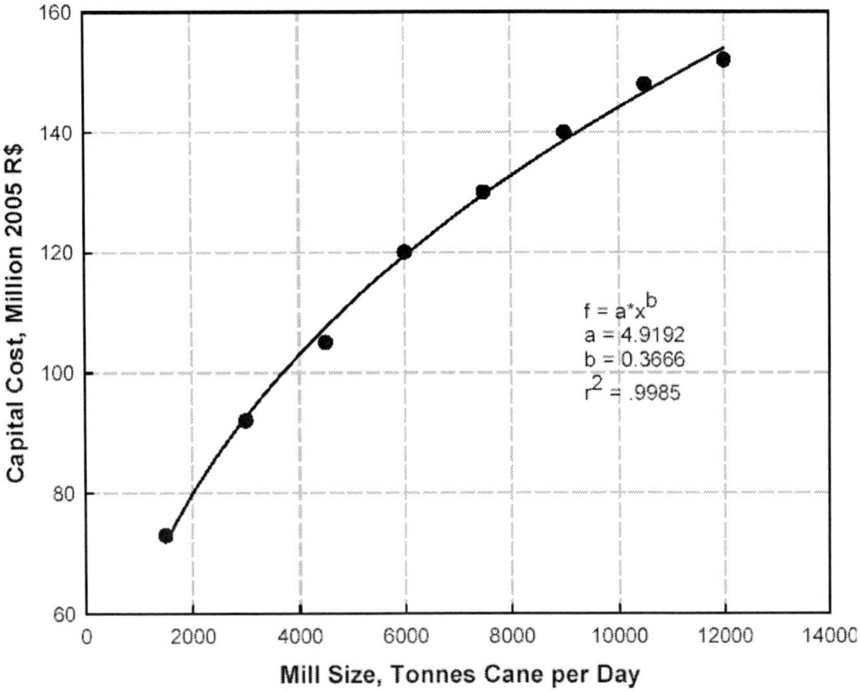


Figure 16. Brazilian sugar mill capital costs, Bohlmann (2006)

A number of operating variables need to be assumed to estimate plant gate prices for sugar and ethanol. Included are cane sugar content, sugar recovery efficiency, ethanol recovery efficiency, ratio of ethanol to sugar production, and plant stream factor. The values selected are those given by Oliverio (2006) for a state-of-the-art Brazilian mill; 97% sugar recovery and 90% sugar to ethanol conversion efficiency. A cane sugar content of 14% (Rosillo-Calle et al. 2000, Bohlmann 2006) and a stream factor of 50% are used for the initial set of comparisons. The ethanol/sugar ratio is set at 60/40, which is about the maximum historical ratio (van den Wall Bake 2006). Additional work is needed to determine capacity factors for individual countries. For example, mill operation in the United States can range from 2 months in some Louisiana mills to 11 months in some Hawaii mills.

The Brazilian base case model was used to evaluate the cost of production versus plant size for the representative feedstock case from the methodology section, and the results are shown in Figure 17. Plant gate costs (10.45 USD/short ton) decrease from 1.27 USD/gallon for a small 4.6 MM GPY plant to 0.84 USD/gallon for a 100 MM GPY plant. As plant size increases, the feedstock cost becomes a larger percentage of plant gate cost, increasing from about 38% for the 4.6 MM GPY plant to 57% for the 100 MM GPY plant.

The ORNL Brazil supply curves were used to estimate plant gate costs as a function of supply for the 2017 low, base, and high growth scenarios. The results are shown in Figure 18. Capital and operating costs estimates for the eight countries are given by Table 32 through Table 39.

Summary cost information for three plant sizes, 25, 50, and 100 MM GPY is given in Table 40- Table 42.

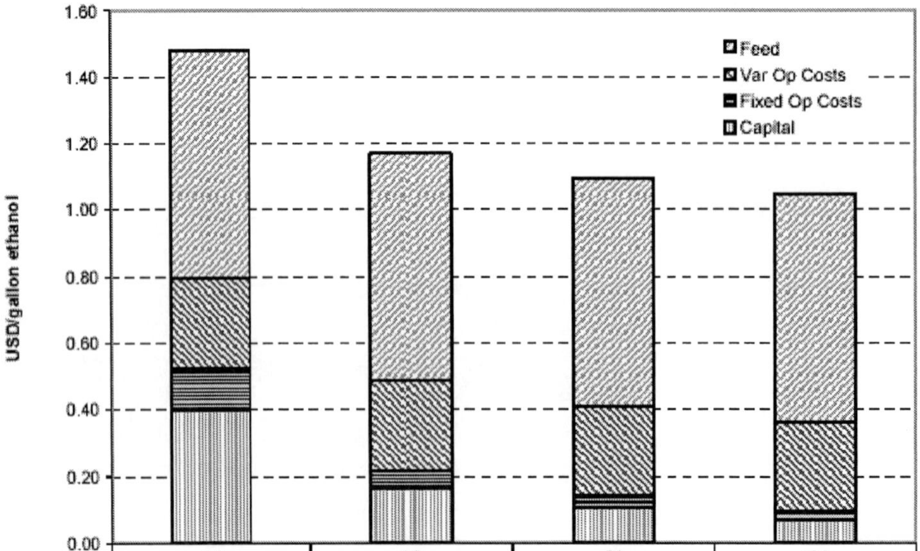


| | 4.6 | 25 | 50 | 100 |
|---|---|---|---|---|
| Feed | 0.68 | 0.68 | 0.68 | 0.68 |
| Var Op Costs | 0.27 | 0.27 | 0.27 | 0.27 |
| Fixed Op Costs | 0.13 | 0.05 | 0.03 | 0.02 |
| Capital | 0.40 | 0.17 | 0.11 | 0.07 |

**Plant Size (MMGPY EtOH) with 60/40 Ethanol/Sucrose**

Figure 17. Brazil sugar cane ethanol cost contributions

## Table 30. Country summary costs, 25 MM GPY plant

| Ethanol-Corn Dry Mill | Units | USA | Argentina | Brazil | Canada | Caribbean Basin | China | Colombia | India | Mexico |
|---|---|---|---|---|---|---|---|---|---|---|
| **Plant Size** | MM gal/yr | 25 | 25 | 25 | 25 | | 25 | | | 25 |
| **Yields** | | | | | | | | | | |
| Primary Product Yield | gal/ton | 101.8 | 101.8 | 101.8 | 101.8 | | 101.8 | | | 101.8 |
| By Product A | | DDG | DDG | DDG | DDG | | DDG | | | DDG |
| Unit | | ton | ton | ton | ton | | ton | | | ton |
| Annual Yield | | 8.10E+04 | 8.10E+04 | 8.10E+04 | 8.10E+04 | | 8.10E+04 | | | 8.10E+04 |
| Price unit | | 80 | 80 | 80 | 80 | | 80 | | | 80 |
| MM USD/yr | | 6.48 | 6.48 | 6.48 | 6.48 | | 6.48 | | | 6.48 |
| By Product B | | | | | | | | | | |
| Unit | | | | | | | | | | |
| Annual Yield | | | | | | | | | | |
| Price unit | | | | | | | | | | |
| Annual Cost | MM SD/yr | | | | | | | | | |
| **Energy Inputs** | | | | | | | | | | |
| Electricity | | | | | | | | | | |
| Cost | $/kWh | 0.05 | 0.05 | 0.05 | 0.05 | | 0.05 | | | 0.05 |
| Amount | kWh/yr | 2.01E+07 | 2.01E+07 | 2.01E+07 | 2.01E+07 | | 2.01E+07 | | | 2.01E+07 |
| Annual Cost | MM SD/yr | 1.01E+00 | 1.01E+00 | 1.01E+00 | 1.01E+00 | | 1.01E+00 | | | 1.01E+00 |
| Fuel | | | | | | | | | | |
| Type | | Nat Gas | Nat Gas | Nat Gas | Nat Gas | | Nat Gas | | | Nat Gas |
| Unit | | GJ | GJ | GJ | GJ | | GJ | | | GJ |
| Amount | | 8.51E+05 | 8.51E+05 | 8.51E+05 | 8.51E+05 | | 8.51E+05 | | | 8.51E+05 |
| Cost | USD/unit | 6.33 | 6.33 | 6.33 | 6.33 | | 6.33 | | | 6.33 |
| Annual Cost | MM USD/yr | 5.39 | 5.39 | 5.39 | 5.39 | | 5.39 | | | 5.39 |
| **Capital & Operating Costs** | | | | | | | | | | |
| Capital | MMUSD | 33.36 | 35.66 | 34.05 | 42.80 | | 36.94 | | | 27.27 |
| Operating w/o feed | MM SD/yr | 10.69 | 10.6 | 10.02 | 11.9 | | 10.95 | | | 9.51 |
| Feed | | | | | | | | | | |

**Table 30. (Continued)**

| Ethanol-Corn Dry Mill | Units | USA | Argentina | Brazil | Canada | Caribbean Basin | China | Colombia | India | Mexico |
|---|---|---|---|---|---|---|---|---|---|---|
| Unit | | ton | | | | | | | | |
| Cost/unit | USD/unit | 107.14 | 127.5 | 134.29 | 80 | | 124.63 | | | 113.57 |
| | USD/bu | 3.00 | 3.57 | 3.76 | 2.24 | | 3.54 | | | 3.18 |
| Rate | unit/day | 642 | 642 | 642 | 642 | | 642 | | | 642 |
| Yearly Cost | MM USD/yr | 26.29 | 31.29 | 32.95 | 19.63 | | 31.02 | | | 31.02 |
| **Product Plant Gate Cost** | | | | | | | | | | |
| Basis | | USTL | 10 SL | 10 SL | 10 SL | | 10 SL | | | 5 SL |
| Value | USD/gal | 1.42 | 1.63 | 1.66 | 1.24 | | 1.63 | | | 1.41 |

**Table 31. Country summary costs, 100 MM GPY plant**

| Ethanol-Corn Dry Mill | Units | USA | Argentina | Brazil | Canada | Caribbean Basin | China | Colombia | India | Mexico |
|---|---|---|---|---|---|---|---|---|---|---|
| **Plant Size** | MM gal/yr | 100 | 100 | 100 | 100 | | 100 | | | 100 |
| **Yields** | | | | | | | | | | |
| Primary Product Yield | gal/ton | 101.8 | 101.8 | 101.8 | 101.8 | | 101.8 | | | 101.8 |
| By Product A | | | | | | | | | | |
| Unit | | DDG | DDG | DDG | DDG | | DDG | | | DDG |
| Yield unit | | ton | ton | ton | ton | | ton | | | ton |
| Annual Yield | | 3.242E+05 | 3.242E+05 | 3.242E+05 | 3.242E+05 | | 3.242E+05 | | | 3.242E+05 |
| Price unit | | 80.00 | 80.00 | 80.00 | 80.00 | | 80.00 | | | 80.00 |
| MM USD/yr | | 25.93 | 25.93 | 25.93 | 25.93 | | 25.93 | | | 25.93 |
| By Product B | | | | | | | | | | |
| Yield unit | | | | | | | | | | |
| Annual Yield | | | | | | | | | | |

**Table 31. (Continued)**

| Ethanol-Corn Dry Mill | Units | USA | Argentina | Brazil | Canada | Caribbean Basin | China | Colombia | India | Mexico |
|---|---|---|---|---|---|---|---|---|---|---|
| Price unit | | | | | | | | | | |
| Annual Cost | MMUSD/yr | | | | | | | | | |
| **Energy Inputs** | | | | | | | | | | |
| Electricity | | | | | | | | | | |
| Cost | $/kWh | 0.05 | 0.05 | 0.05 | 0.05 | | 0.05 | | | 0.05 |
| Amount | kWh/yr | 8.04E+07 | 8.04E+07 | 8.04E+07 | 8.04E+07 | | 8.04E+07 | | | 8.04E+07 |
| Annual Cost | MM USD/yr | 4.02 | 4.02 | 4.02 | 4.02 | | 4.02 | | | 4.02 |
| Fuel | | | | | | | | | | |
| Type | | Nat Gas | Nat Gas | Nat Gas | Nat Gas | | Nat Gas | | | Nat Gas |
| Unit | | GJ/yr | GJ/yr | GJ/yr | GJ/yr | | GJ/yr | | | GJ/yr |
| Amount | | 3.40E+06 | 3.40E+06 | 3.40E+06 | 3.40E+06 | | 3.40E+06 | | | 3.40E+06 |
| Cost | USD/unit | 6.33 | 6.33 | 6.33 | 6.33 | | 6.33 | | | 6.33 |
| Annual Cost | MM USD/yr | 21.67 | 21.67 | 21.67 | 21.67 | | 21.67 | | | 21.67 |
| **Capital & Operating Costs** | | | | | | | | | | |
| Capital | MMUSD | 87.34 | 93.41 | 89.14 | 112.26 | | 95.60 | | | 71.25 |
| Operating w/o feed | MM USD/yr | 37.89 | 37.88 | 36.97 | 40.08 | | 38.42 | | | 32.76 |
| Feed | | | | | | | | | | |
| Unit | | d ton | d ton | d ton | d ton | | d ton | | | d ton |
| Cost/unit | USD/unit | 107.14 | 127.5 | 134.29 | 80 | | 124.63 | | | 113.57 |
| | USD/bu | 3.00 | 3.57 | 5.76 | 2.24 | | 3.54 | | | 3.18 |
| Rate | unit/day | 2570 | 2570 | 2570 | 2570 | | 2570 | | | 2570 |
| Yearly Cost | MM USD/yr | 107.14 | 125.24 | 131.91 | 78.58 | | 124.19 | | | 124.19 |
| **Product Plant Gate Cost** | | | | | | | | | | |
| Basis | | USTL | 10 SL | 10 SL | 10 SL | | 10 SL | | | 5 SL |
| Value | USD/gal | 1.29 | 1.49 | 1.54 | 1.06 | | 1.48 | | | 1.31 |

## Table 32. Capital and operating costs, Brazil sugar-ethanol mill

| Country | Brazil | Code = | 2 | | | | |
|---|---|---|---|---|---|---|---|
| **Cost component** | **Units** | **Cost Factor** | **Scale Factor** | **Brazil Sugar Mill** | **Brazil Sugar Mill** | **Brazil Sugar Mill** | **Brazil Sugar Mill** |
| **Year $** | **$** | | | **2005** | **2005** | **2005** | **2005** |
| Cost Index (2 = M&S, 3 = CE) | | | 3 | 130.93 | 130.93 | 130.93 | 130.93 |
| Plant Size | dry short ton/day | | | 2,206 | 11,118 | 22,236 | 44,473 |
| | dry tonne/day | | | 2,000 | 10,080 | 20,160 | 40,320 |
| Stream Factor | % | | | **50%** | **50%** | **50%** | **50%** |
| Feed | Dry short ton/yr | | | 4.026E+05 | 2.029E+06 | 4.058E+06 | 8.116E+06 |
| | GJ/short ton | | | | | | |
| | GJ/yr | | | 0.000E+00 | 0.000E+00 | 0.000E+00 | 0.000E+00 |
| Feed Cost | USD/dt | | | **$ 10.45** | **$ 10.45** | **$ 10.45** | **$ 10.45** |
| Yield (gal/Dry US Ton) | | | | | | | |
| Ethanol | gal/short ton | | | 12.32 | 12.32 | 12.32 | 12.32 |
| Sugar | ton/short ton | | | 0.057 | 0.057 | 0.057 | 0.057 |
| Process Efficiency - to ethanol | % HHV | | | | | | |
| Process efficiency - overall | % HHV | | | | | | |
| Ethanol | gal/yr | | | 4.960E+06 | 2.500E+07 | 5.000E+07 | 1.000E+08 |
| | Mil Gal/YR | | | **4.96** | **25.00** | **50.00** | **100.00** |
| | tpy | | | 1.639E+04 | 8.263E+04 | 1.653E+05 | 3.305E+05 |
| | GJ/yr | | | 4.401E+05 | 2.218E+06 | 4.436E+06 | 8.873E+06 |
| | gal/Stream day | | | 2.718E+04 | 1.370E+05 | 2.740E+05 | 5.480E+05 |
| | bbl/s stream day | | | 6.472E+02 | 3.262E+03 | 6.523E+03 | 1.305E+04 |
| Sugar, GJ/yr | 14.877 | GJ/yr | | 3.427E+05 | 1.727E+06 | 3.454E+06 | 6.908E+06 |
| Sugar (glucose +fructose) | ton/yr | | | 2.30E+04 | 1.16E+05 | 2.32E+05 | 4.64E+05 |
| Sugar Cane - Sucrose Content | | wt % | | 14.00% | 14.00% | 14.00% | 14.00% |
| Ethanol/sugar Split | | % | | **60%** | **60%** | **60%** | **60%** |
| Sugar extraction efficiency | | % | | 97.0% | 97.0% | 97.0% | 97.0% |
| Ethanol Production efficiency | | % | | 90.0% | 90.0% | 90.0% | 90.0% |
| lb ethanol/lb sucrose to ethanol train | 0.5386 | lb/lb | | 0.4848 | 0.4848 | 0.4848 | 0.4848 |
| **Capital Cost (million USD)** | | | | | | | |

**Table 32. (Continued)**

| Cost component | Units | Cost Factor | Scale Factor | Brazil Sugar Mill | Brazil Sugar Mill | Brazil Sugar Mill | Brazil Sugar Mill |
|---|---|---|---|---|---|---|---|
| Total Capital | | | 0.38 | 15.33 | 33.54 | 43.65 | 56.80 |
| **Direct Fixed Capital (DFC), also called TIC** | | | | 15.33 | 33.54 | 43.65 | 56.80 |
| Engineering | DFC × MF | 0.12 | | 1.84 | 4.03 | 5.24 | 6.82 |
| Construction | DFC × MF | 0.13 | | 1.99 | 4.36 | 5.67 | 7.38 |
| Contractor & Legal | DFC × MF | 0.08 | | 1.23 | 2.68 | 3.49 | 4.54 |
| Process/Project Contingency | DFC × MF | 0.0412 | | 0.63 | 1.38 | 1.80 | 2.34 |
| **Total Plant Cost (TPC)** | | | | 21.02 | 45.99 | 59.85 | 77.89 |
| AFUDC | | | | | | | |
| **Total Plant Investment (TPI)** | | | | 21.02 | 45.99 | 59.85 | 77.89 |
| Land | | | 0.20 | 2.21 | 3.05 | 3.51 | 7.02 |
| Startup | | | | 0.00 | 0.00 | 0.00 | 0.00 |
| **Total Capital Cost (TCC)** | | | | 23.23 | 49.05 | 63.36 | 84.91 |
| **Contingency/TPI** | | | | | | | |
| Working Capital | DFC × MF | 0.05 | | 1.05 | 2.30 | 2.99 | 3.89 |
| **Variable Operating Costs (million USD/yr)** | | | | | | | |
| Feed | | | | 4.21 | 21.20 | 42.41 | 84.82 |
| Utilities | | | | 0.36 | 1.79 | 3.58 | 7.16 |
| Other | | | | 1.18 | 5.96 | 11.93 | 23.86 |
| Catalysts and Chemicals | | | | 0.83 | 4.18 | 8.35 | 16.70 |
| **Total** | | | | 6.57 | 33.13 | 66.27 | 132.53 |
| Fixed Operating Costs (million USD/yr) | | | | | | | |
| Labor (SF from V-R) | DFC × MF | 0.0165 | 0.25 | 0.12 | 0.19 | 0.22 | 0.26 |
| Maintenance (3% of A) | | | 0.03 | 0.46 | 1.01 | 1.31 | 1.70 |
| General Overhead (65% of labor + maint) | | | 0.65 | 0.38 | 0.78 | 1.00 | 1.28 |
| Direct Overhead (45% of Labor) | | | 0.45 | 0.06 | 0.08 | 0.10 | 0.12 |
| Insurance (0.5% of TIC) | | | 0.005 | 0.12 | 0.25 | 0.32 | 0.42 |
| **Total** | | | | 1.14 | 2.30 | 2.94 | 3.79 |

## Table 33. Capital and operating costs, Argentina sugar-ethanol mill

| Country | Argentina | Code = | 5 | | | | |
|---|---|---|---|---|---|---|---|
| **Cost component** | **Units** | **Cost Factor** | **Scale Factor** | **Argentina Sugar Mill** | **Sugar Mill** | **Argentina Sugar Mill** | **Argentina Sugar Mill** |
| **Year $** | **$** | | | **2005** | **2005** | **2005** | **2005** |
| Cost Index (2 = M&S, 3 = CE) | | | 3 | 130.93 | 130.93 | 130.93 | 130.93 |
| Plant Size | dry short ton/day | | | 2,206 | 11,118 | 22,236 | 44,473 |
| | dry tonne/day | | | 2,000 | 10,080 | 20,160 | 40,320 |
| Stream Factor | % | | | **50%** | **50%** | **50%** | **50%** |
| Feed | Dry short ton/yr | | | 4.026E+05 | 2.029E+06 | 4.058E+06 | 8.116E+06 |
| | GJ/short ton | | | | | | |
| | GJ/yr | | | 0.000E+00 | 0.000E+00 | 0.000E+00 | 0.000E+00 |
| Feed Cost | USD/dt | | | **$ 34.30** | **$ 34.30** | **$ 34.30** | **$ 34.30** |
| Yield (gal/Dry US Ton) | | | | | | | |
| Ethanol | gal/short ton | | | 12.32 | 12.32 | 12.32 | 12.32 |
| Sugar | ton/short ton | | | 0.057 | 0.057 | 0.057 | 0.057 |
| Process Efficiency - to ethanol | % HHV | | | | | | |
| Process efficiency - overall | % HHV | | | | | | |
| Ethanol | gal/yr | | | 4.960E+06 | 2.500E+07 | 5.000E+07 | 1.000E+08 |
| | Mil Gal/YR | | | **4.96** | **25.00** | **50.00** | **100.00** |
| | tpy | | | 1.639E+04 | 8.263E+04 | 1.653E+05 | 3.305E+05 |
| | GJ/yr | | | 4.401E+05 | 2.218E+06 | 4.436E+06 | 8.873E+06 |
| | gal/Stream day | | | 2.718E+04 | 1.370E+05 | 2.740E+05 | 5.480E+05 |
| | bbl/s stream day | | | 6.472E+02 | 3.262E+03 | 6.523E+03 | 1.305E+04 |
| Sugar, GJ/yr | 14.877 | GJ/yr | | 3.427E+05 | 1.727E+06 | 3.454E+06 | 6.908E+06 |
| Sugar (glucose +fructose) | ton/yr | | | 2.30E+04 | 1.16E+05 | 2.32E+05 | 4.64E+05 |
| Sugar Cane - Sucrose Content | | wt % | | 14.00% | 14.00% | 14.00% | 14.00% |
| Ethanol/sugar Split | | % | | **60%** | **60%** | **60%** | **60%** |
| Sugar extraction efficiency | | % | | 97.0% | 97.0% | 97.0% | 97.0% |
| Ethanol Production efficiency | | % | | 90.0% | 90.0% | 90.0% | 90.0% |
| lb ethanol/lb sucrose to ethanol train | 0.5386 | lb/lb | | 0.4848 | 0.4848 | 0.4848 | 0.4848 |
| **Capital Cost (million USD)** | | | | | | | |
| Total Capital | | | 0.38 | 16.11 | 35.24 | 45.87 | 59.69 |

**Table 33. (Continued)**

| Country | Argentina | Code = | 5 | | | | |
|---|---|---|---|---|---|---|---|
| **Cost component** | **Units** | **Cost Factor** | **Scale Factor** | **Argentina Sugar Mill** | **Sugar Mill** | **Argentina Sugar Mill** | **Argentina Sugar Mill** |
| **Year $** | **$** | | | **2005** | **2005** | **2005** | **2005** |
| **Direct Fixed Capital (DFC), also called TIC** | | | | 16.11 | 35.24 | 45.87 | 59.69 |
| Engineering | DFC × MF | 0.12 | | 1.93 | 4.23 | 5.50 | 7.16 |
| Construction | DFC × MF | 0.13 | | 2.09 | 4.58 | 5.96 | 7.76 |
| Contractor & Legal | DFC × MF | 0.08 | | 1.29 | 2.82 | 3.67 | 4.77 |
| Process/Project Contingency | DFC × MF | 0.0412 | | 0.66 | 1.45 | 1.89 | 2.46 |
| **Total Plant Cost (TPC)** | | | | 22.08 | 48.33 | 62.89 | 81.84 |
| AFUDC | | | | | | | |
| **Total Plant Investment (TPI)** | | | | 22.08 | 48.33 | 62.89 | 81.84 |
| Land | | | 0.20 | 2.21 | 3.05 | 3.51 | 7.02 |
| Startup | | | | 0.00 | 0.00 | 0.00 | 0.00 |
| **Total Capital Cost (TCC)** | | | | 24.29 | 51.38 | 66.40 | 88.86 |
| **Contingency/TPI** | | | | | | | |
| Working Capital | DFC × MF | 0.05 | | 1.10 | 2.42 | 3.14 | 4.09 |
| **Variable Operating Costs (million USD/yr)** | | | | | | | |
| Feed | | | | 13.81 | 69.60 | 139.19 | 278.39 |
| Utilities | | | | 0.36 | 1.79 | 3.58 | 7.16 |
| Other | | | | 1.18 | 5.96 | 11.93 | 23.86 |
| Catalysts and Chemicals | | | | 0.83 | 4.18 | 8.35 | 16.70 |
| **Total** | | | | 16.18 | 81.53 | 163.05 | 326.11 |
| Fixed Operating Costs (million USD/yr) | | | | | | | |
| Labor (SF from V-R) | DFC × MF | 0.0165 | 0.25 | 0.25 | 0.37 | 0.44 | 0.53 |
| Maintenance (3% of A) | | | 0.03 | 0.48 | 1.06 | 1.38 | 1.79 |
| General Overhead (65% of labor + maint) | | | 0.65 | 0.48 | 0.93 | 1.18 | 1.51 |
| Direct Overhead (45% of Labor) | | | 0.45 | 0.11 | 0.17 | 0.20 | 0.24 |
| Insurance (0.5% of TIC) | | | 0.005 | 0.12 | 0.26 | 0.33 | 0.44 |
| **Total** | | | | 1.44 | 2.78 | 3.53 | 4.50 |

## Table 34. Capital and operating costs, Caribbean basin sugar-ethanol mill

| Country | Caribbean | Code = | 9 | | | | |
|---|---|---|---|---|---|---|---|
| **Cost component** | **Units** | **Cost Factor** | **Scale Factor** | **Caribbean Sugar Mill** | **Caribbean Sugar Mill** | **Caribbean Sugar Mill** | **Caribbean Sugar Mill** |
| **Year $** | **$** | | | **2005** | **2005** | **2005** | **2005** |
| Cost Index (2 = M&S, 3 = CE) | | | 3 | 130.93 | 130.93 | 130.93 | 130.93 |
| Plant Size | dry short ton/day | | | 2,206 | 11,118 | 22,236 | 44,473 |
| | dry tonne/day | | | 2,000 | 10,080 | 20,160 | 40,320 |
| Stream Factor | % | | | **50%** | **50%** | **50%** | **50%** |
| Feed | Dry short ton/yr | | | 4.026E+05 | 2.029E+06 | 4.058E+06 | 8.116E+06 |
| | GJ/short ton | | | | | | |
| | GJ/yr | | | 0.000E+00 | 0.000E+00 | 0.000E+00 | 0.000E+00 |
| Feed Cost | USD/dt | | | **$ 15.00** | **$ 15.00** | **$ 15.00** | **$ 15.00** |
| Yield (gal/Dry US Ton) | | | | | | | |
| Ethanol | gal/short ton | | | 12.32 | 12.32 | 12.32 | 12.32 |
| Sugar | ton/short ton | | | 0.057 | 0.057 | 0.057 | 0.057 |
| Process Efficiency - to ethanol | % HHV | | | | | | |
| Process efficiency - overall | % HHV | | | | | | |
| Ethanol | gal/yr | | | 4.960E+06 | 2.500E+07 | 5.000E+07 | 1.000E+08 |
| | Mil Gal/YR | | | **4.96** | **25.00** | **50.00** | **100.00** |
| | tpy | | | 1.639E+04 | 8.263E+04 | 1.653E+05 | 3.305E+05 |
| | GJ/yr | | | 4.401E+05 | 2.218E+06 | 4.436E+06 | 8.873E+06 |
| | gal/Stream day | | | 2.718E+04 | 1.370E+05 | 2.740E+05 | 5.480E+05 |
| | bbl/s stream day | | | 6.472E+02 | 3.262E+03 | 6.523E+03 | 1.305E+04 |
| Sugar, GJ/yr | 14.877 | GJ/yr | | 3.427E+05 | 1.727E+06 | 3.454E+06 | 6.908E+06 |
| Sugar (glucose +fructose) | ton/yr | | | 2.30E+04 | 1.16E+05 | 2.32E+05 | 4.64E+05 |
| Sugar Cane - Sucrose Content | | wt % | | 14.00% | 14.00% | 14.00% | 14.00% |
| Ethanol/sugar Split | | % | | **60%** | **60%** | **60%** | **60%** |
| Sugar extraction efficiency | | % | | 97.0% | 97.0% | 97.0% | 97.0% |
| Ethanol Production efficiency | | % | | 90.0% | 90.0% | 90.0% | 90.0% |
| lb ethanol/lb sucrose to ethanol train | 0.5386 | lb/lb | | 0.4848 | 0.4848 | 0.4848 | 0.4848 |
| **Capital Cost (million USD)** | | | | | | | |

**Table 34. (Continued)**

| Country | Caribbean | Code = | 9 | | | | |
|---|---|---|---|---|---|---|---|
| **Cost component** | **Units** | **Cost Factor** | **Scale Factor** | **Caribbean Sugar Mill** | **Caribbean Sugar Mill** | **Caribbean Sugar Mill** | **Caribbean Sugar Mill** |
| **Year $** | $ | | | **2005** | **2005** | **2005** | **2005** |
| Total Capital | | | 0.38 | 16.92 | 37.03 | 48.19 | 62.72 |
| **Direct Fixed Capital (DFC), also called TIC** | | | | 16.92 | 37.03 | 48.19 | 62.72 |
| Engineering | DFC × MF | 0.12 | | 2.03 | 4.44 | 5.78 | 7.53 |
| Construction | DFC × MF | 0.13 | | 2.20 | 4.81 | 6.27 | 8.15 |
| Contractor & Legal | DFC × MF | 0.08 | | 1.35 | 2.96 | 3.86 | 5.02 |
| Process/Project Contingency | DFC × MF | 0.0412 | | 0.70 | 1.53 | 1.99 | 2.58 |
| **Total Plant Cost (TPC)** | | | | 23.21 | 50.78 | 66.08 | 86.00 |
| AFUDC | | | | | | | |
| **Total Plant Investment (TPI)** | | | | 23.21 | 50.78 | 66.08 | 86.00 |
| Land | | | 0.20 | 2.21 | 3.05 | 3.51 | 7.02 |
| Startup | | | | 0.00 | 0.00 | 0.00 | 0.00 |
| **Total Capital Cost (TCC)** | | | | 25.42 | 53.84 | 69.59 | 93.02 |
| **Contingency/TPI** | | | | | | | |
| Working Capital | DFC × MF | 0.05 | | 1.16 | 2.54 | 3.30 | 4.30 |
| **Variable Operating Costs (million USD/yr)** | | | | | | | |
| Feed | | | | 6.04 | 30.44 | 60.87 | 121.74 |
| Utilities | | | | 0.36 | 1.79 | 3.58 | 7.16 |
| Other | | | | 1.18 | 5.96 | 11.93 | 23.86 |
| Catalysts and Chemicals | | | | 0.83 | 4.18 | 8.35 | 16.70 |
| **Total** | | | | 8.41 | 42.37 | 84.73 | 169.46 |
| Fixed Operating Costs (million USD/yr) | | | | | | | |
| Labor (SF from V-R) | DFC × MF | 0.0165 | 0.25 | 0.28 | 0.42 | 0.50 | 0.60 |
| Maintenance (3% of A) | | | 0.03 | 0.51 | 1.11 | 1.45 | 1.88 |
| General Overhead (65% of labor + maint) | | | 0.65 | 0.51 | 1.00 | 1.27 | 1.61 |
| Direct Overhead (45% of Labor) | | | 0.45 | 0.13 | 0.19 | 0.23 | 0.27 |
| Insurance (0.5% of TIC) | | | 0.005 | 0.13 | 0.27 | 0.35 | 0.47 |
| **Total** | | | | 1.56 | 2.99 | 3.79 | 4.82 |

**Table 35. Capital and operating costs, China sugar-ethanol mill**

| Country | China | Code = | 3 | | | | |
|---|---|---|---|---|---|---|---|
| **Cost component** | **Units** | **Cost Factor** | **Scale Factor** | **China Sugar Mill** | **China Sugar Mill** | **China Sugar Mill** | **China Sugar Mill** |
| **Year $** | **$** | | | **2005** | **2005** | **2005** | **2005** |
| Cost Index (2 = M&S, 3 = CE) | | | 3 | 130.93 | 130.93 | 130.93 | 130.93 |
| Plant Size | dry short ton/day | | | 2,206 | 11,118 | 22,236 | 44,473 |
| | dry tonne/day | | | 2,000 | 10,080 | 20,160 | 40,320 |
| Stream Factor | % | | | **50%** | **50%** | **50%** | **50%** |
| Feed | Dry short ton/yr | | | 4.026E+05 | 2.029E+06 | 4.058E+06 | 8.116E+06 |
| | GJ/short ton | | | | | | |
| | GJ/yr | | | 0.000E+00 | 0.000E+00 | 0.000E+00 | 0.000E+00 |
| Feed Cost | USD/dt | | | **$ 23.00** | **$ 23.00** | **$ 23.00** | **$ 23.00** |
| Yield (gal/Dry US Ton) | | | | | | | |
| Ethanol | gal/short ton | | | 12.32 | 12.32 | 12.32 | 12.32 |
| Sugar | ton/short ton | | | 0.057 | 0.057 | 0.057 | 0.057 |
| Process Efficiency - to ethanol | % HHV | | | | | | |
| Process efficiency - overall | % HHV | | | | | | |
| Ethanol | gal/yr | | | 4.960E+06 | 2.500E+07 | 5.000E+07 | 1.000E+08 |
| | Mil Gal/YR | | | **4.96** | **25.00** | **50.00** | **100.00** |
| | tpy | | | 1.639E+04 | 8.263E+04 | 1.653E+05 | 3.305E+05 |
| | GJ/yr | | | 4.401E+05 | 2.218E+06 | 4.436E+06 | 8.873E+06 |
| | gal/Stream day | | | 2.718E+04 | 1.370E+05 | 2.740E+05 | 5.480E+05 |
| | bbl/s stream day | | | 6.472E+02 | 3.262E+03 | 6.523E+03 | 1.305E+04 |
| Sugar, GJ/yr | 14.877 | GJ/yr | | 3.427E+05 | 1.727E+06 | 3.454E+06 | 6.908E+06 |
| Sugar (glucose +fructose) | ton/yr | | | 2.30E+04 | 1.16E+05 | 2.32E+05 | 4.64E+05 |
| Sugar Cane - Sucrose Content | | wt % | | 14.00% | 14.00% | 14.00% | 14.00% |
| Ethanol/sugar Split | | % | | **60%** | **60%** | **60%** | **60%** |
| Sugar extraction efficiency | | % | | 97.0% | 97.0% | 97.0% | 97.0% |
| Ethanol Production efficiency | | % | | 90.0% | 90.0% | 90.0% | 90.0% |
| lb ethanol/lb sucrose to ethanol train | 0.5386 | lb/lb | | 0.4848 | 0.4848 | 0.4848 | 0.4848 |
| **Capital Cost (million USD)** | | | | | | | |

**Table 35. (Continued)**

| Country | China | Code = | 3 | | | | |
|---|---|---|---|---|---|---|---|
| **Cost component** | **Units** | **Cost Factor** | **Scale Factor** | **China Sugar Mill** | **China Sugar Mill** | **China Sugar Mill** | **China Sugar Mill** |
| **Year $** | **$** | | | **2005** | **2005** | **2005** | **2005** |
| | | | | | | | |
| Total Capital | | | 0.38 | 16.51 | 36.12 | 47.01 | 61.17 |
| **Direct Fixed Capital (DFC), also called TIC** | | | | 16.51 | 36.12 | 47.01 | 61.17 |
| Engineering | DFC × MF | 0.12 | | 1.98 | 4.33 | 5.64 | 7.34 |
| Construction | DFC × MF | 0.13 | | 2.15 | 4.70 | 6.11 | 7.95 |
| Contractor & Legal | DFC × MF | 0.08 | | 1.32 | 2.89 | 3.76 | 4.89 |
| Process/Project Contingency | DFC × MF | 0.0412 | | 0.68 | 1.49 | 1.94 | 2.52 |
| **Total Plant Cost (TPC)** | | | | 22.63 | 49.53 | 64.45 | 83.88 |
| AFUDC | | | | | | | |
| **Total Plant Investment (TPI)** | | | | 22.63 | 49.53 | 64.45 | 83.88 |
| Land | | | 0.20 | 2.21 | 3.05 | 3.51 | 7.02 |
| Startup | | | | 0.00 | 0.00 | 0.00 | 0.00 |
| **Total Capital Cost (TCC)** | | | | 24.84 | 52.58 | 67.96 | 90.89 |
| **Contingency/TPI** | | | | | | | |
| Working Capital | DFC × MF | 0.05 | | 1.13 | 2.48 | 3.22 | 4.19 |
| **Variable Operating Costs (million USD/yr)** | | | | | | | |
| Feed | | | | 9.26 | 46.67 | 93.34 | 186.68 |
| Utilities | | | | 0.36 | 1.79 | 3.58 | 7.16 |
| Other | | | | 1.18 | 5.96 | 11.93 | 23.86 |
| Catalysts and Chemicals | | | | 0.83 | 4.18 | 8.35 | 16.70 |
| **Total** | | | | 11.63 | 58.60 | 117.20 | 234.39 |
| Fixed Operating Costs (million USD/yr) | | | | | | | |
| Labor (SF from V-R) | DFC × MF | 0.0165 | 0.25 | 0.33 | 0.49 | 0.58 | 0.69 |
| Maintenance (3% of A) | | | 0.03 | 0.50 | 1.08 | 1.41 | 1.84 |
| General Overhead (65% of labor + maint) | | | 0.65 | 0.53 | 1.02 | 1.29 | 1.64 |
| Direct Overhead (45% of Labor) | | | 0.45 | 0.15 | 0.22 | 0.26 | 0.31 |
| Insurance (0.5% of TIC) | | | 0.005 | 0.12 | 0.26 | 0.34 | 0.45 |
| **Total** | | | | 1.63 | 3.08 | 3.89 | 4.93 |

## Table 36. Capital and operating costs, Colombia sugar-ethanol mill

| Country | China | Code = | 6 | | | | |
|---|---|---|---|---|---|---|---|
| Cost component | Units | Cost Factor | Scale Factor | Colombia Sugar Mill | Colombia Sugar Mill | Colombia Sugar Mill | Colombia Sugar Mill |
| **Year $** | $ | | | **2005** | **2005** | **2005** | **2005** |
| Cost Index (2 = M&S, 3 = CE) | | | 3 | 130.93 | 130.93 | 130.93 | 130.93 |
| Plant Size | dry short ton/day | | | 2,206 | 11,118 | 22,236 | 44,473 |
| | dry tonne/day | | | 2,000 | 10,080 | 20,160 | 40,320 |
| Stream Factor | % | | | **50%** | **50%** | **50%** | **50%** |
| Feed | Dry short ton/yr | | | 4.026E+05 | 2.029E+06 | 4.058E+06 | 8.116E+06 |
| | GJ/short ton | | | | | | |
| | GJ/yr | | | 0.000E+00 | | 0.000E+00 | 0.000E+00 |
| Feed Cost | USD/dt | | | **$ 26.85** | **$ 26.85** | **$ 26.85** | **$ 26.85** |
| Yield (gal/Dry US Ton) | | | | | | | |
| Ethanol | gal/short ton | | | 12.32 | 12.32 | 12.32 | 12.32 |
| Sugar | ton/short ton | | | 0.057 | 0.057 | 0.057 | 0.057 |
| Process Efficiency - to ethanol | % HHV | | | | | | |
| Process efficiency - overall | % HHV | | | | | | |
| Ethanol | gal/yr | | | 4.960E+06 | | 5.000E+07 | 1.000E+08 |
| | Mil Gal/YR | | | **4.96** | **25.00** | **50.00** | **100.00** |
| | tpy | | | 1.639E+04 | 048.263E+04 | 1.653E+05 | 3.305E+05 |
| | GJ/yr | | | 4.401E+05 | 2.218E+06 | 4.436E+06 | 8.873E+06 |
| | gal/Stream day | | | 2.718E+04 | 1.370E+05 | 2.740E+05 | 5.480E+05 |
| | bbl/s stream day | | | 6.472E+02 | 3.262E+03 | 6.523E+03 | 1.305E+04 |
| Sugar, GJ/yr | 14.877 | GJ/yr | | 3.427E+05 | 1.727E+06 | 3.454E+06 | 6.908E+06 |
| Sugar (glucose +fructose) | ton/yr | | | 2.30E+04 | 1.16E+05 | 2.32E+05 | 4.64E+05 |
| Sugar Cane - Sucrose Content | | wt % | | 14.00% | 14.00% | 14.00% | 14.00% |
| Ethanol/sugar Split | | % | | **60%** | **60%** | **60%** | **60%** |
| Sugar extraction efficiency | | % | | 97.0% | 97.0% | 97.0% | 97.0% |
| Ethanol Production efficiency | | % | | 90.0% | 90.0% | 90.0% | 90.0% |
| lb ethanol/lb sucrose to ethanol train | 0.5386 | lb/lb | | 0.4848 | 0.4848 | 0.4848 | 0.4848 |
| **Capital Cost (million USD)** | | | | | | | |

Table 36. (Continued)

| Country | China | Code = | 6 | | | | |
|---|---|---|---|---|---|---|---|
| **Cost component** | **Units** | **Cost Factor** | **Scale Factor** | **Colombia Sugar Mill** | **Colombia Sugar Mill** | **Colombia Sugar Mill** | **Colombia Sugar Mill** |
| **Year $** | $ | | | **2005** | **2005** | **2005** | **2005** |
| Total Capital | | | 0.38 | 14.92 | 32.64 | 42.48 | 55.28 |
| **Direct Fixed Capital (DFC), also called TIC** | | | | 14.92 | 32.64 | 42.48 | 55.28 |
| Engineering | DFC × MF | 0.12 | | 1.79 | 3.92 | 5.10 | 6.63 |
| Construction | DFC × MF | 0.13 | | 1.94 | 4.24 | 5.52 | 7.19 |
| Contractor & Legal | DFC × MF | 0.08 | | 1.19 | 2.61 | 3.40 | 4.42 |
| Process/Project Contingency | DFC × MF | 0.0412 | | 0.61 | 1.34 | 1.75 | 2.28 |
| **Total Plant Cost (TPC)** | | | | 20.45 | 44.76 | 58.25 | 75.80 |
| AFUDC | | | | | | | |
| **Total Plant Investment (TPI)** | | | | 20.45 | 44.76 | 58.25 | 75.80 |
| Land | | | 0.20 | 2.21 | 3.05 | 3.51 | 7.02 |
| Startup | | | | 0.00 | 0.00 | 0.00 | 0.00 |
| **Total Capital Cost (TCC)** | | | | 22.66 | 47.81 | 61.75 | 82.81 |
| **Contingency/TPI** | | | | | | | |
| Working Capital | DFC × MF | 0.05 | | 1.02 | 2.24 | 2.91 | 3.79 |
| **Variable Operating Costs (million USD/yr)** | | | | | | | |
| Feed | | | | 10.81 | 54.48 | 108.96 | 217.92 |
| Utilities | | | | 0.36 | 1.79 | 3.58 | 7.16 |
| Other | | | | 1.18 | 5.96 | 11.93 | 23.86 |
| Catalysts and Chemicals | | | | 0.83 | 4.18 | 8.35 | 16.70 |
| **Total** | | | | 13.18 | 66.41 | 132.82 | 265.64 |
| Fixed Operating Costs (million USD/yr) | | | | | | | |
| Labor (SF from V-R) | DFC × MF | 0.0165 | 0.25 | 0.19 | 0.28 | 0.33 | 0.40 |
| Maintenance (3% of A) | | | 0.03 | 0.45 | 0.98 | 1.27 | 1.66 |
| General Overhead (65% of labor + maint) | | | 0.65 | 0.41 | 0.82 | 1.04 | 1.33 |
| Direct Overhead (45% of Labor) | | | 0.45 | 0.08 | 0.13 | 0.15 | 0.18 |
| Insurance (0.5% of TIC) | | | 0.005 | 0.11 | 0.24 | 0.31 | 0.41 |
| **Total** | | | | 1.24 | 2.44 | 3.11 | 3.98 |

### Table 37. Capital and operating costs, India sugar-ethanol mill

| Country | India | Code = | 4 | | | | | |
|---|---|---|---|---|---|---|---|---|
| **Cost component** | **Units** | **Cost Factor** | **Scale Factor** | **Brazil Sugar Mill** | **India Sugar Mill** | **India Sugar Mill** | **India Sugar Mill** | **India Sugar Mill** |
| **Year $** | **$** | | | **2002** | **2005** | **2005** | **2005** | **2005** |
| Cost Index (2 = M&S, 3 = CE) | | | 3 | 110.63 | 130.93 | 130.93 | 130.93 | 130.93 |
| Plant Size | dry short ton/day | | | 2,206 | 2,206 | 11,118 | 22,236 | 44,473 |
| | dry tonne/day | | | 2,000 | 2,000 | 10,080 | 20,160 | 40,320 |
| Stream Factor | % | | | **50%** | **50%** | **50%** | **50%** | **50%** |
| Feed | Dry short ton/yr | | | 4.026E+05 | | 2.029E+06 | | 8.116E+06 |
| | GJ/short ton | | | | | | | |
| | GJ/yr | | | 0.000E+00 | 0.000E+00 | | 0.000E+00 | |
| Feed Cost | USD/dt | | | **$ 15.00** | **$ 15.00** | **$ 15.00** | **$ 15.00** | **$ 15.00** |
| Yield (gal/Dry US Ton) | | | | | | | | |
| Ethanol | gal/short ton | | | 12.32 | 12.32 | 12.32 | 12.32 | 12.32 |
| Sugar | ton/short ton | | | 0.057 | 0.057 | 0.057 | 0.057 | 0.057 |
| Process Efficiency - to ethanol | % HHV | | | | | | | |
| Process efficiency - overall | % HHV | | | | | | | |
| Ethanol | gal/yr | | | 4.960E+06 | 4.960E+06 | | 5.000E+07 | 1.000E+08 |
| | Mil Gal/YR | | | 4.96 | **4.96** | **25.00** | **50.00** | **100.00** |
| | tpy | | | 1.639E+04 | 1.639E+04 | 8.263E+04 | 1.653E+05 | 3.305E+05 |
| | GJ/yr | | | 4.401E+05 | 4.401E+05 | 2.218E+06 | 4.436E+06 | 8.873E+06 |
| | gal/Stream day | | | 2.718E+04 | 2.718E+04 | 1.370E+05 | 2.740E+05 | 5.480E+05 |
| | bbl/s stream day | | | 6.472E+02 | 6.472E+02 | 3.262E+03 | 6.523E+03 | 1.305E+04 |
| Sugar, GJ/yr | 14.877 | GJ/yr | | 3.427E+05 | 3.427E+05 | 1.727E+06 | 3.454E+06 | 6.908E+06 |
| Sugar (glucose +fructose) | ton/yr | | | 2.30E+04 | 2.30E+04 | 1.16E+05 | 2.32E+05 | 4.64E+05 |
| Sugar Cane - Sucrose Content | | wt % | | 14.00% | 14.00% | 14.00% | 14.00% | 14.00% |
| Ethanol/sugar Split | | % | | **60.0%** | **60%** | **60%** | **60%** | **60%** |
| Sugar extraction efficiency | | % | | 97.0% | 97.0% | 97.0% | 97.0% | 97.0% |
| Ethanol Production efficiency | | % | | 90.0% | 90.0% | 90.0% | 90.0% | 90.0% |
| lb ethanol/lb sucrose to ethanol train | 0.5386 | lb/lb | | 0.4848 | 0.4848 | 0.4848 | 0.4848 | 0.4848 |
| **Capital Cost (million USD)** | | | | | | | | |

**Table 37. (Continued)**

| Country | India | Code = | 4 | | | | | |
|---|---|---|---|---|---|---|---|---|
| **Cost component** | **Units** | **Cost Factor** | **Scale Factor** | **Brazil Sugar Mill** | **India Sugar Mill** | **India Sugar Mill** | **India Sugar Mill** | **India Sugar Mill** |
| **Year $** | **$** | | | **2002** | **2005** | **2005** | **2005** | **2005** |
| Total Capital | | | 0.38 | 15.00 | 16.00 | 35.01 | 45.56 | 59.29 |
| **Direct Fixed Capital (DFC), also called TIC** | | | | 15.00 | 16.00 | 35.01 | 45.56 | 59.29 |
| Engineering | DFC × MF | 0.12 | | 1.80 | 1.92 | 4.20 | 5.47 | 7.11 |
| Construction | DFC × MF | 0.13 | | 1.95 | 2.08 | 4.55 | 5.92 | 7.71 |
| Contractor & Legal | DFC × MF | 0.08 | | 1.20 | 1.28 | 2.80 | 3.64 | 4.74 |
| Process/Project Contingency | DFC × MF | 0.0412 | | 0.62 | 0.66 | 1.44 | 1.88 | 2.44 |
| **Total Plant Cost (TPC)** | | | | 20.57 | 21.94 | 48.01 | 62.47 | 81.30 |
| AFUDC | | | | | | | | |
| **Total Plant Investment (TPI)** | | | | 20.57 | 21.94 | 48.01 | 62.47 | 81.30 |
| Land | | | 0.20 | 2.21 | 2.21 | 3.05 | 3.51 | 7.02 |
| Startup | | | | 0.00 | 0.00 | 0.00 | 0.00 | 0.00 |
| **Total Capital Cost (TCC)** | | | | 22.78 | 24.15 | 51.06 | 65.98 | 88.31 |
| **Contingency/TPI** | | | | | | | | |
| Working Capital | DFC × MF | 0.05 | | 1.03 | 1.10 | 2.40 | 3.12 | 4.06 |
| **Variable Operating Costs (million USD/yr)** | | | | | | | | |
| Feed | | | | 6.04 | 6.04 | 30.44 | 60.87 | 121.74 |
| Utilities | | | | 0.30 | 0.36 | 1.79 | 3.58 | 7.16 |
| Other | | | | 1.00 | 1.18 | 5.96 | 11.93 | 23.86 |
| Catalysts and Chemicals | | | | 0.70 | 0.83 | 4.18 | 8.35 | 16.70 |
| **Total** | | | | 8.04 | 8.41 | 42.37 | 84.73 | 169.46 |
| Fixed Operating Costs (million USD/yr) | | | | | | | | |
| Labor (SF from V-R) | DFC × MF | 0.0165 | 0.25 | 0.25 | 0.11 | 0.16 | 0.20 | 0.23 |
| Maintenance (3% of A) | | | 0.03 | 0.45 | 0.48 | 1.05 | 1.37 | 1.78 |
| General Overhead (65% of labor + maint) | | | 0.65 | 0.45 | 0.38 | 0.79 | 1.02 | 1.31 |
| Direct Overhead (45% of Labor) | | | 0.45 | 0.11 | 0.05 | 0.07 | 0.09 | 0.10 |
| Insurance (0.5% of TIC) | | | 0.005 | 0.11 | 0.12 | 0.26 | 0.33 | 0.44 |
| **Total** | | | | 1.38 | 1.14 | 2.33 | 3.00 | 3.86 |

## Table 38. Capital and operating costs, Mexico sugar-ethanol mill

| Country | Mexico | Code = | 8 | | | | |
|---|---|---|---|---|---|---|---|
| **Cost component** | **Units** | **Cost Factor** | **Scale Factor** | **Mexico Sugar Mill** | **Mexico Sugar Mill** | **Mexico Sugar Mill** | **Mexico Sugar Mill** |
| **Year $** | **$** | | | **2005** | **2005** | **2005** | **2005** |
| Cost Index (2 = M&S, 3 = CE) | | | 3 | 130.93 | 130.93 | 130.93 | 130.93 |
| Plant Size | dry short ton/day | | | 2,206 | 11,118 | 22,236 | 44,473 |
| | dry tonne/day | | | 2,000 | 10,080 | 20,160 | 40,320 |
| Stream Factor | % | | | **50%** | **50%** | **50%** | **50%** |
| Feed | Dry short ton/yr | | | 4.026E+05 | 2.029E+06 | 4.058E+06 | 8.116E+06 |
| | GJ/short ton | | | | | | |
| | GJ/yr | | | 0.000E+00 | 0.000E+00 | 0.000E+00 | 0.000E+00 |
| Feed Cost | USD/dt | | | **$ 29.00** | **$ 29.00** | **$ 29.00** | **$ 29.00** |
| Yield (gal/Dry US Ton) | | | | | | | |
| Ethanol | gal/short ton | | | 12.32 | 12.32 | 12.32 | 12.32 |
| Sugar | ton/short ton | | | 0.057 | 0.057 | 0.057 | 0.057 |
| Process Efficiency - to ethanol | % HHV | | | | | | |
| Process efficiency - overall | % HHV | | | | | | |
| Ethanol | gal/yr | | | 4.960E+06 | 2.500E+07 | 5.000E+07 | 1.000E+08 |
| | Mil Gal/YR | | | **4.96** | **25.00** | **50.00** | **100.00** |
| | tpy | | | 1.639E+04 | 8.263E+04 | 1.653E+05 | 3.305E+05 |
| | GJ/yr | | | 4.401E+05 | 2.218E+06 | 4.436E+06 | 8.873E+06 |
| | gal/Stream day | | | 2.718E+04 | 1.370E+05 | 2.740E+05 | 5.480E+05 |
| | bbl/s stream day | | | 6.472E+02 | 3.262E+03 | 6.523E+03 | 1.305E+04 |
| Sugar, GJ/yr | 14.877 | GJ/yr | | 3.427E+05 | 1.727E+06 | 3.454E+06 | 6.908E+06 |
| Sugar (glucose +fructose) | ton/yr | | | 2.30E+04 | 1.16E+05 | 2.32E+05 | 4.64E+05 |
| Sugar Cane - Sucrose Content | | wt % | | 14.00% | 14.00% | 14.00% | 14.00% |
| Ethanol/sugar Split | | % | | **60%** | **60%** | **60%** | **60%** |
| Sugar extraction efficiency | | % | | 97.0% | 97.0% | 97.0% | 97.0% |
| Ethanol Production efficiency | | % | | 90.0% | 90.0% | 90.0% | 90.0% |
| lb ethanol/lb sucrose to ethanol train | 0.5386 | lb/lb | | 0.4848 | 0.4848 | 0.4848 | 0.4848 |
| **Capital Cost (million USD)** | | | | | | | |

**Table 38. (Continued)**

| Country | Mexico | Code = | 8 | | | | |
|---|---|---|---|---|---|---|---|
| **Cost component** | **Units** | **Cost Factor** | **Scale Factor** | **Mexico Sugar Mill** | **Mexico Sugar Mill** | **Mexico Sugar Mill** | **Mexico Sugar Mill** |
| **Year $** | **$** | | | **2005** | **2005** | **2005** | **2005** |
| Total Capital | | | 0.38 | 12.07 | 26.40 | 34.36 | 44.71 |
| **Direct Fixed Capital (DFC), also called TIC** | | | | 12.07 | 26.40 | 34.36 | 44.71 |
| Engineering | DFC × MF | 0.12 | | 1.45 | 3.17 | 4.12 | 5.37 |
| Construction | DFC × MF | 0.13 | | 1.57 | 3.43 | 4.47 | 5.81 |
| Contractor & Legal | DFC × MF | 0.08 | | 0.97 | 2.11 | 2.75 | 3.58 |
| Process/Project Contingency | DFC × MF | 0.0412 | | 0.50 | 1.09 | 1.42 | 1.84 |
| **Total Plant Cost (TPC)** | | | | 16.54 | 36.20 | 47.11 | 61.31 |
| AFUDC | | | | | | | |
| **Total Plant Investment (TPI)** | | | | 16.54 | 36.20 | 47.11 | 61.31 |
| Land | | | 0.20 | 2.21 | 3.05 | 3.51 | 7.02 |
| Startup | | | | 0.00 | 0.00 | 0.00 | 0.00 |
| **Total Capital Cost (TCC)** | | | | 18.75 | 39.26 | 50.62 | 68.32 |
| **Contingency/TPI** | | | | | | | |
| Working Capital | DFC × MF | 0.05 | | 0.83 | 1.81 | 2.36 | 3.07 |
| **Variable Operating Costs (million USD/yr)** | | | | | | | |
| Feed | | | | 11.68 | 58.84 | 117.69 | 235.37 |
| Utilities | | | | 0.36 | 1.79 | 3.58 | 7.16 |
| Other | | | | 1.18 | 5.96 | 11.93 | 23.86 |
| Catalysts and Chemicals | | | | 0.83 | 4.18 | 8.35 | 16.70 |
| **Total** | | | | 14.04 | 70.77 | 141.55 | 283.09 |
| Fixed Operating Costs (million USD/yr) | | | | | | | |
| Labor (SF from V-R) | DFC × MF | 0.0165 | 0.25 | 0.07 | 0.11 | 0.13 | 0.15 |
| Maintenance (3% of A) | | | 0.03 | 0.36 | 0.79 | 1.03 | 1.34 |
| General Overhead (65% of labor + maint) | | | 0.65 | 0.28 | 0.58 | 0.75 | 0.97 |
| Direct Overhead (45% of Labor) | | | 0.45 | 0.03 | 0.05 | 0.06 | 0.07 |
| Insurance (0.5% of TIC) | | | 0.005 | 0.09 | 0.20 | 0.25 | 0.34 |
| Total | | | | 0.84 | 1.73 | 2.22 | 2.87 |

## Table 39. Capital and operating costs, USA sugar-ethanol mill

| Country | USA | Code = | 1 | | | | |
|---|---|---|---|---|---|---|---|
| **Cost component** | **Units** | **Cost Factor** | **Scale Factor** | **USA Sugar Mill** | **USA Sugar Mill** | **USA Sugar Mill** | **USA Sugar Mill** |
| **Year $** | **$** | | | **2005** | **2005** | **2005** | **2005** |
| Cost Index (2 = M&S, 3 = CE) | | | 3 | 130.93 | 130.93 | 130.93 | 130.93 |
| Plant Size | dry short ton/day | | | 2,206 | 11,118 | 22,236 | 44,473 |
| | dry tonne/day | | | 2,000 | 10,080 | 20,160 | 40,320 |
| Stream Factor | % | | | **50%** | **50%** | **50%** | **50%** |
| Feed | Dry short ton/yr | | | 4.026E+05 | 2.029E+06 | 4.058E+06 | 8.116E+06 |
| | GJ/short ton | | | | | | |
| | GJ/yr | | | 0.000E+00 | 0.000E+00 | 0.000E+00 | 0.000E+00 |
| Feed Cost | USD/dt | | | **$ 29.00** | **$ 29.00** | **$ 29.00** | **$ 29.00** |
| Yield (gal/Dry US Ton) | | | | | | | |
| Ethanol | gal/short ton | | | 12.32 | 12.32 | 12.32 | 12.32 |
| Sugar | ton/short ton | | | 0.059 | 0.059 | 0.059 | 0.059 |
| Process Efficiency - to ethanol | % HHV | | | | | | |
| Process efficiency - overall | % HHV | | | | | | |
| Ethanol | gal/yr | | | 4.960E+06 | 2.500E+07 | 5.000E+07 | 1.000E+08 |
| | Mil Gal/YR | | | **4.96** | **25.00** | **50.00** | **100.00** |
| | tpy | | | 1.639E+04 | 8.263E+04 | 1.653E+05 | 3.305E+05 |
| | GJ/yr | | | 4.401E+05 | 2.218E+06 | 4.436E+06 | 8.873E+06 |
| | gal/Stream day | | | 2.718E+04 | 1.370E+05 | 2.740E+05 | 5.480E+05 |
| | bbl/s stream day | | | 6.472E+02 | 3.262E+03 | 6.523E+03 | 1.305E+04 |
| Sugar, GJ/yr | 14.877 | GJ/yr | | 3.532E+05 | 1.780E+06 | 3.561E+06 | 7.121E+06 |
| Sugar (glucose +fructose) | ton/yr | | | 2.37E+04 | 1.20E+05 | 2.39E+05 | 4.79E+05 |
| Sugar Cane - Sucrose Content | | wt % | | 14.00% | 14.00% | 14.00% | 14.00% |
| Ethanol/sugar Split | | % | | **60%** | **60%** | **60%** | **60%** |
| Sugar extraction efficiency | | % | | 97.0% | 97.0% | 97.0% | 97.0% |
| Ethanol Production efficiency | | % | | 90.0% | 90.0% | 90.0% | 90.0% |
| lb ethanol/lb sucrose to ethanol train | 0.5386 | lb/lb | | 0.4848 | 0.4848 | 0.4848 | 0.4848 |
| **Capital Cost (million USD)** | | | | | | | |

**Table 39. (Continued)**

| Country | USA | Code = | 1 | | | | |
|---|---|---|---|---|---|---|---|
| **Cost component** | **Units** | **Cost Factor** | **Scale Factor** | **USA Sugar Mill** | **USA Sugar Mill** | **USA Sugar Mill** | **USA Sugar Mill** |
| **Year $** | **$** | | | **2005** | **2005** | **2005** | **2005** |
| Total Capital | | | 0.38 | 15.00 | 32.82 | 42.72 | 55.59 |
| **Direct Fixed Capital (DFC), also called TIC** | | | | 15.00 | 32.82 | 42.72 | 55.59 |
| Engineering | DFC × MF | | | 1.80 | 3.94 | 5.13 | 6.67 |
| Construction | DFC × MF | | | 1.95 | 4.27 | 5.55 | 7.23 |
| Contractor & Legal | DFC × MF | | | 1.20 | 2.63 | 3.42 | 4.45 |
| Process/Project Contingency | DFC × MF | | | 0.62 | 1.35 | 1.76 | 2.29 |
| **Total Plant Cost (TPC)** | | | | 20.57 | 45.01 | 58.57 | 76.22 |
| AFUDC | | | | | | | |
| **Total Plant Investment (TPI)** | | | | 20.57 | 45.01 | 58.57 | 76.22 |
| Land | | | 0.20 | 2.21 | 3.05 | 3.51 | 7.02 |
| Startup | | | | 0.00 | 0.00 | 0.00 | 0.00 |
| **Total Capital Cost (TCC)** | | | | 22.78 | 48.06 | 62.08 | 83.24 |
| **Contingency/TPI** | | | | | | | |
| Working Capital | DFC × MF | | | 1.03 | 2.25 | 2.93 | 3.81 |
| **Variable Operating Costs (million USD/yr)** | | | | | | | |
| Feed | | | | 11.68 | 58.84 | 117.69 | 235.37 |
| Utilities | | | | 0.36 | 1.79 | 3.58 | 7.16 |
| Other | | | | 1.18 | 5.96 | 11.93 | 23.86 |
| Catalysts and Chemicals | | | | 0.83 | 4.18 | 8.35 | 16.70 |
| **Total** | | | | 14.04 | 70.77 | 141.55 | 283.09 |
| Fixed Operating Costs (million USD/yr) | | | | | | | |
| Labor (SF from V-R) | DFC ×MF | | 0.25 | 0.29 | 0.44 | 0.52 | 0.62 |
| Maintenance (3% of A) | | | 0.03 | 0.45 | 0.98 | 1.28 | 1.67 |
| General Overhead (65% of labor + maint) | | | 0.65 | 0.48 | 0.93 | 1.17 | 1.49 |
| Direct Overhead (45% of Labor) | | | 0.45 | 0.13 | 0.20 | 0.23 | 0.28 |
| Insurance (0.5% of TIC) | | | 0.005 | 0.11 | 0.24 | 0.31 | 0.42 |
| **Total** | | | | 1.47 | 2.79 | 3.52 | 4.47 |

**Table 40. Sugar cane ethanol: summary yields, costs, and plant gate prices, 25 MM GPY ethanol**

| Sugar Cane | Units | USA | Argentina | Brazil | Canada | Caribbean Basin | China | Colombia | India | Mexico |
|---|---|---|---|---|---|---|---|---|---|---|
| **Plant Size** | MM gal/yr | 25 | | 25 | | 25 | 25 | 25 | 25 | 25 |
| **Yields** | | | | | | | | | | |
| Primary Product Yield | gal/ton | 12.32 | | 12.32 | | 12.32 | 12.32 | 12.32 | 12.32 | 12.32 |
| By Product A | | Sugar | | Sugar | | Sugar | Sugar | Sugar | Sugar | Sugar |
| Yield unit | ton | | | | | | | | | |
| Annual Yield | ton/yr | 1.16E+05 | | 1.16E+05 | | 1.16E+05 | 1.16E+05 | 1.16E+05 | 1.16E+05 | 1.16E+05 |
| Price unit | | 300.60 | | 161.36 | | 201.52 | 262.21 | 157.79 | 194.48 | 295.76 |
| MM USD/yr | | 34.87 | | 18.72 | | 23.38 | 30.42 | 18.30 | 22.56 | 34.31 |
| By Product B | | | | | | | | | | |
| Yield unit | | | | | | | | | | |
| Annual Yield | | | | | | | | | | |
| Price unit | | | | | | | | | | |
| Annual Cost | MM USD/yr | | | | | | | | | |
| **Energy Inputs** | | | | | | | | | | |
| Electricity | | | | | | | | | | |
| Cost | $/kWh | | | | | | | | | |
| Amount | kWh/yr | N/A | | N/A | | N/A | N/A | N/A | N/A | N/A |
| Annual Cost | MM USD/yr | | | | | | | | | |
| Fuel | | | | | | | | | | |
| Type | | | | | | | | | | |
| Unit | | | | | | | | | | |
| Amount | | N/A | | N/A | | N/A | N/A | N/A | N/A | N/A |
| Cost | USD/unit | | | | | | | | | |
| Annual Cost | MM USD/yr | | | | | | | | | |
| **Capital & Operating Costs** | | | | | | | | | | |
| Capital | MMUSD | 48.06 | | 49.05 | | 53.84 | 52.58 | 47.81 | 51.06 | 39.26 |
| Operating w/o feed | MM USD/yr | 14.72 | | 14.23 | | 14.92 | 15.01 | 14.37 | 14.26 | 13.66 |
| Feed | | | | | | | | | | |
| Unit | | d ton | | d ton | | d ton | d ton | d ton | d ton | d ton |

**Table 40. (Continued)**

| Sugar Cane | Units | USA | Argentina | Brazil | Canada | Caribbean Basin | China | Colombia | India | Mexico |
|---|---|---|---|---|---|---|---|---|---|---|
| Cost/unit | USD/unit | 29 | | 10.45 | | 15 | 23 | 26.85 | 15 | 29 |
| Rate | unit/day | 10080 | | 10080 | | 10080 | 10080 | 10080 | 10080 | 10080 |
| Yearly Cost | MM USD/yr | 58.84 | | 21.2 | | 30.44 | 46.67 | 54.48 | 30.44 | 58.84 |
| **Product Plant Gate Cost** | | | | | | | | | | |
| Basis | | | | 10 SL | | 10 SL | 10 SL | 10 SL | 10 SL | 5 SL |
| Value | USD/gal | 1.79 | | 1.17 | | 1.2 | 1.56 | 1.71 | 1.18 | 1.76 |

**Table 41. Sugar cane ethanol: summary yields, costs, and plant gate prices, 50 MM GPY ethanol**

| Sugar Cane | Units | USA | Argentina | Brazil | Canada | Caribbean Basin | China | Colombia | India | Mexico |
|---|---|---|---|---|---|---|---|---|---|---|
| **Plant Size** | MM gal/yr | 50 | | 50 | | 50 | 50 | 50 | 50 | 50 |
| **Yields** | | | | | | | | | | |
| Primary Product Yield | gal/ton | 12.32 | | 12.32 | | 12.32 | 12.32 | 12.32 | 12.32 | 12.32 |
| By Product A | | Sugar | | Sugar | | Sugar | Sugar | Sugar | Sugar | Sugar |
| Yield unit | | ton | | ton | | ton | ton | ton | ton | ton |
| Annual Yield | | 2.32E+05 | | 2.32E+05 | | 2.32E+05 | 2.32E+05 | 2.32E+05 | 2.32E+05 | 2.32E+05 |
| Price unit | | 287.22 | | 148.4 | | 186.58 | 247.42 | 144.83 | 184.6 | 285.49 |
| MM USD/yr | | 66.64 | | 34.43 | | 43.29 | 57.40 | 33.60 | 42.83 | 66.23 |
| By Product B | | | | | | | | | | |
| Yield unit | | | | | | | | | | |
| Annual Yield | | | | | | | | | | |
| Price unit | | | | | | | | | | |
| Annual Cost | MM USD/yr | | | | | | | | | |
| **Energy Inputs** | | | | | | | | | | |
| Electricity | | | | | | | | | | |
| Cost | $/kWh | N/A | | N/A | | N/A | N/A | N/A | N/A | N/A |
| Amount | kWh/yr | | | | | | | | | |
| Annual Cost | MM USD/yr | | | | | | | | | |
| | | | | | | | | | | |

**Table 41. (Continued)**

| Sugar Cane | Units | USA | Argentina | Brazil | Canada | Caribbean Basin | China | Colombia | India | Mexico |
|---|---|---|---|---|---|---|---|---|---|---|
| Fuel | | | | | | | | | | |
| Type | | N/A | | N/A | | N/A | N/A | N/A | N/A | N/A |
| Unit | | | | | | | | | | |
| Amount | | | | | | | | | | |
| Cost | USD/unit | | | | | | | | | |
| Annual Cost | MM USD/yr | | | | | | | | | |
| **Capital & Operating Costs** | | | | | | | | | | |
| Capital | MMUSD | 62.08 | | 63.36 | | 69.59 | 67.96 | 61.75 | 65.98 | 50.62 |
| Operating w/o feed | MM USD/yr | 27.38 | | 26.8 | | 27.65 | 27.75 | 26.97 | 26.85 | 26.08 |
| Feed | | | | | | | | | | |
| Unit | | d ton | | d ton | | d ton | d ton | d ton | d ton | d ton |
| Cost/unit | USD/unit | | | 10.45 | | 15 | 23 | 26.85 | 15 | 29 |
| Rate | unit/day | 20160 | | 20160 | | 20160 | 20160 | 20160 | 20160 | 20160 |
| Yearly Cost | MM USD/yr | 117.69 | | 42.41 | | 60.87 | 93.34 | 108.96 | 60.87 | 117.69 |
| **Product Plant Gate Cost** | | | | | | | | | | |
| Basis | | | | 10-SL | | 10-SL | 10-SL | 10-SL | 10-SL | 5-SL |
| Value | USD/gal | 1.71 | | 1.09 | | 1.11 | 1.48 | 1.63 | 1.10 | 1.7 |

**Table 42. Sugar cane ethanol: summary yields, costs, and plant gate prices, 100 MM GPY ethanol**

| Sugar Cane | Units | USA | Argentina | Brazil | Canada | Caribbean Basin | China | Colombia | India | Mexico |
|---|---|---|---|---|---|---|---|---|---|---|
| **Plant Size** | MM gal/yr | 100 | | 100 | | 100 | 100 | 100 | 100 | 100 |
| **Yields** | | | | | | | | | | |
| Primary Product Yield | gal/ton | 12.32 | | 12.32 | | 12.32 | 12.32 | 12.32 | 12.32 | 12.32 |
| By Product A | | Sugar | | Sugar | | Sugar | Sugar | Sugar | Sugar | Sugar |
| Yield unit | | ton | | ton | | ton | ton | ton | ton | ton |
| Annual Yield | | 4.64E+05 | | 4.64E+05 | | 4.64E+05 | 4.64E+05 | 4.64E+05 | 4.64E+05 | 4.64E+05 |
| Price unit | | 279.16 | | 140.57 | | 177.51 | 238.46 | 137.02 | 176.29 | 179.31 |
| MM USD/yr | | 1.30E+02 | | 6.52E+01 | | 8.24E+01 | 1.11E+02 | 6.36E+01 | 8.18E+01 | 8.32E+01 |

**Table 42. (Continued)**

| Sugar Cane | Units | USA | Argentina | Brazil | Canada | Caribbean Basin | China | Colombia | India | Mexico |
|---|---|---|---|---|---|---|---|---|---|---|
| By Product B | | | | | | | | | | |
| Yield unit | | | | | | | | | | |
| Annual Yield | | | | | | | | | | |
| Price unit | | | | | | | | | | |
| Annual Cost | MM USD/yr | | | | | | | | | |
| **Energy Inputs** | | | | | | | | | | |
| Electricity | | N/A | | N/A | | N/A | N/A | N/A | N/A | N/A |
| Cost | $/kWh | | | | | | | | | |
| Amount | kWh/yr | | | | | | | | | |
| Annual Cost | MM USD/yr | | | | | | | | | |
| Fuel | | N/A | | N/A | | N/A | N/A | N/A | N/A | N/A |
| Type | | | | | | | | | | |
| Unit | | | | | | | | | | |
| Amount | | | | | | | | | | |
| Cost | USD/unit | | | | | | | | | |
| Annual Cost | MM USD/yr | | | | | | | | | |
| **Capital & Operating Costs** | | | | | | | | | | |
| Capital | MMUSD | 83.24 | | 84.91 | | 93.02 | 90.89 | 82.81 | 88.31 | 68.32 |
| Operating w/o feed | MM USD/yr | 52.19 | | 51.51 | | 52.54 | 52.65 | 51.70 | 51.58 | 50.59 |
| Feed | | | | | | | | | | |
| Unit | | d ton | | d ton | | d ton | d ton | d ton | d ton | d ton |
| Cost/unit | USD/unit | 29 | | 10.45 | | 15 | 23 | 26.85 | 15 | 29 |
| Rate | unit/day | 40320 | | 40320 | | 40320 | 40320 | 40320 | 40320 | 40320 |
| Yearly Cost | MM USD/yr | 235.37 | | 84.82 | | 121.74 | 186.68 | 217.92 | 121.74 | 235.37 |
| **Product Plant Gate Cost** | | | | | | | | | | |
| Basis | | | | 10-SL | | 10-SL | 10-SL | 10-SL | 10-SL | 5-SL |
| Value | USD/gal | 1.66 | | 1.05 | | 1.06 | 1.42 | 1.59 | 1.05 | 1.67 |

**Table 43. USA cellulosic ethanol (bc) capital and operating costs**

| Country | USA | Code = | 1 | | | | | |
|---|---|---|---|---|---|---|---|---|
| **Cost component** | **Units** | **Cost Factor** | **Scale Factor** | **USA Cellulosic Ethanol (TC)** | **USA Cellulosic Ethanol (TC)** | **USA Cellulosic Ethanol (TC)** | **USA Cellulosic Ethanol (TC)** | **USA Cellulosic Ethanol (TC)** |
| **Year $** | **$** | | | **2002** | **2005** | **2005** | **2005** | **2005** |
| Cost Index (2 = M&S, 3 = CE) | | | 3 | 110.63 | 130.93 | 130.93 | 130.93 | 130.93 |
| Plant Size | dry short ton/day | | | 2,206 | **1,608** | **804** | **3,215** | **9,645** |
| | dry tonne/day | | | 2,000 | 1,457 | 729 | 2,915 | 8,744 |
| Stream Factor | % | | | 95% | 95% | 95% | 95% | 95% |
| Feed | Dry short ton/yr | | | 7.649E+05 | 5.574E+05 | 2.787E+05 | 1.115E+06 | 063.344E+06 |
| | GJ/short ton | | | 1.794E+01 | 1.794E+01 | 1.794E+01 | 1.794E+01 | 1.794E+01 |
| | GJ/yr | | | 1.372E+07 | 9.997E+06 | 4.999E+06 | 1.999E+07 | 5.998E+07 |
| Feed Cost | USD/dt | | | **$ 35.00** | **$ 35.00** | **$ 35.00** | **$ 35.00** | **$ 35.00** |
| Yield (gal/Dry US Ton) | | | | | | | | |
| Ethanol | gal/short ton | | | 89.7 | 89.7 | 89.7 | 89.70 | 89.7 |
| Electricity | kWh/gal EtOH | | | 2.18 | 2.18 | 2.18 | 2.18 | 2.18 |
| Process Efficiency - to ethanol | % HHV | | | 44.4% | 44.4% | 44.4% | 44.4% | 44.4% |
| Process efficiency - overall | % HHV | | | | | | | |
| Ethanol | gal/yr | | | 6.861E+07 | 5.000E+07 | 2.500E+07 | | 3.000E+08 |
| | Mil Gal/YR | | | 68.61 | **50.00** | **25.00** | **100.00** | **300.00** |
| | tpy | | | 2.275E+05 | 1.657E+05 | 8.287E+04 | 3.315E+05 | 9.945E+05 |
| | GJ/yr | | | 6.088E+06 | 4.436E+06 | 2.218E+06 | 8.872E+06 | 2.662E+07 |
| | gal/Stream day | | | 1.979E+05 | 1.442E+05 | 7.210E+04 | 2.884E+05 | 8.652E+05 |
| | bbl/s stream day | | | 4.711E+03 | 3.433E+03 | 1.717E+03 | 6.866E+03 | 2.060E+04 |
| Electricity | kWh/yr | | | 1.50E+08 | 1.09E+08 | 5.45E+07 | 2.18E+08 | 6.54E+08 |

Table 43. (Continued)

| Country | USA | Code = | 1 | | | | | |
|---|---|---|---|---|---|---|---|---|
| **Cost component** | **Units** | **Cost Factor** | **Scale Factor** | **USA Cellulosic Ethanol (TC)** | **USA Cellulosic Ethanol (TC)** | **USA Cellulosic Ethanol (TC)** | **USA Cellulosic Ethanol (TC)** | **USA Cellulosic Ethanol (TC)** |
| **Year $** | **$** | | | **2002** | **2005** | **2005** | **2005** | **2005** |
| **Capital Cost (million USD)** | | | | | | | | |
| Feed Handling | | | 0.70 | 7.50 | 8.88 | 6.47 | 17.07 | 36.82 |
| Pretreatment | | | 0.80 | 19.00 | 22.49 | 12.92 | 39.15 | 94.29 |
| Neutralization/Conditioning | | | 0.70 | 7.90 | 9.35 | 5.76 | 15.19 | 32.77 |
| Saccharification & Fermentation | | | 0.85 | 9.40 | 11.13 | 6.17 | 20.05 | 51.02 |
| Distillation & Solids Recovery | | | 0.65 | 21.90 | 25.92 | 16.52 | 40.67 | 83.06 |
| Wastewater Treatment | | | 0.65 | 3.10 | 3.67 | 2.34 | 5.76 | 11.76 |
| Storage | | | 0.65 | 2.10 | 2.49 | 1.58 | 3.90 | 7.97 |
| Boiler/Turbogenerator | | | 0.70 | 38.60 | 45.68 | 28.12 | 74.21 | 160.13 |
| Utilities | | | 0.70 | 4.60 | 5.44 | 3.35 | 8.84 | 19.08 |
| **Direct Fixed Capital (DFC), also called TIC** | | | | 114.10 | 135.04 | 83.22 | 224.84 | 496.90 |
| Engineering | DFC × MF | 0.12 | | 13.69 | 16.20 | 9.99 | 26.98 | 59.63 |
| Construction | DFC × MF | 0.13 | | 14.83 | 17.56 | 10.82 | 29.23 | 64.60 |
| Contractor & Legal | DFC × MF | 0.08 | | 9.13 | 10.80 | 6.66 | 17.99 | 39.75 |
| Process/Project Contingency | DFC × MF | 0.0412 | | 4.70 | 5.56 | 3.43 | 9.26 | 20.47 |
| **Total Plant Cost (TPC)** | | | | 156.45 | 185.17 | 114.11 | 308.30 | 681.35 |
| AFUDC | | | | | | | | |
| **Total Plant Investment (TPI)** | | | | 156.45 | 185.17 | 114.11 | 308.30 | 681.35 |
| Land | | | 0.60 | 2.21 | 2.21 | 2.21 | 5.08 | 15.23 |
| Startup | | | | 0 | 0 | 0.00 | 0.00 | 0.00 |

**Table 43. (Continued)**

| **Country** | **USA** | **Code =** | **1** | | | | | |
|---|---|---|---|---|---|---|---|---|
| **Cost component** | **Units** | **Cost Factor** | **Scale Factor** | **USA Cellulosic Ethanol (TC)** | **USA Cellulosic Ethanol (TC)** | **USA Cellulosic Ethanol (TC)** | **USA Cellulosic Ethanol (TC)** | **USA Cellulosic Ethanol (TC)** |
| | | | | | | | | |
| **Total Capital Cost (TCC)** | | | | 158.66 | 187.38 | 116.32 | 313.38 | 696.58 |
| **Contingency/TPI** | | | | | | | | |
| Working Capital | DFC × MF | 0.05 | | 7.82 | 9.26 | 5.71 | 15.42 | 34.07 |
| **Variable Operating Costs (million USD/yr)** | | | | | | | | |
| Feed | | | | 26.77 | 19.51 | 9.75 | 39.02 | 117.06 |
| Utilities | | | | 0.00 | 0.00 | 0.00 | 0.00 | 0.00 |
| Waste Disposal | | | | 2.50 | 1.82 | 0.91 | 3.64 | 10.93 |
| Catalysts and Chemicals | Major comp is cellulase | | | 12.70 | 9.25 | 4.63 | 18.51 | 55.53 |
| **Total** | | | | 41.97 | 30.59 | 15.29 | 61.17 | 183.51 |
| Fixed Operating Costs (million USD/yr) | | | | | | | | |
| Labor (SF from V-R) | DFC × MF | 0.0165 | 0.25 | 1.88 | 2.23 | 1.73 | 2.45 | 3.22 |
| Maintenance (3% of A) | | | 0.03 | 3.42 | 4.05 | 2.50 | 6.75 | 14.91 |
| General Overhead (65% of labor + maint) | | | 0.65 | 3.45 | 4.08 | 2.75 | 5.98 | 11.78 |
| Direct Overhead (45% of Labor) | | | 0.45 | 0.85 | 1.00 | 0.78 | 1.10 | 1.45 |
| Insurance (0.5% of TIC) | | | 0.005 | 0.79 | 0.94 | 0.58 | 1.57 | 3.48 |
| **Total** | | | | 10.39 | 12.30 | 8.34 | 17.84 | 34.85 |
| | | | | | | | | |

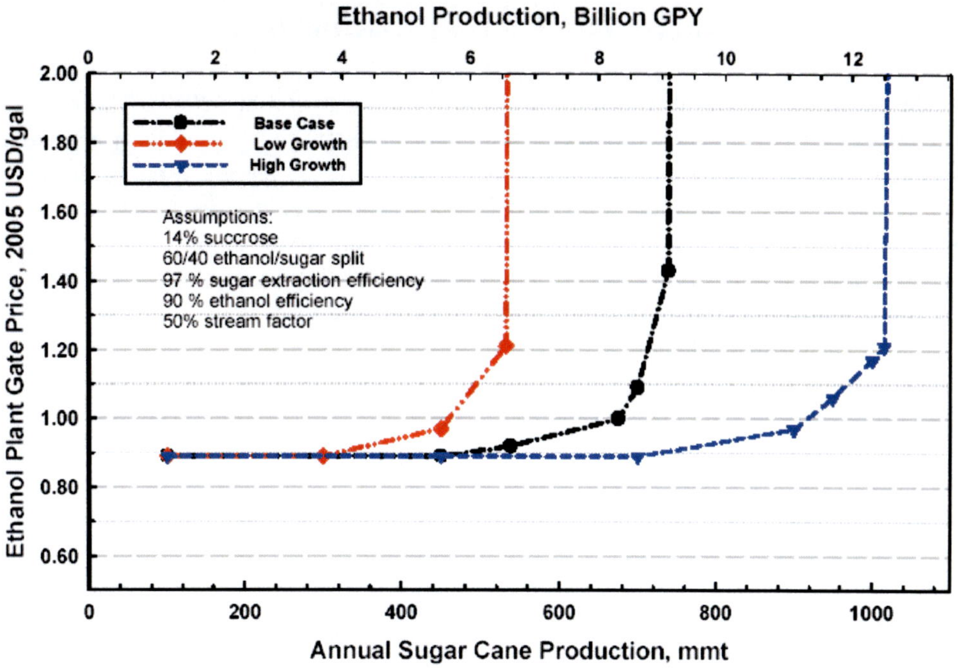


Figure 18. Brazil ethanol potential (based on 07-25-2007 ORNL Supply Curves Table 1-3)

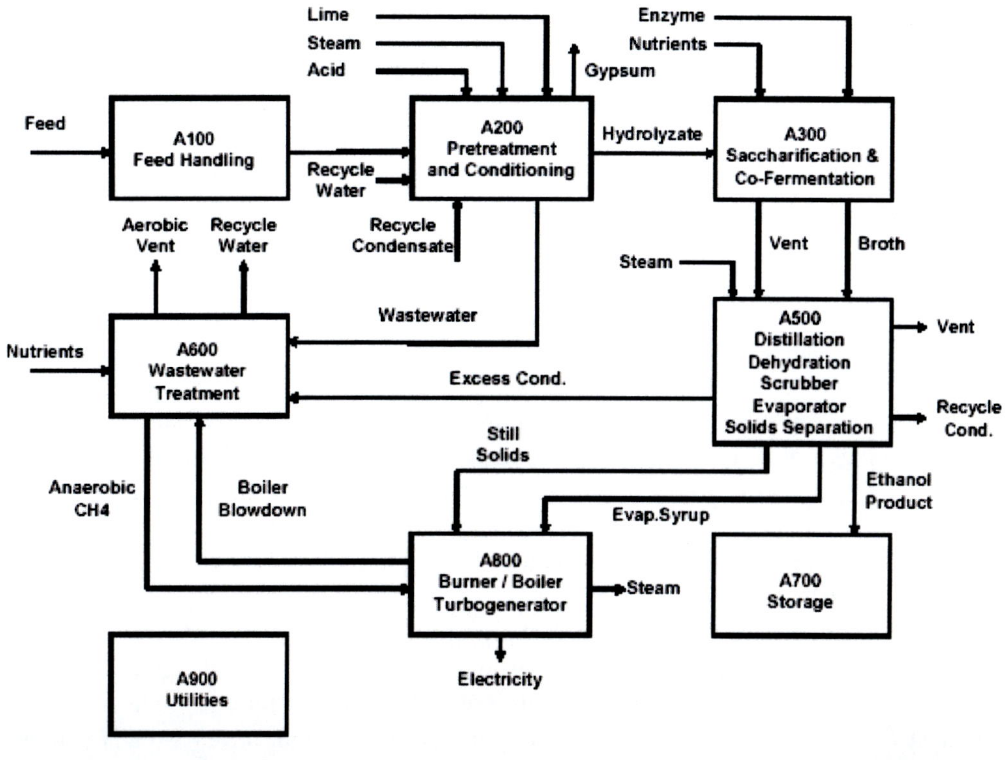


Figure 19. Overall biochemical process flow diagram

## 5. CELLULOSIC ETHANOL (BIOCHEMICAL)

The process overview discussion is excerpted from Aden et al. (2002). The process can be described as using co-current dilute acid prehydrolysis of the lignocellulosic biomass with enzymatic saccharification of the remaining cellulose and co-fermentation of the resulting glucose and xylose to ethanol. The process design also includes feedstock handling and storage, product purification, wastewater treatment, lignin combustion, product storage, and all other required utilities. In all, the process is divided into eight areas, see Figure 19.

The feedstock, in this case corn stover (comprised of leaves, cobs, and husks), is delivered to the feedstock handling area (A100) for storage and size reduction. From there the biomass is conveyed to pretreatment and detoxification (A200). In this area, the biomass is treated with dilute sulfuric acid catalyst for a short time, liberating the hemicellulose sugars and other compounds. Separation with washing removes the acid from the solids for neutralization. Overliming is required to remove compounds liberated in the pretreatment that are toxic to the fermenting organism. Detoxification is applied only to the liquid portion of the hydrolysis stream.

Enzymatic hydrolysis (or saccharification) coupled with co-fermentation (A300) of the detoxified hydrolyzate slurry is carried out in continuous hydrolysis tanks and anaerobic fermentation tanks in series. A purchased cellulase enzyme preparation is added to the hydrolyzate in the hydrolysis tanks that are maintained at a temperature to optimize the enzyme's activity. The fermenting organism *Zymomonas mobilis* is first grown in a series of progressively larger batch anaerobic fermentations to make enough cells to inoculate the main fermentors. The inoculum, along with other nutrients, is added to the first ethanol fermentor along with the partially saccharified slurry at a reduced temperature. The cellulose will continue to be hydrolyzed, although at a slower rate, at the lower temperature. After several days of separate and combined saccharification and co-fermentation, most of the cellulose and xylose will have been converted to ethanol. The resulting beer is sent to product recovery.

Product recovery (A500) involves distilling the beer to separate ethanol from the water and residual solids. A mixture of nearly azeotropic water and ethanol is purified to pure ethanol using a vapor-phase molecular sieve. Solids from the distillation bottoms are separated and sent to the boiler. Concentration of the distillation bottoms liquid is performed by evaporation, using waste heat. The evaporated condensate is returned to the process and the concentrated syrup is sent to the combustor.

Part of the evaporator condensate, along with other wastewater, is treated by anaerobic and aerobic digestion (A600). The biogas (high in methane) from the anaerobic digestion is sent to the combustor for energy recovery. The treated water is suitable for recycling and is returned to the process.

The solids from distillation, the concentrated syrup from the evaporator, and biogas form anaerobic digestion are combusted in a fluidized bed combustor (A800) to produce high-pressure steam for electricity production and process heat. The majority of the process steam demand is in the pretreatment reactor and distillation areas. Generally, the process produces excess steam that is converted to electricity for use in the plant and for sale to the grid.

Capital and operating costs for the USA base case are given in Table 43. Figure 20 gives the cost breakdown for the USA base case as a function of plant size using 35 USD/ton feed.

Cellulosic feedstock costs for all countries are given in Table 19 (in the methodology section). Capital and operating costs for other countries are given in Table 44 through Table 51.

Table 52 and Table 53 summarize the bagasse capital/operating costs and plant gate costs for 25 MM GPY and 100 MM GPY plants, respectively. Table 54 – Table 57 summarize the same information for agricultural residues and wood/perennial feedsrespectively. Table 58 summarizes the cellulosic ethanol (BC) plant gate price estimates.

## 6. CELLULOSIC ETHANOL (THERMOCHEMICAL)

The cellulosic ethanol process for the country comparisons is a conceptual nth plant process that is used by the Office of the Biomass Program as a benchmark for defining research and development goals and performance metrics. Portions of the Phillips et al. (2007) design report are excerpted below to provide background information and a brief discussion of the technology.

This assessment directly builds upon an initial analysis for the TC production of ethanol and other alcohol co-products (Aden et al. 2006), which, in turn, was based upon a detailed design and economic analysis for the production of hydrogen from biomass (Spath et al). This design report is complementary to other studies being funded by the DOE OBP, including the RBAEF (Role of Biomass in America's Energy Future) study. However, the RBAEF study differs in many ways from this study. For example, RBAEF is designed for a further time horizon than 2012. It is based on a different feedstock, switchgrass, and it considers a variety of thermochemical product options, including ethanol, power and Fischer-Tropsch liquids (Larson et al, 2006).

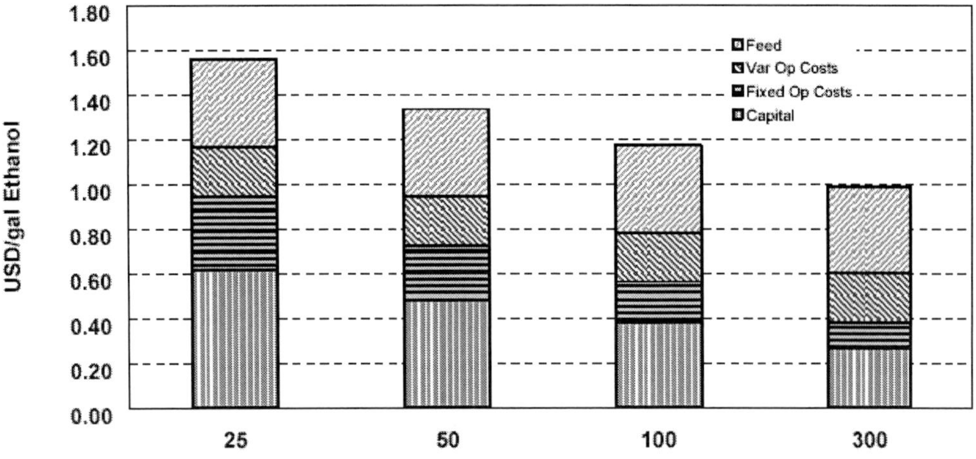


| | 25 | 50 | 100 | 300 |
|---|---|---|---|---|
| Feed | 0.39 | 0.39 | 0.39 | 0.39 |
| Var Op Costs | 0.22 | 0.22 | 0.22 | 0.22 |
| Fixed Op Costs | 0.33 | 0.25 | 0.18 | 0.12 |
| Capital | $0.62 | $0.48 | $0.38 | $0.26 |

Plant Size (mil gal/yr)

Figure 20. USA base case cellulosic ethanol (biochemical) plant gate prices

## Table 44. Argentina cellulosic ethanol (bc) capital and operating costs, bagasse

| Country | Argentina | Code = | 5 | | | | |
|---|---|---|---|---|---|---|---|
| **Cost component** | **Units** | **Cost Factor** | **Scale Factor** | **Argentina Cellulosic Ethanol (TC)** | **Argentina Cellulosic Ethanol (TC)** | **Argentina Cellulosic Ethanol (TC)** | **Argentina Cellulosic Ethanol (TC)** |
| Year $ | $ | | | 2005 | 2005 | 2005 | 2005 |
| Cost Index (2 = M&S, 3 = CE) | | | 3 | 130.93 | 130.93 | 130.93 | 130.93 |
| Plant Size | dry short ton/day | | | **1,608** | **804** | **3,215** | **9,645** |
| | dry tonne/day | | | 1,457 | 729 | 2,915 | 8,744 |
| Stream Factor | % | | | 95% | 95% | 95% | 95% |
| Feed | Dry short ton/yr | | | 5.574E+05 | 2.787E+05 | 1.115E+06 | 3.344E+06 |
| | GJ/short ton | | | 1.794E+01 | 1.794E+01 | 1.794E+01 | 1.794E+01 |
| | GJ/yr | | | 9.997E+06 | 4.999E+06 | 1.999E+07 | 5.998E+07 |
| Feed Cost | USD/dt | | | **$ 9.10** | **$ 9.10** | **$ 9.10** | **$ 9.10** |
| Yield (gal/Dry US Ton) | | | | | | | |
| Ethanol | gal/short ton | | | 89.7 | 89.7 | 89.70 | 89.7 |
| Electricity | kWh/gal EtOH | | | 2.18 | 2.18 | 2.18 | 2.18 |
| Process Efficiency - to ethanol | % HHV | | | 44.4% | 44.4% | 44.4% | 44.4% |
| Process efficiency - overall | % HHV | | | | | | |
| Ethanol | gal/yr | | | 5.000E+07 | 2.500E+07 | 1.000E+08 | 3.000E+08 |
| | Mil Gal/YR | | | **50.00** | **25.00** | **100.00** | **300.00** |
| | tpy | | | 1.657E+05 | 8.287E+04 | 3.315E+05 | 9.945E+05 |
| | GJ/yr | | | 4.436E+06 | 2.218E+06 | 8.872E+06 | 2.662E+07 |
| | gal/Stream day | | | 1.442E+05 | 7.210E+04 | 2.884E+05 | 8.652E+05 |
| | bbl/s stream day | | | 3.433E+03 | 1.717E+03 | 6.866E+03 | 2.060E+04 |
| Electricity | kWh/yr | | | 1.09E+08 | 5.45E+07 | 2.18E+08 | 6.54E+08 |
| **Capital Cost (million USD)** | | | | | | | |
| Feed Handling | | | 0.70 | 9.53 | 6.94 | 18.32 | 39.54 |
| Pretreatment | | | 0.80 | 24.15 | 13.87 | 42.04 | 101.24 |
| Neutralization/Conditioning | | | 0.70 | 10.04 | 6.18 | 16.31 | 35.19 |
| Saccharification & Fermentation | | | 0.85 | 11.95 | 6.63 | 21.53 | 54.78 |
| Distillation & Solids Recovery | | | 0.65 | 27.83 | 17.74 | 43.67 | 89.19 |
| Wastewater Treatment | | | 0.65 | 3.94 | 2.51 | 6.18 | 12.63 |
| Storage | | | 0.65 | 2.67 | 1.70 | 4.19 | 8.55 |
| Boiler/Turbogenerator | | | 0.70 | 49.05 | 30.20 | 79.68 | 171.94 |
| Utilities | | | 0.70 | 5.85 | 3.60 | 9.50 | 20.49 |
| **Direct Fixed Capital (DFC), also called TIC** | | | | 145.00 | 89.36 | 241.42 | 533.55 |
| Engineering | DFC x MF | 0.12 | | 17.40 | 10.72 | 28.97 | 64.03 |
| Construction | DFC x MF | 0.13 | | 18.85 | 11.62 | 31.38 | 69.36 |
| Contractor & Legal | DFC x MF | 0.08 | | 11.60 | 7.15 | 19.31 | 42.68 |
| Process/Project Contingency | DFC x MF | 0.0412 | | 5.97 | 3.68 | 9.95 | 21.98 |
| **Total Plant Cost (TPC)** | | | | 198.82 | 122.53 | 331.04 | 731.60 |
| AFUDC | | | | | | | |
| **Total Plant Investment (TPI)** | | | | 198.82 | 122.53 | 331.04 | 731.60 |
| Land | | | 0.60 | 2.21 | 2.21 | 5.08 | 15.23 |
| Startup | | | | 0 | 0.00 | 0.00 | 0.00 |
| **Total Capital Cost (TCC)** | | | | 201.03 | **124.74** | **336.11** | 746.83 |
| **Contingency/TPI** | | | | | | | |
| Working Capital | DFC x MF | 0.05 | | 9.94 | 6.13 | 16.55 | 36.58 |
| **Variable Operating Costs (million USD/yr)** | | | | | | | |
| Feed | | | | 5.07 | 2.54 | 10.14 | 30.43 |
| Utilities | | | | 0.00 | 0.00 | 0.00 | 0.00 |
| Waste Disposal | | | | 1.82 | 0.91 | 3.64 | 10.93 |
| Catalysts and Chemicals | Major comp is cellulase | | | 9.25 | 4.63 | 18.51 | 55.53 |
| **Total** | | | | 16.15 | 8.07 | 32.30 | 96.89 |
| Fixed Operating Costs (million USD/yr) | | | | | | | |
| Labor (SF from V-R) | DFC x MF | 0.0165 | 0.25 | 1.89 | 1.47 | 2.08 | 2.73 |
| Maintenance (3% of A) | | | 0.03 | 4.35 | 2.68 | 7.24 | 16.01 |
| General Overhead (65% of labor + maint) | | | 0.65 | 4.06 | 2.70 | 6.06 | 12.18 |
| Direct Overhead (45% of Labor) | | | 0.45 | 0.85 | 0.66 | 0.93 | 1.23 |
| Insurance (0.5% of TIC) | | | 0.005 | 1.01 | 0.62 | 1.68 | 3.73 |
| **Total** | | | | 12.15 | 8.13 | 17.99 | 35.89 |

**Table 45. Brazil cellulosic ethanol (bc) capital and operating costs, bagasse**

| Country | Brazil | Code = | 2 | | | | |
|---|---|---|---|---|---|---|---|
| **Cost component** | **Units** | **Cost Factor** | **Scale Factor** | **Brazil Cellulosic Ethanol (TC)** | **Brazil ellulosic Ethanol (TC)** | **Brazil Cellulosic Ethanol (TC)** | **Brazil Cellulosic Ethanol (TC)** |
| Year $ | $ | | | 2005 | 2005 | 2005 | 2005 |
| Cost Index (2 = M&S, 3 = CE) | | | 3 | 130.93 | 130.93 | 130.93 | 130.93 |
| Plant Size | dry short ton/day | | | **1,608** | **804** | **3,215** | **9,645** |
| | dry tonne/day | | | 1,457 | 729 | 2,915 | 8,744 |
| | | | | | | | |
| Stream Factor | % | | | 95% | 95% | 95% | 95% |
| Feed | Dry short ton/yr | | | 5.574E+05 | 2.787E+05 | 1.115E+06 | 3.344E+06 |
| | GJ/short ton | | | 1.794E+01 | 1.794E+01 | 1.794E+01 | 1.794E+01 |
| | GJ/yr | | | 9.997E+06 | 4.999E+06 | 1.999E+07 | 5.998E+07 |
| Feed Cost | USD/dt | | | **$ 7.60** | **$ 7.60** | **$ 7.60** | **$ 7.60** |
| Yield (gal/Dry US Ton) | | | | | | | |
| Ethanol | gal/short ton | | | 89.7 | 89.7 | 89.70 | 89.7 |
| Electricity | kWh/gal EtOH | | | 2.18 | 2.18 | 2.18 | 2.18 |
| Process Efficiency - to ethanol | % HHV | | | 44.4% | 44.4% | 44.4% | 44.4% |
| Process efficiency - overall | % HHV | | | | | | |
| Ethanol | gal/yr | | | 5.000E+07 | 2.500E+07 | 1.000E+08 | 3.000E+08 |
| | Mil Gal/YR | | | **50.00** | **25.00** | **100.00** | **300.00** |
| | tpy | | | 1.657E+05 | 8.287E+04 | 3.315E+05 | 9.945E+05 |
| | GJ/yr | | | 4.436E+06 | 2.218E+06 | 8.872E+06 | 2.662E+07 |
| | gal/Stream day | | | 1.442E+05 | 7.210E+04 | 2.884E+05 | 8.652E+05 |
| | bbl/s stream day | | | 3.433E+03 | 1.717E+03 | 6.866E+03 | 2.060E+04 |
| Electricity | kWh/yr | | | 1.09E+08 | 5.45E+07 | 2.18E+08 | 6.54E+08 |
| | | | | | | | |
| **Capital Cost (million USD)** | | | | | | | |
| Feed Handling | | | 0.70 | 9.07 | 6.61 | 17.44 | 37.63 |
| Pretreatment | | | 0.80 | 22.98 | 13.20 | 40.01 | 96.35 |
| Neutralization/Conditioning | | | 0.70 | 9.55 | 5.88 | 15.52 | 33.49 |
| Saccharification & Fermentation | | | 0.85 | 11.37 | 6.31 | 20.49 | 52.14 |
| Distillation & Solids Recovery | | | 0.65 | 26.49 | 16.88 | 41.56 | 84.89 |
| Wastewater Treatment | | | 0.65 | 3.75 | 2.39 | 5.88 | 12.02 |
| Storage | | | 0.65 | 2.54 | 1.62 | 3.99 | 8.14 |
| Boiler/Turbogenerator | | | 0.70 | 46.68 | 28.74 | 75.84 | 163.64 |
| Utilities | | | 0.70 | 5.56 | 3.42 | 9.04 | 19.50 |
| **Direct Fixed Capital (DFC), also called TIC** | | | | 138.00 | 85.05 | 229.77 | 507.79 |
| Engineering | DFC x MF | 0.12 | | 16.56 | 10.21 | 27.57 | 60.93 |
| Construction | DFC x MF | 0.13 | | 17.94 | 11.06 | 29.87 | 66.01 |
| Contractor & Legal | DFC x MF | 0.08 | | 11.04 | 6.80 | 18.38 | 40.62 |
| Process/Project Contingency | DFC x MF | 0.0412 | | 5.69 | 3.50 | 9.47 | 20.92 |
| **Total Plant Cost (TPC)** | | | | 189.22 | 116.61 | 315.06 | 696.28 |
| AFUDC | | | | | | | |
| **Total Plant Investment (TPI)** | | | | 189.22 | 116.61 | 315.06 | 696.28 |
| Land | | | 0.60 | 2.21 | 2.21 | 5.08 | 15.23 |
| Startup | | | | 0 | 0.00 | 0.00 | 0.00 |
| **Total Capital Cost (TCC)** | | | | 191.43 | **118.82** | **320.13** | 711.51 |
| **Contingency/TPI** | | | | | | | |
| Working Capital | DFC x MF | 0.05 | | 9.46 | 5.83 | 15.75 | 34.81 |
| **Variable Operating Costs (million USD/yr)** | | | | | | | |
| Feed | | | | 4.24 | 2.12 | 8.47 | 25.42 |
| Utilities | | | | 0.00 | 0.00 | 0.00 | 0.00 |
| Waste Disposal | | | | 1.82 | 0.91 | 3.64 | 10.93 |
| Catalysts and Chemicals | Major comp is cellulase | | | 9.25 | 4.63 | 18.51 | 55.53 |
| **Total** | | | | 15.31 | 7.66 | 30.62 | 91.88 |
| Fixed Operating Costs (million USD/yr) | | | | | | | |
| Labor (SF from V-R) | DFC x MF | 0.0165 | 0.25 | 0.95 | 0.74 | 1.04 | 1.37 |
| Maintenance (3% of A) | | | 0.03 | 4.14 | 2.55 | 6.89 | 15.23 |
| General Overhead (65% of labor + maint) | | | 0.65 | 3.31 | 2.14 | 5.16 | 10.79 |
| Direct Overhead (45% of Labor) | | | 0.45 | 0.43 | 0.33 | 0.47 | 0.62 |
| Insurance (0.5% of TIC) | | | 0.005 | 0.96 | 0.59 | 1.60 | 3.56 |
| **Total** | | | | 9.78 | 6.35 | 15.17 | 31.58 |

## Table 46. Canada cellulosic ethanol (bc) capital and operating costs, ag residues

| Country | Canada | Code = | 7 | | | | |
|---|---|---|---|---|---|---|---|
| Cost component | Units | Cost Factor | Scale Factor | Canada Cellulosic Ethanol (BC) | Canada Cellulosic Ethanol (BC) | Canada Cellulosic Ethanol (BC) | Canada Cellulosic Ethanol (BC) |
| Year $ | $ | | | 2005 | 2005 | 2005 | 2005 |
| Cost Index (2 = M&S, 3 = CE) | | | 3 | 130.93 | 130.93 | 130.93 | 130.93 |
| Plant Size | dry short ton/day | | | **1,608** | **804** | **3,215** | **9,645** |
| | dry tonne/day | | | 1,457 | 729 | 2,915 | 8,744 |
| Stream Factor | % | | | 95% | 95% | 95% | 95% |
| Feed | Dry short ton/yr | | | 5.574E+05 | 2.787E+05 | 1.115E+06 | 3.344E+06 |
| | GJ/short ton | | | 1.794E+01 | 1.794E+01 | 1.794E+01 | 1.794E+01 |
| | GJ/yr | | | 9.997E+06 | 4.999E+06 | 1.999E+07 | 5.998E+07 |
| Feed Cost | USD/dt | | | **$ 10.00** | **$ 10.00** | **$ 10.00** | **$ 10.00** |
| Yield (gal/Dry US Ton) | | | | | | | |
| Ethanol | gal/short ton | | | 89.7 | 89.7 | 89.70 | 89.7 |
| Electricity | kWh/gal EtOH | | | 2.18 | 2.18 | 2.18 | 2.18 |
| Process Efficiency - to ethanol | % HHV | | | 44.4% | 44.4% | 44.4% | 44.4% |
| Process efficiency - overall | % HHV | | | | | | |
| Ethanol | gal/yr | | | 5.000E+07 | 2.500E+07 | 1.000E+08 | 3.000E+08 |
| | Mil Gal/YR | | | **50.00** | **25.00** | **100.00** | **300.00** |
| | tpy | | | 1.657E+05 | 8.287E+04 | 3.315E+05 | 9.945E+05 |
| | GJ/yr | | | 4.436E+06 | 2.218E+06 | 8.872E+06 | 2.662E+07 |
| | gal/Stream day | | | 1.442E+05 | 7.210E+04 | 2.884E+05 | 8.652E+05 |
| | bbl/s stream day | | | 3.433E+03 | 1.717E+03 | 6.866E+03 | 2.060E+04 |
| Electricity | kWh/yr | | | 1.09E+08 | 5.45E+07 | 2.18E+08 | 6.54E+08 |
| **Capital Cost (million USD)** | | | | | | | |
| Feed Handling | | | 0.70 | 11.57 | 8.43 | 22.24 | 47.98 |
| Pretreatment | | | 0.80 | 29.30 | 16.83 | 51.01 | 122.85 |
| Neutralization/Conditioning | | | 0.70 | 12.18 | 7.50 | 19.79 | 42.70 |
| Saccharification & Fermentation | | | 0.85 | 14.50 | 8.04 | 26.13 | 66.48 |
| Distillation & Solids Recovery | | | 0.65 | 33.77 | 21.52 | 52.99 | 108.23 |
| Wastewater Treatment | | | 0.65 | 4.78 | 3.05 | 7.50 | 15.32 |
| Storage | | | 0.65 | 3.24 | 2.06 | 5.08 | 10.38 |
| Boiler/Turbogenerator | | | 0.70 | 59.52 | 36.64 | 96.69 | 208.64 |
| Utilities | | | 0.70 | 7.09 | 4.37 | 11.52 | 24.86 |
| **Direct Fixed Capital (DFC), also called TIC** | | | | 175.95 | 108.43 | 292.96 | 647.44 |
| Engineering | DFC x MF | 0.12 | | 21.11 | 13.01 | 35.15 | 77.69 |
| Construction | DFC x MF | 0.13 | | 22.87 | 14.10 | 38.08 | 84.17 |
| Contractor & Legal | DFC x MF | 0.08 | | 14.08 | 8.67 | 23.44 | 51.79 |
| Process/Project Contingency | DFC x MF | 0.0412 | | 7.25 | 4.47 | 12.07 | 26.67 |
| **Total Plant Cost (TPC)** | | | | 241.26 | 148.69 | 401.70 | 887.76 |
| AFUDC | | | | | | | |
| **Total Plant Investment (TPI)** | | | | 241.26 | 148.69 | 401.70 | 887.76 |
| Land | | | 0.60 | 2.21 | 2.21 | 5.08 | 15.23 |
| Startup | | | | 0 | 0.00 | 0.00 | 0.00 |
| **Total Capital Cost (TCC)** | | | | 243.47 | **150.90** | **406.78** | 902.99 |
| **Contingency/TPI** | | | | | | | |
| Working Capital | DFC x MF | 0.05 | | 12.06 | 7.43 | 20.09 | 44.39 |
| **Variable Operating Costs (million USD/yr)** | | | | | | | |
| Feed | | | | 5.57 | 2.79 | 11.15 | 33.44 |
| Utilities | | | | 0.00 | 0.00 | 0.00 | 0.00 |
| Waste Disposal | | | | 1.82 | 0.91 | 3.64 | 10.93 |
| Catalysts and Chemicals | Major comp is cellulase | | | 9.25 | 4.63 | 18.51 | 55.53 |
| **Total** | | | | 16.65 | 8.33 | 33.30 | 99.90 |
| Fixed Operating Costs (million USD/yr) | | | | | | | |
| Labor (SF from V-R) | DFC x MF | 0.0165 | 0.25 | 3.74 | 2.91 | 4.11 | 5.41 |
| Maintenance (3% of A) | | | 0.03 | 5.28 | 3.25 | 8.79 | 19.42 |
| General Overhead (65% of labor + maint) | | | 0.65 | 5.86 | 4.00 | 8.38 | 16.14 |
| Direct Overhead (45% of Labor) | | | 0.45 | 1.68 | 1.31 | 1.85 | 2.43 |
| Insurance (0.5% of TIC) | | | 0.005 | 1.22 | 0.75 | 2.03 | 4.51 |
| **Total** | | | | 17.78 | 12.22 | 25.16 | 47.92 |

**Table 47. Caribbean basin cellulosic ethanol (bc) capital and operating costs, bagasse**

| Country | Caribbean | Code = | 9 | | | | |
|---|---|---|---|---|---|---|---|
| **Cost component** | **Units** | **Cost Factor** | **Scale Factor** | **Caribbean Cellulosic Ethanol (TC)** | **Caribbean Cellulosic Ethanol (TC)** | **Caribbean Cellulosic Ethanol (TC)** | **Caribbean Cellulosic Ethanol (TC)** |
| Year $ | $ | | | 2005 | 2005 | 2005 | 2005 |
| Cost Index (2 = M&S, 3 = CE) | | | 3 | 130.93 | 130.93 | 130.93 | 130.93 |
| Plant Size | dry short ton/day | | | **1,608** | **804** | **3,215** | **9,645** |
| | dry tonne/day | | | 1,457 | 729 | 2,915 | 8,744 |
| Stream Factor | % | | | 95% | 95% | 95% | 95% |
| Feed | Dry short ton/yr | | | 5.574E+05 | 2.787E+05 | 1.115E+06 | 3.344E+06 |
| | GJ/short ton | | | 1.794E+01 | 1.794E+01 | 1.794E+01 | 1.794E+01 |
| | GJ/yr | | | 9.997E+06 | 4.999E+06 | 1.999E+07 | 5.998E+07 |
| Feed Cost | USD/dt | | | **$ 7.60** | **$ 7.60** | **$ 7.60** | **$ 7.60** |
| Yield (gal/Dry US Ton) | | | | | | | |
| Ethanol | gal/short ton | | | 89.7 | 89.7 | 89.70 | 89.7 |
| Electricity | kWh/gal EtOH | | | 2.18 | 2.18 | 2.18 | 2.18 |
| Process Efficiency - to ethanol | % HHV | | | 44.4% | 44.4% | 44.4% | 44.4% |
| Process efficiency - overall | % HHV | | | | | | |
| Ethanol | gal/yr | | | 5.000E+07 | 2.500E+07 | 1.000E+08 | 3.000E+08 |
| | Mil Gal/YR | | | **50.00** | **25.00** | **100.00** | **300.00** |
| | tpy | | | 1.657E+05 | 8.287E+04 | 3.315E+05 | 9.945E+05 |
| | GJ/yr | | | 4.436E+06 | 2.218E+06 | 8.872E+06 | 2.662E+07 |
| | gal/Stream day | | | 1.442E+05 | 7.210E+04 | 2.884E+05 | 8.652E+05 |
| | bbl/s stream day | | | 3.433E+03 | 1.717E+03 | 6.866E+03 | 2.060E+04 |
| Electricity | kWh/yr | | | 1.09E+08 | 5.45E+07 | 2.18E+08 | 6.54E+08 |
| **Capital Cost (million USD)** | | | | | | | |
| Feed Handling | | | 0.70 | 10.02 | 7.30 | 19.25 | 41.55 |
| Pretreatment | | | 0.80 | 25.37 | 14.57 | 44.17 | 106.38 |
| Neutralization/Conditioning | | | 0.70 | 10.55 | 6.49 | 17.14 | 36.98 |
| Saccharification & Fermentation | | | 0.85 | 12.55 | 6.96 | 22.62 | 57.56 |
| Distillation & Solids Recovery | | | 0.65 | 29.24 | 18.64 | 45.89 | 93.72 |
| Wastewater Treatment | | | 0.65 | 4.14 | 2.64 | 6.50 | 13.27 |
| Storage | | | 0.65 | 2.80 | 1.79 | 4.40 | 8.99 |
| Boiler/Turbogenerator | | | 0.70 | 51.54 | 31.73 | 83.73 | 180.67 |
| Utilities | | | 0.70 | 6.14 | 3.76 | 9.98 | 21.53 |
| **Direct Fixed Capital (DFC), also called TIC** | | | | 152.36 | 93.90 | 253.68 | 560.65 |
| Engineering | DFC x MF | 0.12 | | 18.28 | 11.27 | 30.44 | 67.28 |
| Construction | DFC x MF | 0.13 | | 19.81 | 12.21 | 32.98 | 72.88 |
| Contractor & Legal | DFC x MF | 0.08 | | 12.19 | 7.51 | 20.29 | 44.85 |
| Process/Project Contingency | DFC x MF | 0.0412 | | 6.28 | 3.87 | 10.45 | 23.10 |
| **Total Plant Cost (TPC)** | | | | 208.92 | 128.75 | 347.85 | 768.76 |
| AFUDC | | | | | | | |
| **Total Plant Investment (TPI)** | | | | 208.92 | 128.75 | 347.85 | 768.76 |
| Land | | | 0.60 | 2.21 | 2.21 | 5.08 | 15.23 |
| Startup | | | | 0 | 0.00 | 0.00 | 0.00 |
| **Total Capital Cost (TCC)** | | | | 211.13 | **130.96** | **352.93** | 783.99 |
| **Contingency/TPI** | | | | | | | |
| Working Capital | DFC x MF | 0.05 | | 10.45 | 6.44 | 17.39 | 38.44 |
| **Variable Operating Costs (million USD/yr)** | | | | | | | |
| Feed | | | | 4.24 | 2.12 | 8.47 | 25.42 |
| Utilities | | | | 0.00 | 0.00 | 0.00 | 0.00 |
| Waste Disposal | | | | 1.82 | 0.91 | 3.64 | 10.93 |
| Catalysts and Chemicals | Major comp is cellulase | | | 9.25 | 4.63 | 18.51 | 55.53 |
| **Total** | | | | 15.31 | 7.66 | 30.62 | 91.88 |
| Fixed Operating Costs (million USD/yr) | | | | | | | |
| Labor (SF from V-R) | DFC x MF | 0.0165 | 0.25 | 2.14 | 1.66 | 2.35 | 3.09 |
| Maintenance (3% of A) | | | 0.03 | 4.57 | 2.82 | 7.61 | 16.82 |
| General Overhead (65% of labor + maint) | | | 0.65 | 4.36 | 2.91 | 6.47 | 12.94 |
| Direct Overhead (45% of Labor) | | | 0.45 | 0.96 | 0.75 | 1.06 | 1.39 |
| Insurance (0.5% of TIC) | | | 0.005 | 1.06 | 0.65 | 1.76 | 3.92 |
| **Total** | | | | 13.09 | 8.79 | 19.26 | 38.17 |

## Table 48. China cellulosic ethanol capital (bc) and operating costs, bagasse

Country China code = 3

| Cost component | Units | Cost Factor | Scale Factor | China Cellulosic Ethanol (TC) | China ellulosic Ethanol (TC) | China Cellulosic Ethanol (TC) | China Cellulosic Ethanol (TC) |
|---|---|---|---|---|---|---|---|
| Year $ | $ | | | 2005 | 2005 | 2005 | 2005 |
| Cost Index (2 = M&S, 3 = CE) | | | 3 | 130.93 | 130.93 | 130.93 | 130.93 |
| Plant Size | dry short ton/day | | | **1,608** | **804** | **3,215** | **9,645** |
| | dry tonne/day | | | 1,457 | 729 | 2,915 | 8,744 |
| Stream Factor | % | | | 95% | 95% | 95% | 95% |
| Feed | Dry short ton/yr | | | 5.574E+05 | 2.787E+05 | 1.115E+06 | 3.344E+06 |
| | GJ/short ton | | | 1.794E+01 | 1.794E+01 | 1.794E+01 | 1.794E+01 |
| | GJ/yr | | | 9.997E+06 | 4.999E+06 | 1.999E+07 | 5.998E+07 |
| Feed Cost | USD/dt | | | **$ 9.10** | **$ 9.10** | **$ 9.10** | **$ 9.10** |
| Yield (gal/Dry US Ton) | | | | | | | |
| Ethanol | gal/short ton | | | 89.7 | 89.7 | 89.70 | 89.7 |
| Electricity | kWh/gal EtOH | | | 2.18 | 2.18 | 2.18 | 2.18 |
| Process Efficiency - to ethanol | % HHV | | | 44.4% | 44.4% | 44.4% | 44.4% |
| Process efficiency - overall | % HHV | | | | | | |
| Ethanol | gal/yr | | | 5.000E+07 | 2.500E+07 | 1.000E+08 | 3.000E+08 |
| | Mil Gal/YR | | | **50.00** | **25.00** | **100.00** | **300.00** |
| | tpy | | | 1.657E+05 | 8.287E+04 | 3.315E+05 | 9.945E+05 |
| | GJ/yr | | | 4.436E+06 | 2.218E+06 | 8.872E+06 | 2.662E+07 |
| | gal/Stream day | | | 1.442E+05 | 7.210E+04 | 2.884E+05 | 8.652E+05 |
| | bbl/s stream day | | | 3.433E+03 | 1.717E+03 | 6.866E+03 | 2.060E+04 |
| Electricity | kWh/yr | | | 1.09E+08 | 5.45E+07 | 2.18E+08 | 6.54E+08 |
| **Capital Cost (million USD)** | | | | | | | |
| Feed Handling | | | 0.70 | 9.77 | 7.12 | 18.78 | 40.52 |
| Pretreatment | | | 0.80 | 24.75 | 14.21 | 43.08 | 103.76 |
| Neutralization/Conditioning | | | 0.70 | 10.29 | 6.33 | 16.71 | 36.06 |
| Saccharification & Fermentation | | | 0.85 | 12.24 | 6.79 | 22.07 | 56.14 |
| Distillation & Solids Recovery | | | 0.65 | 28.52 | 18.18 | 44.76 | 91.41 |
| Wastewater Treatment | | | 0.65 | 4.04 | 2.57 | 6.34 | 12.94 |
| Storage | | | 0.65 | 2.74 | 1.74 | 4.29 | 8.77 |
| Boiler/Turbogenerator | | | 0.70 | 50.27 | 30.95 | 81.67 | 176.21 |
| Utilities | | | 0.70 | 5.99 | 3.69 | 9.73 | 21.00 |
| **Direct Fixed Capital (DFC), also called TIC** | | | | 148.60 | 91.58 | 247.43 | 546.81 |
| Engineering | DFC x MF | 0.12 | | 17.83 | 10.99 | 29.69 | 65.62 |
| Construction | DFC x MF | 0.13 | | 19.32 | 11.91 | 32.17 | 71.09 |
| Contractor & Legal | DFC x MF | 0.08 | | 11.89 | 7.33 | 19.79 | 43.75 |
| Process/Project Contingency | DFC x MF | 0.0412 | | 6.12 | 3.77 | 10.19 | 22.53 |
| **Total Plant Cost (TPC)** | | | | 203.77 | 125.58 | 339.27 | 749.79 |
| AFUDC | | | | | | | |
| **Total Plant Investment (TPI)** | | | | 203.77 | 125.58 | 339.27 | 749.79 |
| Land | | | 0.60 | 2.21 | 2.21 | 5.08 | 15.23 |
| Startup | | | | 0 | 0.00 | 0.00 | 0.00 |
| **Total Capital Cost (TCC)** | | | | 205.98 | **127.79** | **344.35** | 765.02 |
| **Contingency/TPI** | | | | | | | |
| Working Capital | DFC x MF | 0.05 | | 10.19 | 6.28 | 16.96 | 37.49 |
| **Variable Operating Costs (million USD/yr)** | | | | | | | |
| Feed | | | | 5.07 | 2.54 | 10.14 | 30.43 |
| Utilities | | | | 0.00 | 0.00 | 0.00 | 0.00 |
| Waste Disposal | | | | 1.82 | 0.91 | 3.64 | 10.93 |
| Catalysts and Chemicals | Major comp is cellulase | | | 9.25 | 4.63 | 18.51 | 55.53 |
| **Total** | | | | 16.15 | 8.07 | 32.30 | 96.89 |
| Fixed Operating Costs (million USD/yr) | | | | | | | |
| Labor (SF from V-R) | DFC x MF | 0.0165 | 0.25 | 2.48 | 1.92 | 2.72 | 3.58 |
| Maintenance (3% of A) | | | 0.03 | 4.46 | 2.75 | 7.42 | 16.40 |
| General Overhead (65% of labor + maint) | | | 0.65 | 4.51 | 3.04 | 6.59 | 12.99 |
| Direct Overhead (45% of Labor) | | | 0.45 | 1.11 | 0.87 | 1.22 | 1.61 |
| Insurance (0.5% of TIC) | | | 0.005 | 1.03 | 0.64 | 1.72 | 3.83 |
| **Total** | | | | 13.59 | 9.21 | 19.69 | 38.42 |

## Table 49. Colombia cellulosic ethanol (bc) capital and operating costs, bagasse

| Country | Colombia | Code = | 6 | | | | | |
|---|---|---|---|---|---|---|---|---|
| **Cost component** | **Units** | **Cost Factor** | **Scale Factor** | **USA Cellulosic Ethanol (TC)** | **Colombia Cellulosic Ethanol (TC)** | **Colombia Cellulosic Ethanol (TC)** | **Colombia Cellulosic Ethanol (TC)** | **Colombia Cellulosic Ethanol (TC)** |
| Year $ | $ | | | 2002 | 2005 | 2005 | 2005 | 2005 |
| Cost Index (2 = M&S, 3 = CE) | | | 3 | 110.63 | 130.93 | 130.93 | 130.93 | 130.93 |
| Plant Size | dry short ton/day | | | 2,206 | **1,608** | **804** | **3,215** | **9,645** |
| | dry tonne/day | | | 2,000 | 1,457 | 729 | 2,915 | 8,744 |
| Stream Factor | % | | | 95% | 95% | 95% | 95% | 95% |
| Feed | Dry short ton/yr | | | 7.649E+05 | 5.574E+05 | 2.787E+05 | 1.115E+06 | 3.344E+06 |
| | GJ/short ton | | | 1.794E+01 | 1.794E+01 | 1.794E+01 | 1.794E+01 | 1.794E+01 |
| | GJ/yr | | | 1.372E+07 | 9.997E+06 | 4.999E+06 | 1.999E+07 | 5.998E+07 |
| Feed Cost | USD/dt | | | **$ 35.00** | **$ 14.50** | **$ 14.50** | **$ 14.50** | **$ 14.50** |
| Yield (gal/Dry US Ton) | | | | | | | | |
| Ethanol | gal/short ton | | | 89.7 | 89.7 | 89.7 | 89.70 | 89.7 |
| Electricity | kWh/gal EtOH | | | 2.18 | 2.18 | 2.18 | 2.18 | 2.18 |
| Process Efficiency - to ethanol | % HHV | | | 44.4% | 44.4% | 44.4% | 44.4% | 44.4% |
| Process efficiency - overall | % HHV | | | | | | | |
| Ethanol | gal/yr | | | 6.861E+07 | 5.000E+07 | 2.500E+07 | 1.000E+08 | 3.000E+08 |
| | Mil Gal/YR | | | 68.61 | **50.00** | **25.00** | **100.00** | **300.00** |
| | tpy | | | 2.275E+05 | 1.657E+05 | 8.287E+04 | 3.315E+05 | 9.945E+05 |
| | GJ/yr | | | 6.088E+06 | 4.436E+06 | 2.218E+06 | 8.872E+06 | 2.662E+07 |
| | gal/Stream day | | | 1.979E+05 | 1.442E+05 | 7.210E+04 | 2.884E+05 | 8.652E+05 |
| | bbl/s stream day | | | 4.711E+03 | 3.433E+03 | 1.717E+03 | 6.866E+03 | 2.060E+04 |
| Electricity | kWh/yr | | | 1.50E+08 | 1.09E+08 | 5.45E+07 | 2.18E+08 | 6.54E+08 |
| **Capital Cost (million USD)** | | | | | | | | |
| Feed Handling | | | 0.70 | 7.50 | 8.83 | 6.43 | 16.97 | 36.62 |
| Pretreatment | | | 0.80 | 19.00 | 22.36 | 12.84 | 38.93 | 93.76 |
| Neutralization/Conditioning | | | 0.70 | 7.90 | 9.30 | 5.72 | 15.10 | 32.59 |
| Saccharification & Fermentation | | | 0.85 | 9.40 | 11.06 | 6.14 | 19.94 | 50.74 |
| Distillation & Solids Recovery | | | 0.65 | 21.90 | 25.77 | 16.43 | 40.44 | 82.60 |
| Wastewater Treatment | | | 0.65 | 3.10 | 3.65 | 2.33 | 5.72 | 11.69 |
| Storage | | | 0.65 | 2.10 | 2.47 | 1.58 | 3.88 | 7.92 |
| Boiler/Turbogenerator | | | 0.70 | 38.60 | 45.43 | 27.97 | 73.80 | 159.24 |
| Utilities | | | 0.70 | 4.60 | 5.41 | 3.33 | 8.79 | 18.98 |
| **Direct Fixed Capital (DFC), also called TIC** | | | | 114.10 | 134.29 | 82.76 | 223.59 | 494.14 |
| Engineering | DFC x MF | 0.12 | | 13.69 | 16.11 | 9.93 | 26.83 | 59.30 |
| Construction | DFC x MF | 0.13 | | 14.83 | 17.46 | 10.76 | 29.07 | 64.24 |
| Contractor & Legal | DFC x MF | 0.08 | | 9.13 | 10.74 | 6.62 | 17.89 | 39.53 |
| Process/Project Contingency | DFC x MF | 0.0412 | | 4.70 | 5.53 | 3.41 | 9.21 | 20.36 |
| **Total Plant Cost (TPC)** | | | | 156.45 | 184.14 | 113.48 | 306.59 | 677.56 |
| AFUDC | | | | | | | | |
| **Total Plant Investment (TPI)** | | | | 156.45 | 184.14 | 113.48 | 306.59 | 677.56 |
| Land | | | 0.60 | 2.21 | 2.21 | 2.21 | 5.08 | 15.23 |
| Startup | | | | 0 | 0 | 0.00 | 0.00 | 0.00 |
| **Total Capital Cost (TCC)** | | | | 158.66 | 186.35 | **115.69** | **311.66** | 692.79 |
| **Contingency/TPI** | | | | | | | | |
| Working Capital | DFC x MF | 0.05 | | 7.82 | 9.21 | 5.67 | 15.33 | 33.88 |
| **Variable Operating Costs (million USD/yr)** | | | | | | | | |
| Feed | | | | 26.77 | 8.08 | 4.04 | 16.16 | 48.49 |
| Utilities | | | | 0.00 | 0.00 | 0.00 | 0.00 | 0.00 |
| Waste Disposal | | | | 2.50 | 1.82 | 0.91 | 3.64 | 10.93 |
| Catalysts and Chemicals | Major comp is cellulase | | | 12.70 | 9.25 | 4.63 | 18.51 | 55.53 |
| **Total** | | | | 41.97 | 19.16 | 9.58 | 38.32 | 114.95 |
| Fixed Operating Costs (million USD/yr) | | | | | | | | |
| Labor (SF from V-R) | DFC x MF | 0.0165 | 0.25 | 1.88 | 1.42 | 1.10 | 1.56 | 2.05 |
| Maintenance (3% of A) | | | 0.03 | 3.42 | 4.03 | 2.48 | 6.71 | 14.82 |
| General Overhead (65% of labor + maint) | | | 0.65 | 3.45 | 3.54 | 2.33 | 5.37 | 10.97 |
| Direct Overhead (45% of Labor) | | | 0.45 | 0.85 | 0.64 | 0.50 | 0.70 | 0.92 |
| Insurance (0.5% of TIC) | | | 0.005 | 0.79 | 0.93 | 0.58 | 1.56 | 3.46 |
| **Total** | | | | 10.39 | 10.56 | 6.99 | 15.90 | 32.23 |

## Table 50. India cellulosic ethanol (bc) capital and operating costs, bagasse

Country India code = 4

| Cost component | Units | Cost Factor | Scale Factor | India Cellulosic Ethanol (TC) | India Cellulosic Ethanol (TC) | India Cellulosic Ethanol (TC) | India Cellulosic Ethanol (TC) |
|---|---|---|---|---|---|---|---|
| Year $ | $ | | | 2005 | 2005 | 2005 | 2005 |
| Cost Index (2 = M&S, 3 = CE) | | | 3 | 130.93 | 130.93 | 130.93 | 130.93 |
| Plant Size | dry short ton/day | | | **1,608** | **804** | **3,215** | **9,645** |
| | dry tonne/day | | | 1,457 | 729 | 2,915 | 8,744 |
| Stream Factor | % | | | 95% | 95% | 95% | 95% |
| Feed | Dry short ton/yr | | | 5.574E+05 | 2.787E+05 | 1.115E+06 | 3.344E+06 |
| | GJ/short ton | | | 1.794E+01 | 1.794E+01 | 1.794E+01 | 1.794E+01 |
| | GJ/yr | | | 9.997E+06 | 4.999E+06 | 1.999E+07 | 5.998E+07 |
| Feed Cost | USD/dt | | | **$ 7.30** | **$ 7.30** | **$ 7.30** | **$ 7.30** |
| Yield (gal/Dry US Ton) | | | | | | | |
| Ethanol | gal/short ton | | | 89.7 | 89.7 | 89.70 | 89.7 |
| Electricity | kWh/gal EtOH | | | 2.18 | 2.18 | 2.18 | 2.18 |
| Process Efficiency - to ethanol | % HHV | | | 44.4% | 44.4% | 44.4% | 44.4% |
| Process efficiency - overall | % HHV | | | | | | |
| Ethanol | gal/yr | | | 5.000E+07 | 2.500E+07 | 1.000E+08 | 3.000E+08 |
| | Mil Gal/YR | | | **50.00** | **25.00** | **100.00** | **300.00** |
| | tpy | | | 1.657E+05 | 8.287E+04 | 3.315E+05 | 9.945E+05 |
| | GJ/yr | | | 4.436E+06 | 2.218E+06 | 8.872E+06 | 2.662E+07 |
| | gal/Stream day | | | 1.442E+05 | 7.210E+04 | 2.884E+05 | 8.652E+05 |
| | bbl/s stream day | | | 3.433E+03 | 1.717E+03 | 6.866E+03 | 2.060E+04 |
| Electricity | kWh/yr | | | 1.09E+08 | 5.45E+07 | 2.18E+08 | 6.54E+08 |
| **Capital Cost (million USD)** | | | | | | | |
| Feed Handling | | | 0.70 | 9.47 | 6.90 | 18.20 | 39.28 |
| Pretreatment | | | 0.80 | 23.98 | 13.78 | 41.76 | 100.57 |
| Neutralization/Conditioning | | | 0.70 | 9.97 | 6.14 | 16.20 | 34.96 |
| Saccharification & Fermentation | | | 0.85 | 11.87 | 6.58 | 21.39 | 54.42 |
| Distillation & Solids Recovery | | | 0.65 | 27.65 | 17.62 | 43.38 | 88.60 |
| Wastewater Treatment | | | 0.65 | 3.91 | 2.49 | 6.14 | 12.54 |
| Storage | | | 0.65 | 2.65 | 1.69 | 4.16 | 8.50 |
| Boiler/Turbogenerator | | | 0.70 | 48.73 | 29.99 | 79.16 | 170.79 |
| Utilities | | | 0.70 | 5.81 | 3.57 | 9.43 | 20.35 |
| **Direct Fixed Capital (DFC), also called TIC** | | | | 144.03 | 88.77 | 239.82 | 530.00 |
| Engineering | DFC x MF | 0.12 | | 17.28 | 10.65 | 28.78 | 63.60 |
| Construction | DFC x MF | 0.13 | | 18.72 | 11.54 | 31.18 | 68.90 |
| Contractor & Legal | DFC x MF | 0.08 | | 11.52 | 7.10 | 19.19 | 42.40 |
| Process/Project Contingency | DFC x MF | 0.0412 | | 5.93 | 3.66 | 9.88 | 21.84 |
| **Total Plant Cost (TPC)** | | | | 197.50 | 121.72 | 328.84 | 726.74 |
| AFUDC | | | | | | | |
| **Total Plant Investment (TPI)** | | | | 197.50 | 121.72 | 328.84 | 726.74 |
| Land | | | 0.60 | 2.21 | 2.21 | 5.08 | 15.23 |
| Startup | | | | 0 | 0.00 | 0.00 | 0.00 |
| **Total Capital Cost (TCC)** | | | | 199.71 | **123.93** | **333.91** | 741.97 |
| **Contingency/TPI** | | | | | | | |
| Working Capital | DFC x MF | 0.05 | | 9.88 | 6.09 | 16.44 | 36.34 |
| **Variable Operating Costs (million USD/yr)** | | | | | | | |
| Feed | | | | 4.07 | 2.03 | 8.14 | 24.41 |
| Utilities | | | | 0.00 | 0.00 | 0.00 | 0.00 |
| Waste Disposal | | | | 1.82 | 0.91 | 3.64 | 10.93 |
| Catalysts and Chemicals | Major comp is cellulase | | | 9.25 | 4.63 | 18.51 | 55.53 |
| **Total** | | | | 15.15 | 7.57 | 30.29 | 90.87 |
| Fixed Operating Costs (million USD/yr) | | | | | | | |
| Labor (SF from V-R) | DFC x MF | 0.0165 | 0.25 | 0.83 | 0.65 | 0.92 | 1.21 |
| Maintenance (3% of A) | | | 0.03 | 4.32 | 2.66 | 7.19 | 15.90 |
| General Overhead (65% of labor + maint) | | | 0.65 | 3.35 | 2.15 | 5.27 | 11.12 |
| Direct Overhead (45% of Labor) | | | 0.45 | 0.38 | 0.29 | 0.41 | 0.54 |
| Insurance (0.5% of TIC) | | | 0.005 | 1.00 | 0.62 | 1.67 | 3.71 |
| **Total** | | | | 9.88 | 6.37 | 15.46 | 32.48 |

## Table 51. Mexico cellulosic ethanol (bc) capital and operating costs, bagasse

| Country | Mexico | Code = | 8 | | | | |
|---|---|---|---|---|---|---|---|
| **Cost component** | **Units** | **Cost Factor** | **Scale Factor** | **Mexico Cellulosic Ethanol (TC)** | **Mexico Cellulosic Ethanol (TC)** | **Mexico Cellulosic Ethanol (TC)** | **Mexico Cellulosic Ethanol (TC)** |
| Year $ | $ | | | 2005 | 2005 | 2005 | 2005 |
| Cost Index (2 = M&S, 3 = CE) | | | 3 | 130.93 | 130.93 | 130.93 | 130.93 |
| Plant Size | dry short ton/day | | | **1,608** | **804** | **3,215** | **9,645** |
| | dry tonne/day | | | 1,457 | 729 | 2,915 | 8,744 |
| Stream Factor | % | | | 95% | 95% | 95% | 95% |
| Feed | Dry short ton/yr | | | 5.574E+05 | 2.787E+05 | 1.115E+06 | 3.344E+06 |
| | GJ/short ton | | | 1.794E+01 | 1.794E+01 | 1.794E+01 | 1.794E+01 |
| | GJ/yr | | | 9.997E+06 | 4.999E+06 | 1.999E+07 | 5.998E+07 |
| Feed Cost | USD/dt | | | **$ 15.40** | **$ 15.40** | **$ 15.40** | **$ 15.40** |
| Yield (gal/Dry US Ton) | | | | | | | |
| Ethanol | gal/short ton | | | 89.7 | 89.7 | 89.70 | 89.7 |
| Electricity | kWh/gal EtOH | | | 2.18 | 2.18 | 2.18 | 2.18 |
| Process Efficiency - to ethanol | % HHV | | | 44.4% | 44.4% | 44.4% | 44.4% |
| Process efficiency - overall | % HHV | | | | | | |
| Ethanol | gal/yr | | | 5.000E+07 | 2.500E+07 | 1.000E+08 | 3.000E+08 |
| | Mil Gal/YR | | | **50.00** | **25.00** | **100.00** | **300.00** |
| | tpy | | | 1.657E+05 | 8.287E+04 | 3.315E+05 | 9.945E+05 |
| | GJ/yr | | | 4.436E+06 | 2.218E+06 | 8.872E+06 | 2.662E+07 |
| | gal/Stream day | | | 1.442E+05 | 7.210E+04 | 2.884E+05 | 8.652E+05 |
| | bbl/s stream day | | | 3.433E+03 | 1.717E+03 | 6.866E+03 | 2.060E+04 |
| Electricity | kWh/yr | | | 1.09E+08 | 5.45E+07 | 2.18E+08 | 6.54E+08 |
| **Capital Cost (million USD)** | | | | | | | |
| Feed Handling | | | 0.70 | 7.14 | 5.20 | 13.73 | 29.62 |
| Pretreatment | | | 0.80 | 18.09 | 10.39 | 31.49 | 75.84 |
| Neutralization/Conditioning | | | 0.70 | 7.52 | 4.63 | 12.22 | 26.36 |
| Saccharification & Fermentation | | | 0.85 | 8.95 | 4.96 | 16.13 | 41.04 |
| Distillation & Solids Recovery | | | 0.65 | 20.85 | 13.29 | 32.71 | 66.81 |
| Wastewater Treatment | | | 0.65 | 2.95 | 1.88 | 4.63 | 9.46 |
| Storage | | | 0.65 | 2.00 | 1.27 | 3.14 | 6.41 |
| Boiler/Turbogenerator | | | 0.70 | 36.75 | 22.62 | 59.69 | 128.80 |
| Utilities | | | 0.70 | 4.38 | 2.70 | 7.11 | 15.35 |
| **Direct Fixed Capital (DFC), also called TIC** | | | | 108.62 | 66.94 | 180.85 | 399.68 |
| Engineering | DFC x MF | 0.12 | | 13.03 | 8.03 | 21.70 | 47.96 |
| Construction | DFC x MF | 0.13 | | 14.12 | 8.70 | 23.51 | 51.96 |
| Contractor & Legal | DFC x MF | 0.08 | | 8.69 | 5.36 | 14.47 | 31.97 |
| Process/Project Contingency | DFC x MF | 0.0412 | | 4.48 | 2.76 | 7.45 | 16.47 |
| **Total Plant Cost (TPC)** | | | | 148.94 | 91.79 | 247.98 | 548.04 |
| AFUDC | | | | | | | |
| **Total Plant Investment (TPI)** | | | | 148.94 | 91.79 | 247.98 | 548.04 |
| Land | | | 0.60 | 2.21 | 2.21 | 5.08 | 15.23 |
| Startup | | | | 0 | 0.00 | 0.00 | 0.00 |
| **Total Capital Cost (TCC)** | | | | 151.15 | **94.00** | **253.06** | 563.27 |
| **Contingency/TPI** | | | | | | | |
| Working Capital | DFC x MF | 0.05 | | 7.45 | 4.59 | 12.40 | 27.40 |
| **Variable Operating Costs (million USD/yr)** | | | | | | | |
| Feed | | | | 8.58 | 4.29 | 17.17 | 51.50 |
| Utilities | | | | 0.00 | 0.00 | 0.00 | 0.00 |
| Waste Disposal | | | | 1.82 | 0.91 | 3.64 | 10.93 |
| Catalysts and Chemicals | Major comp is cellulase | | | 9.25 | 4.63 | 18.51 | 55.53 |
| **Total** | | | | 19.66 | 9.83 | 39.32 | 117.96 |
| Fixed Operating Costs (million USD/yr) | | | | | | | |
| Labor (SF from V-R) | DFC x MF | 0.0165 | 0.25 | 0.54 | 0.42 | 0.59 | 0.78 |
| Maintenance (3% of A) | | | 0.03 | 3.26 | 2.01 | 5.43 | 11.99 |
| General Overhead (65% of labor + maint) | | | 0.65 | 2.47 | 1.58 | 3.91 | 8.30 |
| Direct Overhead (45% of Labor) | | | 0.45 | 0.24 | 0.19 | 0.27 | 0.35 |
| Insurance (0.5% of TIC) | | | 0.005 | 0.76 | 0.47 | 1.27 | 2.82 |
| **Total** | | | | 7.26 | 4.66 | 11.46 | 24.23 |

## Table 52. Cellulosic ethanol (biochemical) summary yields, costs, and plant gate prices, 25 MM GPY, bagasse

| Technology | Units | USA | Argentina | Brazil | Canada | Caribbean Basin | China | Colombia | India | Mexico |
|---|---|---|---|---|---|---|---|---|---|---|
| **Plant Size** | MM gal/yr | 25 | 25 | 25 | 25 | 25 | 25 | 25 | 25 | 25 |
| **Yields** | | | | | | | | | | |
| Primary Product Yield | gal/ton | 89.7 | 89.7 | 89.7 | | 89.7 | 89.7 | 89.7 | 89.7 | 89.7 |
| By Product A | | | | | | | | | | |
| Unit | | electricity | electricity | electricity | | electricity | electricity | electricity | electricity | electricity |
| Yield unit | | kWh | kWh | kWh | | kWh | kWh | kWh | kWh | kWh |
| Annual Yield | | 5.45E+07 | 5.45E+07 | 5.45E+07 | | 5.45E+07 | 5.45E+07 | 5.45E+07 | 5.45E+07 | 5.45E+07 |
| Price unit | | 0.04 | 0.04 | 0.04 | | 0.04 | 0.04 | 0.04 | 0.04 | 0.04 |
| MM USD/yr | | 2.18 | 2.18 | 2.18 | | 2.18 | 2.18 | 2.18 | 2.18 | 2.18 |
| By Product B | | | | | | | | | | |
| Unit | | | | | | | | | | |
| Yield unit | | | | | | | | | | |
| Annual Yield | | | | | | | | | | |
| Price unit | | | | | | | | | | |
| Annual Cost | MM USD/yr | | | | | | | | | |
| **Energy Inputs** | | | | | | | | | | |
| Electricity | | | | | | | | | | |
| Cost | $/kWh | N/A | N/A | N/A | | N/A | N/A | N/A | N/A | N/A |
| Amount | kWh/yr | | | | | | | | | |
| Annual Cost | MM USD/yr | | | | | | | | | |
| Fuel | | | | | | | | | | |
| Type | | N/A | N/A | N/A | | N/A | N/A | N/A | N/A | N/A |
| Unit | | | | | | | | | | |
| Amount | | | | | | | | | | |
| Cost | USD/unit | | | | | | | | | |
| Annual Cost | MM USD/yr | | | | | | | | | |
| **Capital & Operating Costs** | | | | | | | | | | |
| Capital | MMUSD | 116.32 | 124.74 | 118.82 | | 130.96 | 127.79 | 115.69 | 123.93 | 94.00 |
| Operating w/o feed | MM USD/yr | 13.87 | 13.67 | 11.89 | | 14.33 | 14.75 | 12.53 | 11.91 | 10.2 |
| Feed | | | | | | | | | | |
| Unit | | ton | | | | | | | | |
| Cost/unit | USD/unit | 35 | 9.1 | 7.6 | | 7.6 | 9.1 | 14.5 | 7.3 | 15.4 |
| Rate | unit/day | 804 | 804 | 804 | | 804 | 804 | 804 | 804 | 804 |
| Yearly Cost | MM USD/yr | 9.75 | 2.54 | 2.12 | | 2.12 | 2.54 | 4.04 | 2.03 | 4.29 |
| **Product Plant Gate Cost** | | | | | | | | | | |
| Basis | | | | | | | | | | |
| Value | USD/gal | 1.56 | 1.29 | 1.15 | | 1.30 | 1.35 | 1.24 | 1.18 | 1.05 |

## Table 53. Cellulosic ethanol (biochemical) summary yields, costs, and plant gate prices, 100 MM GPY, bagasse

| Technology | Units | USA | Argentina | Brazil | Canada | Caribbean Basin | China | Colombia | India | Mexico |
|---|---|---|---|---|---|---|---|---|---|---|
| **Plant Size** | MM gal/yr | 100 | 100 | 100 | 100 | 100 | 100 | 100 | 100 | 100 |
| **Yields** | | | | | | | | | | |
| Primary Product Yield | gal/ton | 89.7 | 89.7 | 89.7 | | 89.7 | 89.7 | 89.7 | 89.7 | 89.7 |
| By Product A | | | | | | | | | | |
| Unit | | electricity | electricity | electricity | | electricity | electricity | electricity | electricity | electricity |
| Yield unit | | kWh | kWh | kWh | | kWh | kWh | kWh | kWh | kWh |
| Annual Yield | | 2.18E+08 | 2.18E+08 | 2.18E+08 | | 2.18E+08 | 2.18E+08 | 2.18E+08 | 2.18E+08 | 2.18E+08 |
| Price unit | | 0.04 | 0.04 | 0.04 | | 0.04 | 0.04 | 0.04 | 0.04 | 0.04 |
| MM USD/yr | | 8.72 | 8.72 | 8.72 | | 8.72 | 8.72 | 8.72 | 8.72 | 8.72 |
| By Product B | | | | | | | | | | |
| Unit | | | | | | | | | | |
| Yield unit | | | | | | | | | | |
| Annual Yield | | | | | | | | | | |
| Price unit | | | | | | | | | | |
| Annual Cost | MM USD/yr | | | | | | | | | |
| **Energy Inputs** | | | | | | | | | | |
| Electricity | | N/A | N/A | N/A | | N/A | N/A | N/A | N/A | N/A |
| Cost | $/kWh | | | | | | | | | |
| Amount | kWh/yr | | | | | | | | | |
| Annual Cost | MM USD/yr | | | | | | | | | |
| Fuel | | | | | | | | | | |
| Type | | N/A | N/A | N/A | | N/A | N/A | N/A | N/A | N/A |
| Unit | | | | | | | | | | |
| Amount | | | | | | | | | | |
| Cost | USD/unit | | | | | | | | | |
| Annual Cost | MM USD/yr | | | | | | | | | |
| **Capital & Operating Costs** | | | | | | | | | | |
| Capital | MMUSD | 313.38 | 336.11 | 320.13 | | 352.93 | 344.35 | 311.66 | 333.91 | 253.06 |
| Operating w/o feed | MM USD/yr | 39.99 | 40.15 | 37.32 | | 41.41 | 41.84 | 38.05 | 37.62 | 33.61 |
| Feed | | | | | | | | | | |
| Unit | | ton | | | | | | | | |
| Cost/unit | USD/unit | 35 | 9.1 | 7.6 | | 7.6 | 9.1 | 14.5 | 7.3 | 15.4 |
| Rate | unit/day | 3125 | 3125 | 3125 | | 3125 | 3125 | 3125 | 3125 | 3125 |
| Yearly Cost | MM USD/yr | 39.03 | 10.14 | 8.47 | | 8.47 | 10.14 | 16.16 | 8.14 | 17.17 |
| **Product Plant Gate Cost** | | | | | | | | | | |
| Basis | | | | | | | | | | |
| Value | USD/gal | 1.17 | 0.90 | 0.83 | | 0.90 | 0.92 | 0.90 | 0.81 | 0.81 |

## Table 54. Cellulosic ethanol (biochemical) summary yields, costs, and plant gate prices, 25 MM GPY, ag residue

| Technology | Units | USA | Argentina | Brazil | Canada | Caribbean Basin | China | Colombia | India | Mexico |
|---|---|---|---|---|---|---|---|---|---|---|
| **Plant Size** | MM gal/yr | 25 | 25 | 25 | 25 | 25 | 25 | 25 | 25 | 25 |
| **Yields** | | | | | | | | | | |
| Primary Product Yield | gal/ton | 89.7 | 89.7 | 89.7 | 89.7 | 89.7 | 89.7 | 89.7 | 89.7 | 89.7 |
| By Product A | | | | | | | | | | |
| Unit | | electricity | electricity | electricity | electricity | electricity | electricity | electricity | electricity | electricity |
| Yield unit | | kWh | kWh | kWh | kWh | kWh | kWh | kWh | kWh | kWh |
| Annual Yield | | 5.45E+07 | 5.45E+07 | 5.45E+07 | 5.45E+07 | 5.45E+07 | 5.45E+07 | 5.45E+07 | 5.45E+07 | 5.45E+07 |
| Price unit | | 0.04 | 0.04 | 0.04 | 0.04 | 0.04 | 0.04 | 0.04 | 0.04 | 0.04 |
| MM USD/yr | | 2.18 | 2.18 | 2.18 | 2.18 | 2.18 | 2.18 | 2.18 | 2.18 | 2.18 |
| By Product B | | | | | | | | | | |
| Unit | | | | | | | | | | |
| Yield unit | | | | | | | | | | |
| Annual Yield | | | | | | | | | | |
| Price unit | | | | | | | | | | |
| Annual Cost | MM USD/yr | | | | | | | | | |
| **Energy Inputs** | | | | | | | | | | |
| Electricity | | | | | | | | | | |
| Cost | $/kWh | N/A | N/A | N/A | N/A | N/A | N/A | N/A | N/A | N/A |
| Amount | kWh/yr | | | | | | | | | |
| Annual Cost | MM USD/yr | | | | | | | | | |
| Fuel | | | | | | | | | | |
| Type | | N/A | N/A | N/A | N/A | N/A | N/A | N/A | N/A | N/A |
| Unit | | | | | | | | | | |
| Amount | | | | | | | | | | |
| Cost | USD/unit | | | | | | | | | |
| Annual Cost | MM USD/yr | | | | | | | | | |
| **Capital & Operating Costs** | | | | | | | | | | |
| Capital | MMUSD | 116.32 | 124.74 | 118.82 | 150.9 | 130.96 | 127.79 | 115.69 | 123.93 | 94.00 |
| Operating w/o feed | MM USD/yr | 13.87 | 13.67 | 11.89 | 17.76 | 14.33 | 14.75 | 12.53 | 11.91 | 10.2 |
| Feed | | | | | | | | | | |
| Unit | | | | | | | | | | |
| Cost/unit | USD/unit | 35 | 47.2 | 30.8 | 10 | 14.5 | 28.1 | 14.5 | 7.3 | 47.2 |
| Rate | unit/day | 804 | 804 | 804 | 804 | 804 | 804 | 804 | 804 | 804 |
| Yearly Cost | MM USD/yr | 9.75 | 26.31 | 8.58 | 2.79 | 4.04 | 7.83 | 4.04 | 2.03 | 13.15 |
| **Product Plant Gate Cost** | | | | | | | | | | |
| Basis | | | | | | | | | | |
| Value | USD/gal | 1.56 | 1.70 | 1.41 | 1.57 | 1.37 | 1.56 | 1.24 | 1.18 | 1.4 |

## Table 55. Cellulosic ethanol (biochemical) summary yields, costs, and plant gate prices, 100 MM GPY, ag residues

| Technology | Units | USA | Argentina | Brazil | Canada | Caribbean Basin | China | Colombia | India | Mexico |
|---|---|---|---|---|---|---|---|---|---|---|
| **Plant Size** | MM gal/yr | 100 | 100 | 100 | 100 | 100 | 100 | 100 | 100 | 100 |
| **Yields** | | | | | | | | | | |
| Primary Product Yield | gal/ton | 89.7 | 89.7 | 89.7 | 89.7 | 89.7 | 89.7 | 89.7 | 89.7 | 89.7 |
| By Product A | | | | | | | | | | |
| Unit | | electricity | electricity | electricity | electricity | electricity | electricity | electricity | electricity | electricity |
| Yield unit | | kWh | kWh | kWh | kWh | kWh | kWh | kWh | kWh | kWh |
| Annual Yield | | 2.18E+08 | 2.18E+08 | 2.18E+08 | 2.18E+08 | 2.18E+08 | 2.18E+08 | 2.18E+08 | 2.18E+08 | 2.18E+08 |
| Price unit | | 0.04 | 0.04 | 0.04 | 0.04 | 0.04 | 0.04 | 0.04 | 0.04 | 0.04 |
| MM USD/yr | | 8.72 | 8.72 | 8.72 | 8.72 | 8.72 | 8.72 | 8.72 | 8.72 | 8.72 |
| By Product B | | | | | | | | | | |
| Unit | | | | | | | | | | |
| Yield unit | | | | | | | | | | |
| Annual Yield | | | | | | | | | | |
| Price unit | | | | | | | | | | |
| Annual Cost | MM USD/yr | | | | | | | | | |
| **Energy Inputs** | | | | | | | | | | |
| Electricity | | N/A | N/A | N/A | N/A | N/A | N/A | N/A | N/A | N/A |
| Cost | $/kWh | | | | | | | | | |
| Amount | kWh/yr | | | | | | | | | |
| Annual Cost | MM USD/yr | | | | | | | | | |
| Fuel | | | | | | | | | | |
| Type | | N/A | N/A | N/A | N/A | N/A | N/A | N/A | N/A | N/A |
| Unit | | | | | | | | | | |
| Amount | | | | | | | | | | |
| Cost | USD/unit | | | | | | | | | |
| Annual Cost | MM USD/yr | | | | | | | | | |
| **Capital & Operating Costs** | | | | | | | | | | |
| Capital | MMUSD | 313.38 | 336.11 | 320.13 | 406.78 | 352.93 | 344.35 | 311.66 | 333.91 | 253.06 |
| Operating w/o feed | MM USD/yr | 39.99 | 40.15 | 37.32 | 47.31 | 41.41 | 41.84 | 38.05 | 37.62 | 33.61 |
| Feed | | | | | | | | | | |
| Unit | | ton | | | | | | | | |
| Cost/unit | USD/unit | 35 | 47.2 | 30.8 | 10 | 14.5 | 28.1 | 14.5 | 7.3 | 47.2 |
| Rate | unit/day | 3125 | 3125 | 3125 | 3125 | 3125 | 3125 | 3125 | 3125 | 3125 |
| Yearly Cost | MM USD/yr | 39.03 | 52.62 | 34.33 | 4.11 | 16.16 | 31.33 | 16.16 | 8.14 | 52.62 |
| **Product Plant Gate Cost** | | | | | | | | | | |
| Basis | | | | | | | | | | |
| Value | USD/gal | 1.17 | 1.32 | 1.09 | 1.06 | 0.98 | 1.14 | 0.90 | 0.81 | 1.17 |

**Table 56. Cellulosic ethanol (biochemical) summary yields, costs, and plant gate prices, 25 MM GPY, wood/perennial**

| Technology | Units | USA | Argentina | Brazil | Canada | Caribbean Basin | China | Colombia | India | Mexico |
|---|---|---|---|---|---|---|---|---|---|---|
| **Plant Size** | MM gal/yr | 25 | 25 | 25 | 25 | 25 | 25 | 25 | 25 | 25 |
| **Yields** | | | | | | | | | | |
| Primary Product Yield | gal/ton | 89.7 | 89.7 | 89.7 | 89.7 | 89.7 | 89.7 | 89.7 | 89.7 | 89.7 |
| By Product A | | | | | | | | | | |
| Unit | | electricity | electricity | electricity | electricity | electricity | electricity | electricity | electricity | electricity |
| Yield unit | | kWh | kWh | kWh | kWh | kWh | kWh | kWh | kWh | kWh |
| Annual Yield | | 5.45E+07 | 5.45E+07 | 5.45E+07 | 5.45E+07 | 5.45E+07 | 5.45E+07 | 5.45E+07 | 5.45E+07 | 5.45E+07 |
| Price unit | | 0.04 | 0.04 | 0.04 | 0.04 | 0.04 | 0.04 | 0.04 | 0.04 | 0.04 |
| MM USD/yr | | 2.18 | 2.18 | 2.18 | 2.18 | 2.18 | 2.18 | 2.18 | 2.18 | 2.18 |
| By Product B | | | | | | | | | | |
| Unit | | | | | | | | | | |
| Yield unit | | | | | | | | | | |
| Annual Yield | | | | | | | | | | |
| Price unit | | | | | | | | | | |
| Annual Cost | MM USD/yr | | | | | | | | | |
| **Energy Inputs** | | | | | | | | | | |
| Electricity | | | | | | | | | | |
| Cost | $/kWh | N/A | N/A | N/A | N/A | N/A | N/A | N/A | N/A | N/A |
| Amount | kWh/yr | | | | | | | | | |
| Annual Cost | MM USD/yr | | | | | | | | | |
| Fuel | | | | | | | | | | |
| Type | | N/A | N/A | N/A | N/A | N/A | N/A | N/A | N/A | N/A |
| Unit | | | | | | | | | | |
| Amount | | | | | | | | | | |
| Cost | USD/unit | | | | | | | | | |
| Annual Cost | MM USD/yr | | | | | | | | | |
| **Capital & Operating Costs** | | | | | | | | | | |
| Capital | MMUSD | 116.32 | 124.74 | 118.82 | 150.9 | 130.96 | 127.79 | 115.69 | 123.93 | 94.00 |
| Operating w/o feed | MM USD/yr | 13.87 | 13.67 | 11.89 | 17.76 | 14.33 | 14.75 | 12.53 | 11.91 | 10.2 |
| Feed | | | | | | | | | | |
| Unit | | | | | | | | | | |
| Cost/unit | USD/unit | 35 | 47.2 | 47.2 | 31.8 | 47.2 | 47.2 | 47.2 | 47.2 | 47.2 |
| Rate | unit/day | 804 | 804 | 804 | 804 | 804 | 804 | 804 | 804 | 804 |
| Yearly Cost | MM USD/yr | 9.75 | 13.15 | 13.15 | 8.86 | 13.15 | 13.15 | 13.15 | 13.15 | 13.15 |
| **Product Plant Gate Cost** | | | | | | | | | | |
| Basis | | | | | | | | | | |
| Value | USD/gal | 1.56 | 1.70 | 1.59 | 1.81 | 1.74 | 1.77 | 1.40 | 1.64 | 1.4 |

**Table 57. Cellulosic ethanol (biochemical) summary yields, costs, and plant gate prices, 100 MM GPY, wood/perennial**

| Technology | Units | USA | Argentina | Brazil | Canada | Caribbean Basin | China | Colombia | India | Mexico |
|---|---|---|---|---|---|---|---|---|---|---|
| **Plant Size** | MM gal/yr | 100 | 100 | 100 | 100 | 100 | 100 | 100 | 100 | 100 |
| **Yields** | | | | | | | | | | |
| Primary Product Yield | gal/ton | 89.7 | 89.7 | 89.7 | 89.7 | 89.7 | 89.7 | 89.7 | 89.7 | 89.7 |
| By Product A | | | | | | | | | | |
| Unit | | electricity | electricity | electricity | electricity | electricity | electricity | electricity | electricity | electricity |
| Yield unit | | kWh | kWh | kWh | kWh | kWh | kWh | kWh | kWh | kWh |
| Annual Yield | | 2.18E+08 | 2.18E+08 | 2.18E+08 | 2.18E+08 | 2.18E+08 | 2.18E+08 | 2.18E+08 | 2.18E+08 | 2.18E+08 |
| Price unit | | 0.04 | 0.04 | 0.04 | 0.04 | 0.04 | 0.04 | 0.04 | 0.04 | 0.04 |
| MM USD/yr | | 8.72 | 8.72 | 8.72 | 8.72 | 8.72 | 8.72 | 8.72 | 8.72 | 8.72 |
| By Product B | | | | | | | | | | |
| Unit | | | | | | | | | | |
| Yield unit | | | | | | | | | | |
| Annual Yield | | | | | | | | | | |
| Price unit | | | | | | | | | | |
| Annual Cost | MM USD/yr | | | | | | | | | |
| **Energy Inputs** | | | | | | | | | | |
| Electricity | | N/A | N/A | N/A | N/A | N/A | N/A | N/A | N/A | N/A |
| Cost | $/kWh | | | | | | | | | |
| Amount | kWh/yr | | | | | | | | | |
| Annual Cost | MM USD/yr | | | | | | | | | |
| Fuel | | | | | | | | | | |
| Type | | N/A | N/A | N/A | N/A | N/A | N/A | N/A | N/A | N/A |
| Unit | | | | | | | | | | |
| Amount | | | | | | | | | | |
| Cost | USD/unit | | | | | | | | | |
| Annual Cost | MM USD/yr | | | | | | | | | |
| **Capital & Operating Costs** | | | | | | | | | | |
| Capital | MMUSD | 313.38 | 336.11 | 320.13 | 406.78 | 352.93 | 344.35 | 311.66 | 333.91 | 253.06 |
| Operating w/o feed | MM USD/yr | 39.99 | 40.15 | 37.32 | 47.31 | 41.41 | 41.84 | 38.05 | 37.62 | 33.61 |
| Feed | | | | | | | | | | |
| Unit | | ton | | | | | | | | |
| Cost/unit | USD/unit | 35 | 47.2 | 47.2 | 31.8 | 47.2 | 47.2 | 47.2 | 47.2 | 47.2 |
| Rate | unit/day | 3125 | 3125 | 3125 | 3125 | 3125 | 3125 | 3125 | 3125 | 3125 |
| Yearly Cost | MM USD/yr | 39.03 | 52.62 | 52.62 | 35.45 | 52.62 | 52.62 | 52.62 | 52.62 | 52.62 |
| **Product Plant Gate Cost** | | | | | | | | | | |
| Basis | | | | | | | | | | |
| Value | USD/gal | 1.17 | 1.32 | 1.27 | 1.30 | 1.34 | 1.35 | 1.27 | 1.28 | 1.17 |

**Table 58. Cellulosic ethanol (biochemical) plant gate prices**

| Plant Size, MM GPY | 25 | 100 | 25 | 100 | 25 | 100 |
|---|---|---|---|---|---|---|
| Feed | Bagasse | | Ag Residues | | Wood /Perennials | |
| Country | Ethanol Plant Gate Cost (USD/gal ethanol) | | | | | |
| USA | 1.56 | 1.17 | 1.56 | 1.17 | 1.56 | 1.17 |
| Argentina | 1.29 | 0.90 | 1.70 | 1.32 | 1.70 | 1.32 |
| Brazil | 1.15 | 0.83 | 1.41 | 1.09 | 1.59 | 1.27 |
| Canada | -- | -- | 1.57 | 1.06 | 1.81 | 1.30 |
| Caribbean Basin | 1.30 | 0.90 | 1.37 | 0.98 | 1.74 | 1.34 |
| China | 1.35 | 0.92 | 1.56 | 1.14 | 1.77 | 1.35 |
| Colombia | 1.24 | 0.90 | 1.24 | 0.90 | 1.60 | 1.27 |
| India | 1.18 | 0.81 | 1.18 | 0.81 | 1.64 | 1.28 |
| Mexico | 1.05 | 0.81 | 1.40 | 1.17 | 1.40 | 1.17 |

Indirect steam gasification was chosen as the technology around which this process was developed based upon previous technoeconomic studies for the production of methanol and hydrogen from biomass (Bain et al. 2000). The sub-process operations for ethanol production are very similar to those for methanol production (although the specific process configuration will be different). The general process areas include: feedstock preparation, gasification, gas cleanup and conditioning, and alcohol synthesis and purification.

Gasification involves the devolatilization and conversion of biomass in an atmosphere of steam and/or oxygen to produce a medium-calorific value gas. There are two general classes of gasifiers. *Partial oxidation (POX)* gasifiers (directly-heated gasifiers) use the exothermic reaction between oxygen and organics to provide the heat necessary to devolatilize biomass and to convert residual carbon-rich chars. In POX gasifiers, the heat to drive the process is generated internally within the gasifier. A disadvantage of POX gasifiers is that oxygen production is expensive and typically requires large plant sizes to improve economics.

The second general class, *steam gasifiers* (indirectly-heated gasifiers), accomplish biomass heating and gasification through heat transfer from a hot solid or through a heat transfer surface. Either byproduct char and/or a portion of the product gas can be combusted with air (external to the gasifier itself) to provide the energy required for gasification. Steam gasifiers have the advantage of not requiring oxygen; but since most operate at low pressure they require product gas compression for downstream purification and synthesis unit operations.

A number of POX and steam gasifiers are under development and have the potential to produce a synthesis gas suitable for liquid fuel synthesis. These gasifiers have been operated in the 4 to 350 ton per day scale. The decision as to which type of gasifier (POX or steam) will be the most economic depends upon the entire process, not just the cost for the gasifier itself. One indicator for comparing processes is "capital intensity," the capital cost required on a per unit product basis. Figure 21 shows the capital intensity of methanol processes (Wyman et al. 1993; Chem Systems 1989; Feldmann et al. 1988; Wan and Fraser 1989; Hamelinck and Faaij 2001; Williams et al. 1995) based on indirect steam gasification and direct POX gasification. This figure shows that steam gasification capital intensity is comparable or lower than POX gasification. The estimates indicate that both steam gasification and POX gasification processes should be evaluated, but if the processes need to be evaluated sequentially, choosing steam gasification for the first evaluation is reasonable.

Another philosophy applied to the process development was the idea of making the process "electrical energy neutral." It was recognized that the heat and power requirements of the process could not be met just with char combustion and would require additional fuel. Several options were considered. Additional biomass could be added as fuel directly to the heat and power system; however, this would increase the process beyond 2,000 tonnes/day. Fossil fuels (coal or natural gas) could also be added directly to provide the additional fuel. Alternately syngas could be diverted from liquid fuel production to heat and power production. This option makes the design more energy self-sufficient, but also lowers the overall process yield of alcohols.

It was decided that (1) no additional fuel would be used for heat and power and (2) only enough syngas would be diverted so that the internal heat and power requirements would be exactly met. Thus, there would neither be electricity sales to the grid nor electricity purchases. The only exception to this would be if other operating specifications were such that syngas could no longer be backed out of the heat and power system but there is still excess electricity (that could then be sold to the grid for a co-product credit).

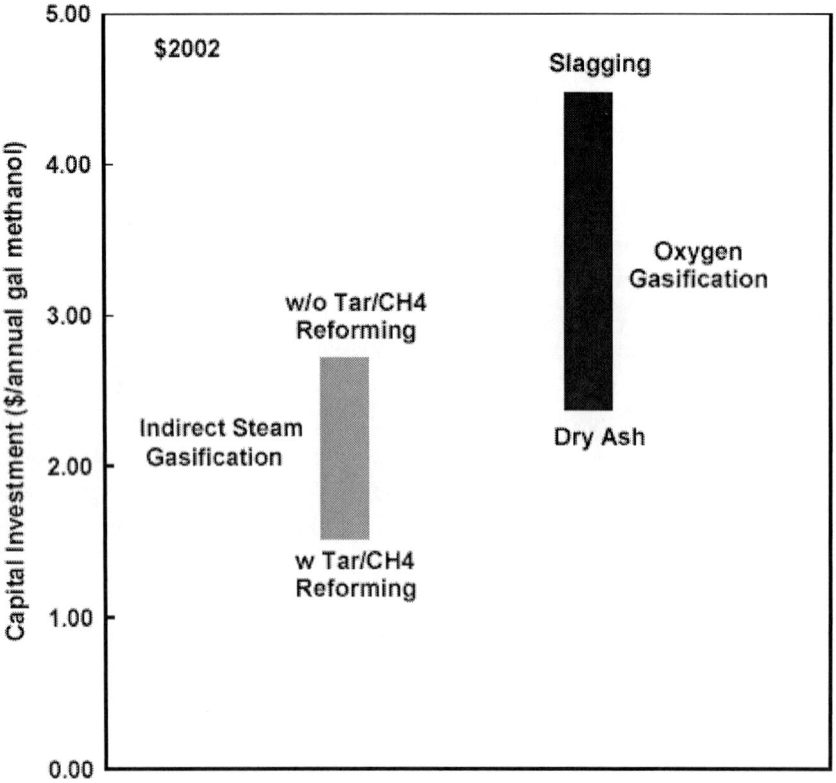


Figure 21. Estimated capital intensities for 2000 tpd biomass-to-methanol processes (Wyman et al. 1993, Williams et al. 1995, Hamelinck and Faaij 2001)

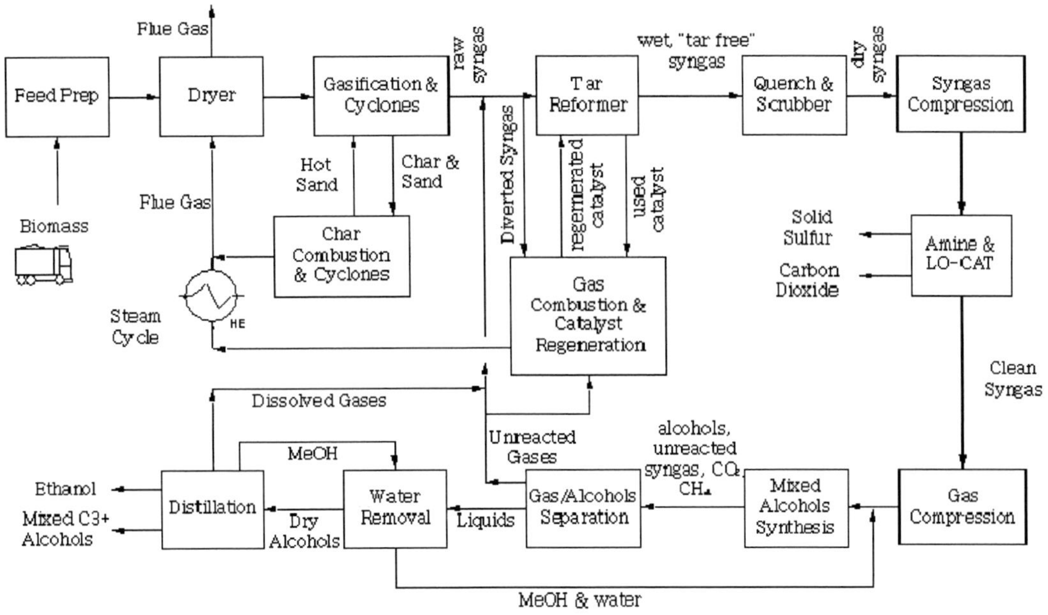


Figure 22. Process flow diagram, cellulosic ethanol (thermochemical)

The capital costs were developed from a variety of sources. For some sub-processes that are well known technology and can be purchased as modular packages (i.e. amine treatment, acid gas removal), an overall cost for the package unit was used. Many of the common equipment items (tanks, pumps, simple heat exchangers) were costed using the Aspen Icarus™ Questimate® costing software. Other more specific unit operations (gasifier, molecular sieve, etc) used cost estimates from other studies and/or from vendor quotes. As documented in the hydrogen design report (Spath et al. 2005) capital costs were developed using general plant-wide factors. The installation costs incorporated cost contributions for not only the actual installation of the purchased equipment but also instrumentation and controls, piping, electrical systems, buildings, yard improvements, etc. A simple block flow diagram of the design is depicted in Figure 22.

The process has the following steps:

- *Feedstock Handling and Preparation.* The biomass feedstock is dried from the as-received moisture to that required for proper feeding into the gasifier using flue gases from the char combustor and tar reformer catalyst regenerator.

- *Gasification.* Indirect gasification is considered in this assessment. Heat for the endothermic gasification reactions is supplied by circulating hot synthetic olivine between the gasifier and the char combustor. Conveyors and hoppers are used to feedstock the biomass to the low-pressure indirectly-heated entrained flow gasifier. Steam is injected into the gasifier to aid in stabilizing the entrained flow of biomass and sand through the gasifier. The biomass chemically converts to a mixture of syngas components ($CO$, $H_2$, $CO_2$, $CH_4$, etc.), tars, and a solid “char” that is mainly the fixed carbon residual from the biomass plus carbon (coke) deposited on the sand.

Cyclones at the exit of the gasifier separate the char and sand from the syngas. These solids flow by gravity from the cyclones into the char combustor. Air is introduced to the bottom of the reactor and serves as a carrier gas for the fluidized bed plus the oxidant for burning the char and coke. The heat of combustion heats the sand to over 1800°F. The hot sand and residual ash from the char is carried out of the combustor by the combustion gases and separated from the hot gases using another pair of cyclones. The first cyclone is designed to capture mostly sand while the smaller ash particles remain entrained in the gas exiting the cyclone. The second cyclone is designed to capture the ash and any sand passing through the first cyclone. The hot sand captured by the first cyclone flows by gravity back into the gasifier to provide the heat for the gasification reaction. Ash and sand particles captured in the second cyclone are cooled, moistened to minimize dust and sent to a land fill for disposal.

- *Gas Cleanup and Conditioning.* This consists of multiple operations: reforming of tars and other hydrocarbons to CO and $H_2$; syngas cooling/quench; and acid gas ($CO_2$ and $H_2S$) removal with subsequent reduction of $H_2S$ to sulfur. Tar reforming is envisioned to occur in an isothermal fluidized bed reactor; de-activated reforming catalyst is separated from the effluent syngas and regenerated on-line. The hot syngas is cooled through heat exchange with the steam cycle and additional cooling via water scrubbing. The scrubber also removes impurities such as particulates and ammonia along with any residual tars. The excess scrubber water is sent off-site to a waste-water treatment facility. The cooled syngas enters an amine unit to remove the $CO_2$ and $H_2S$. The $H_2S$ is reduced to elemental sulfur and stockpiled for disposal. The $CO_2$ is vented to the atmosphere in this design.

- *Alcohol Synthesis.* The cleaned and conditioned syngas is converted to alcohols in a fixed bed reactor. The mixture of alcohol and unconverted syngas is cooled through heat exchange with the steam cycle and other process streams. The liquid alcohols are separated by condensing them away from the unconverted syngas. Though the unconverted syngas has the potential to be recycled back to the entrance of the alcohol synthesis reactor, this recycle is not done in this process design. Instead it is recycled to the Gas Cleanup and Conditioning section, mostly as feed to the tar reformer.

- *Alcohol Separation.* The alcohol stream from the Alcohol Synthesis section is depressurized in preparation of dehydration and separation. Another rough separation is performed in a flash separator; the evolved syngas is recycled to the Gas Cleanup and Conditioning section, mostly as feed to the tar reformer. The depressurized alcohol stream is dehydrated using vapor-phase molecular sieves. The dehydrated alcohol stream is introduced to the main alcohol separation column that splits methanol and ethanol from the higher molecular weight alcohols. The overheads are topped in a second column to remove the methanol to ASTM sales specifications. The methanol leaving in the overheads is used to flush the adsorbed water from the molecular sieves. This methanol/water mixture is recycled back to the entrance of the alcohol synthesis reactor.

- *Heat and Power.* A conventional steam cycle produces heat (as steam) for the gasifier and reformer operations and electricity for internal power requirements (with the possibility of exporting excess electricity as a co-product). The steam cycle is integrated with the biomass conversion process. Pre-heaters, steam generators, and super-heaters are integrated within the process design to create the steam. The steam will run through turbines to drive compressors, generate electricity or be withdrawn at various pressure levels for injection into the process. The condensate will be sent back to the steam cycle, de-gassed, and combined with make-up water.

A cooling water system is also included in the model to determine the requirements of each cooling water heat exchanger within the biomass conversion process as well as the requirements of the cooling tower.

Capital and operating costs for the USA base case are given in Table 59. Figure 23 shows the impact of plant size for the USA base case. Table 19 (in the methodology section) gives cellulosic feedstock costs from the ORNL feedstock report (Draft, August 2007); Table 60 gives the corresponding plant gate costs. Representative capital and operating costs for the other countries are given by Table 61 through Table 68.

Table 69 and Table 70 summarize the capital and operating costs, and plant gate prices for 25 MM GPY and 100 MM GPY plants, respectively, using bagasse feedstock. Similarly Table 71 and Table 72 are for agricultural residues, and Table 73 and Table 74 are for wood/perennial feeds.

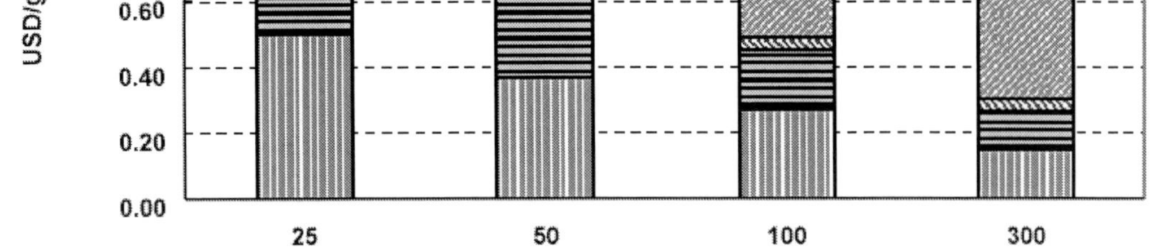


| | 25 | 50 | 100 | 300 |
|---|---|---|---|---|
| Feed | 0.44 | 0.44 | 0.44 | 0.44 |
| Var Op Costs | 0.04 | 0.04 | 0.04 | 0.04 |
| Fixed Op Costs | 0.35 | 0.25 | 0.18 | 0.12 |
| Capital | $0.50 | $0.37 | $0.27 | $0.15 |

Plant Size (Mil gal/yr)

Figure 23. Cellulosic Ethanol (TC) Plant Gate Price

# Table 59. USA cellulosic ethanol (TC) capital and operating costs

| Country | USA | Code = | 1 | | | | |
|---|---|---|---|---|---|---|---|
| Cost component | Units | Cost Factor | Scale Factor | USA Cellulosic Ethanol (TC) | USA Cellulosic Ethanol (TC) | USA Cellulosic Ethanol (TC) | USA Cellulosic Ethanol (TC) |
| Year $ | $ | | | 2005 | 2005 | 2005 | 2005 |
| Cost Index (2 = M&S, 3 = CE) | | | 3 | 130.93 | 130.93 | 130.93 | 130.93 |
| Plant Size | dry short ton/day | | | 1,800 | 900 | 3,601 | 10,801 |
| | dry tonne/day | | | 1,632 | 816 | 3,264 | 9,793 |
| Stream Factor | % | | | 95% | 95% | 95% | 95% |
| Feed | Dry short ton/yr | | | 6.242E+05 | 3.121E+05 | 1.248E+06 | 3.745E+06 |
| | GJ/short ton | | | 1.794E+01 | 1.794E+01 | 1.794E+01 | 1.794E+01 |
| | GJ/yr | | | 1.119E+07 | 5.598E+06 | 2.239E+07 | 6.717E+07 |
| Feed Cost | USD/dt | | | **$ 35.00** | **$ 35.00** | **$ 35.00** | **$ 35.00** |
| Yield (gal/Dry US Ton) | | | | | | | |
| Ethanol | gal/short ton | | | 80.1 | 80.1 | 80.1 | 80.1 |
| Mixed alcohols | gal/short ton | | | 94.1 | 94.1 | 94.1 | 94.1 |
| Process Efficiency - to ethanol | % HHV | | | 39.6% | 39.6% | 39.6% | 39.6% |
| Process efficiency - overall | % HHV | | | 47.4% | 47.4% | 47.4% | 47.4% |
| Ethanol | gal/yr | | | 5.000E+07 | 2.500E+07 | 1.000E+08 | 3.000E+08 |
| | Mil Gal/YR | | | **50.00** | **25.00** | **100.00** | **300.00** |
| | tpy | | | 1.657E+05 | 8.288E+04 | 3.315E+05 | 9.945E+05 |
| | GJ/yr | | | 4.436E+06 | 2.218E+06 | 8.873E+06 | 2.662E+07 |
| | gal/Stream day | | | 1.442E+05 | 7.210E+04 | 2.884E+05 | 8.652E+05 |
| | bbl/s stream day | | | 3.433E+03 | 1.717E+03 | 6.867E+03 | 2.060E+04 |
| Higher Alcohols | | | | | | | |
| | gal/ton | | | 14.0 | 14.0 | 14.0 | 14.0 |
| | gal/yr | | | 8.74E+06 | 4.37E+06 | 1.75E+07 | 5.24E+07 |
| | GJ/gal | | | | | | |
| **Capital Cost (million USD)** | | | | | | | |
| Feed Handling & Drying | | | 0.70 | 23.20 | 14.28 | 37.69 | 81.32 |
| Gasification | | | 0.80 | 12.90 | 7.41 | 22.46 | 54.09 |
| Tar Reforming & Quench | | | 0.70 | 38.40 | 23.64 | 62.38 | 134.60 |
| Acid Gas & Sulfur Removal | | | 0.85 | 14.50 | 8.05 | 26.14 | 66.50 |
| Alcohol synthesis - Compression | | | 0.65 | 16.00 | 10.20 | 25.11 | 51.28 |
| Alcohol Synthesis - Other | | | 0.65 | 4.60 | 2.93 | 7.22 | 14.74 |
| Alcohol Separation | | | 0.65 | 7.20 | 4.59 | 11.30 | 23.08 |
| Steam System & Power Generation | | | 0.70 | 16.80 | 10.34 | 27.29 | 58.89 |
| Cooling Water & Other Utilities | | | 0.70 | 3.60 | 2.22 | 5.85 | 12.62 |
| **Direct Fixed Capital (DFC), also called TIC** | | | | 137.20 | 83.65 | 225.44 | 497.12 |
| Engineering | DFC x MF | 0.12 | | 16.46 | 10.04 | 27.05 | 59.65 |
| Construction | DFC x MF | 0.13 | | 17.84 | 10.88 | 29.31 | 64.63 |
| Contractor & Legal | DFC x MF | 0.08 | | 10.98 | 6.69 | 18.04 | 39.77 |
| Process/Project Contingency | DFC x MF | 0.0412 | | 5.65 | 3.45 | 9.29 | 20.48 |
| **Total Plant Cost (TPC)** | | | | 188.13 | 114.71 | 309.13 | 681.65 |
| AFUDC | | | | | | | |
| **Total Plant Investment (TPI)** | | | | 188.13 | 114.71 | 309.13 | 681.65 |
| Land | | | 0.60 | 2.21 | 2.21 | 5.08 | 15.23 |
| Startup | | | | 0.00 | 0.00 | 0.00 | 0.00 |
| **Total Capital Cost (TCC)** | | | | 190.34 | 116.92 | 314.21 | 696.88 |
| **Contingency/TPI** | | | | | | | |
| Working Capital | DFC x MF | 0.05 | | 9.41 | 5.74 | 15.46 | 34.08 |
| **Variable Operating Costs (million USD/yr)** | | | | | | | |
| Feed | | | | 21.85 | 10.92 | 43.70 | 131.09 |
| Utilities | | | | 0.30 | 0.12 | 0.49 | 1.47 |
| Other | | | | 1.00 | 0.50 | 2.00 | 6.00 |
| Catalysts and Chemicals | | | | 0.70 | 0.35 | 1.40 | 4.20 |
| **Total** | | | | 23.85 | 11.90 | 47.59 | 142.75 |
| Fixed Operating Costs (million USD/yr) | | | | | | | |
| Labor (SF from V-R) | DFC x MF | 0.0165 | 0.25 | 2.26 | 1.90 | 2.69 | 3.54 |
| Maintenance (3% of A) | | | 0.03 | 4.12 | 2.51 | 6.76 | 14.91 |
| General Overhead (65% of labor + maint) | | | 0.65 | 4.15 | 2.87 | 6.15 | 12.00 |
| Direct Overhead (45% of Labor) | | | 0.45 | 1.02 | 0.86 | 1.21 | 1.59 |
| Insurance (0.5% of TIC) | | | 0.005 | 0.95 | 0.58 | 1.57 | 3.48 |
| **Total** | | | | 12.50 | 8.72 | 18.38 | 35.53 |

## Table 60. Cellulosic ethanol (TC) plant gate costs

| Plant Size, MM GPY | 25 | 100 | 25 | 100 | 25 | 100 |
|---|---|---|---|---|---|---|
| Feed | Bagasse | | Ag Residues | | Wood /Perennials | |
| Country | Ethanol Plant Gate Cost (USD/gal ethanol) | | | | | |
| USA | 1.33 | 0.93 | 1.33 | 0.93 | 1.33 | 0.93 |
| Argentina | 1.02 | 0.62 | 1.49 | 1.10 | 1.49 | 1.10 |
| Brazil | 0.88 | 0.54 | 1.17 | 0.83 | 1.37 | 1.04 |
| Canada | -- | -- | 1.32 | 0.79 | 1.60 | 1.06 |
| Caribbean Basin | 1.03 | 0.62 | 1.12 | 0.71 | 1.53 | 1.12 |
| China | 1.07 | 0.65 | 1.31 | 0.88 | 1.59 | 1.13 |
| Colombia | 1.01 | 0.63 | 1.01 | 0.63 | 1.39 | 1.04 |
| India | 0.89 | 0.57 | 0.89 | 0.57 | 1.39 | 1.05 |
| Mexico | 0.80 | 0.54 | 1.20 | 0.93 | 1.20 | 0.93 |

## Table 61. Brazil cellulosic ethanol (TC) capital and operating costs, bagasse

| Country | Brazil | Code = | 2 | | | | |
|---|---|---|---|---|---|---|---|
| Cost component | Units | Cost Factor | Scale Factor | Brazil Cellulosic Ethanol (TC) | Brazil Cellulosic Ethanol (TC) | Brazil Cellulosic Ethanol (TC) | Brazil Cellulosic Ethanol (TC) |
| Year $ | $ | | | 2005 | 2005 | 2005 | 2005 |
| Cost Index (2 = M&S, 3 = CE) | | | 3 | 130.93 | 130.93 | 130.93 | 130.93 |
| Plant Size | dry short ton/day | | | 1,800 | 900 | 3,601 | 10,801 |
| | dry tonne/day | | | 1,632 | 816 | 3,264 | 9,793 |
| Stream Factor | % | | | 95% | 95% | 95% | 95% |
| Feed | Dry short ton/yr | | | .242E+05 | 3.121E+05 | .248E+06 | 3.745E+06 |
| | GJ/short ton | | | .794E+01 | 1.794E+01 | .794E+01 | 1.794E+01 |
| | GJ/yr | | | .119E+07 | 5.598E+06 | 2.239E+07 | 6.717E+07 |
| Feed Cost | USD/dt | | | **7.60** | **7.60** | **7.60** | **7.60** |
| Yield (gal/Dry US Ton) | | | | | | | |
| Ethanol | gal/short ton | | | 80.1 | 80.1 | 80.1 | 80.1 |
| Mixed alcohols | gal/short ton | | | 94.1 | 94.1 | 94.1 | 94.1 |
| Process Efficiency - to ethanol | % HHV | | | 39.6% | 39.6% | 39.6% | 39.6% |
| Process efficiency - overall | % HHV | | | 47.4% | 47.4% | 47.4% | 47.4% |
| Ethanol | gal/yr | | | .000E+07 | 2.500E+07 | .000E+08 | 3.000E+08 |
| | Mil Gal/YR | | | **50.00** | **25.00** | **100.00** | **300.00** |
| | tpy | | | .657E+05 | 8.288E+04 | 3.315E+05 | 9.945E+05 |
| | GJ/yr | | | .436E+06 | 2.218E+06 | 8.873E+06 | 2.662E+07 |
| | gal/Stream day | | | .442E+05 | 7.210E+04 | 2.884E+05 | 8.652E+05 |
| | bbl/s stream day | | | .433E+03 | 1.717E+03 | 6.867E+03 | 2.060E+04 |
| Higher Alcohols | | | | | | | |
| | gal/ton | | | 14.0 | 14.0 | 14.0 | 14.0 |
| | gal/yr | | | 8.74E+06 | 4.37E+06 | 1.75E+07 | 5.24E+07 |
| | GJ/gal | | | | | | |
| **Capital Cost (million USD)** | | | | | | | |
| Feed Handling & Drying | | | 0.70 | 23.71 | 14.60 | 38.52 | 83.10 |
| Gasification | | | 0.80 | 13.18 | 7.57 | 22.95 | 55.28 |
| Tar Reforming & Quench | | | 0.70 | 39.24 | 24.16 | 63.75 | 137.55 |
| Acid Gas & Sulfur Removal | | | 0.85 | 14.82 | 8.22 | 26.71 | 67.96 |
| Alcohol synthesis - Compression | | | 0.65 | 16.35 | 10.42 | 25.66 | 52.40 |
| Alcohol Synthesis - Other | | | 0.65 | 4.70 | 3.00 | 7.38 | 15.07 |
| Alcohol Separation | | | 0.65 | 7.36 | 4.69 | 11.56 | 23.58 |
| Steam System & Power Generation | | | 0.70 | 17.17 | 10.57 | 27.89 | 60.18 |
| Cooling Water & Other Utilities | | | 0.70 | 3.68 | 2.26 | 5.98 | 12.90 |
| **Direct Fixed Capital (DFC), also called TIC** | | | | 140.21 | 85.49 | 230.38 | 508.01 |
| Engineering | DFC x MF | 0.12 | | 16.82 | 10.26 | 27.65 | 60.96 |
| Construction | DFC x MF | 0.13 | | 18.23 | 11.11 | 29.95 | 66.04 |
| Contractor & Legal | DFC x MF | 0.08 | | 11.22 | 6.84 | 18.43 | 40.64 |
| Process/Project Contingency | DFC x MF | 0.0412 | | 5.78 | 3.52 | 9.49 | 20.93 |
| **Total Plant Cost (TPC)** | | | | 192.25 | 117.22 | 315.90 | 696.59 |
| AFUDC | | | | | | | |
| **Total Plant Investment (TPI)** | | | | 192.25 | 117.22 | 315.90 | 696.59 |
| Land | | | 0.60 | 2.21 | 2.21 | 5.08 | 15.23 |
| Startup | | | | 0.00 | 0.00 | 0.00 | 0.00 |
| **Total Capital Cost (TCC)** | | | | 194.46 | 119.43 | 320.98 | 711.82 |
| **Contingency/TPI** | | | | | | | |
| Working Capital | DFC x MF | 0.05 | | 9.61 | 5.86 | 15.80 | 34.83 |
| **Variable Operating Costs (million USD/yr)** | | | | | | | |
| Feed | | | | 4.74 | 2.37 | 9.49 | 28.46 |
| Utilities | | | | 0.30 | 0.12 | 0.49 | 1.47 |
| Other | | | | 1.00 | 0.50 | 2.00 | 6.00 |
| Catalysts and Chemicals | | | | 0.70 | 0.35 | 1.40 | 4.20 |
| **Total** | | | | 6.74 | 3.34 | 13.38 | 40.13 |
| Fixed Operating Costs (million USD/yr) | | | | | | | |
| Labor (SF from V-R) | DFC x MF | 0.0165 | 0.25 | 0.96 | 0.81 | 1.15 | 1.51 |
| Maintenance (3% of A) | | | 0.03 | 4.21 | 2.56 | 6.91 | 15.24 |
| General Overhead (65% of labor + maint) | | | 0.65 | 3.36 | 2.19 | 5.24 | 10.89 |
| Direct Overhead (45% of Labor) | | | 0.45 | 0.43 | 0.37 | 0.52 | 0.68 |
| Insurance (0.5% of TIC) | | | 0.005 | 0.97 | 0.60 | 1.60 | 3.56 |
| **Total** | | | | 9.94 | 6.53 | 15.42 | 31.88 |

## Table 62. Argentina cellulosic ethanol (TC) capital and operating costs, bagasse

| Cost component | Units | Cost Factor | Scale Factor | Argentina Cellulosic Ethanol (TC) | Argentina Cellulosic Ethanol (TC) | Argentina Cellulosic Ethanol (TC) | Argentina Cellulosic Ethanol (TC) |
|---|---|---|---|---|---|---|---|
| Country | Argentina | Code = | 5 | | | | |
| Year $ | $ | | | 2005 | 2005 | 2005 | 2005 |
| Cost Index (2 = M&S, 3 = CE) | | | 3 | 130.93 | 130.93 | 130.93 | 130.93 |
| Plant Size | dry short ton/day | | | 1,800 | 900 | 3,601 | 10,801 |
| | dry tonne/day | | | 1,632 | 816 | 3,264 | 9,793 |
| Stream Factor | % | | | 95% | 95% | 95% | 95% |
| Feed | Dry short ton/yr | | | 6.242E+05 | 3.121E+05 | 1.248E+06 | 3.745E+06 |
| | GJ/short ton | | | 1.794E+01 | 1.794E+01 | 1.794E+01 | 1.794E+01 |
| | GJ/yr | | | 1.119E+07 | 5.598E+06 | 2.239E+07 | 6.717E+07 |
| Feed Cost | USD/dt | | | **$ 9.10** | **$ 9.10** | **$ 9.10** | **$ 9.10** |
| Yield (gal/Dry US Ton) | | | | | | | |
| Ethanol | gal/short ton | | | 80.1 | 80.1 | 80.1 | 80.1 |
| Mixed alcohols | gal/short ton | | | 94.1 | 94.1 | 94.1 | 94.1 |
| Process Efficiency - to ethanol | % HHV | | | 39.6% | 39.6% | 39.6% | 39.6% |
| Process efficiency - overall | % HHV | | | 47.4% | 47.4% | 47.4% | 47.4% |
| Ethanol | gal/yr | | | 5.000E+07 | 2.500E+07 | 1.000E+08 | 3.000E+08 |
| | Mil Gal/YR | | | **50.00** | **25.00** | **100.00** | **300.00** |
| | tpy | | | 1.657E+05 | 8.288E+04 | 3.315E+05 | 9.945E+05 |
| | GJ/yr | | | 4.436E+06 | 2.218E+06 | 8.873E+06 | 2.662E+07 |
| | gal/Stream day | | | 1.442E+05 | 7.210E+04 | 2.884E+05 | 8.652E+05 |
| | bbl/s stream day | | | 3.433E+03 | 1.717E+03 | 6.867E+03 | 2.060E+04 |
| Higher Alcohols | | | | | | | |
| | gal/ton | | | 14.0 | 14.0 | 14.0 | 14.0 |
| | gal/yr | | | 8.74E+06 | 4.37E+06 | 1.75E+07 | 5.24E+07 |
| **Capital Cost (million USD)** | | | | | | | |
| Feed Handling & Drying | | | 0.70 | 24.91 | 15.34 | 40.47 | 87.32 |
| Gasification | | | 0.80 | 13.85 | 7.96 | 24.12 | 58.08 |
| Tar Reforming & Quench | | | 0.70 | 41.23 | 25.38 | 66.99 | 144.53 |
| Acid Gas & Sulfur Removal | | | 0.85 | 15.57 | 8.64 | 28.07 | 71.40 |
| Alcohol synthesis - Compression | | | 0.65 | 17.18 | 10.95 | 26.96 | 55.06 |
| Alcohol Synthesis - Other | | | 0.65 | 4.94 | 3.15 | 7.75 | 15.83 |
| Alcohol Separation | | | 0.65 | 7.73 | 4.93 | 12.13 | 24.78 |
| Steam System & Power Generation | | | 0.70 | 18.04 | 11.11 | 29.31 | 63.23 |
| Cooling Water & Other Utilities | | | 0.70 | 3.87 | 2.38 | 6.28 | 13.55 |
| **Direct Fixed Capital (DFC), also called TIC** | | | | 147.32 | 89.82 | 242.07 | 533.78 |
| Engineering | DFC x MF | 0.12 | | 17.68 | 10.78 | 29.05 | 64.05 |
| Construction | DFC x MF | 0.13 | | 19.15 | 11.68 | 31.47 | 69.39 |
| Contractor & Legal | DFC x MF | 0.08 | | 11.79 | 7.19 | 19.37 | 42.70 |
| Process/Project Contingency | DFC x MF | 0.0412 | | 6.07 | 3.70 | 9.97 | 21.99 |
| **Total Plant Cost (TPC)** | | | | 202.00 | 123.17 | 331.93 | 731.92 |
| AFUDC | | | | | | | |
| **Total Plant Investment (TPI)** | | | | 202.00 | 123.17 | 331.93 | 731.92 |
| Land | | | 0.60 | 2.21 | 2.21 | 5.08 | 15.23 |
| Startup | | | | 0.00 | 0.00 | 0.00 | 0.00 |
| **Total Capital Cost (TCC)** | | | | 204.21 | 125.38 | 337.00 | 747.15 |
| **Contingency/TPI** | | | | | | | |
| Working Capital | DFC x MF | 0.05 | | 10.10 | 6.16 | 16.60 | 36.60 |
| **Variable Operating Costs (million USD/yr)** | | | | | | | |
| Feed | | | | 5.68 | 2.84 | 11.36 | 34.08 |
| Utilities | | | | 0.30 | 0.12 | 0.49 | 1.47 |
| Other | | | | 1.00 | 0.50 | 2.00 | 6.00 |
| Catalysts and Chemicals | | | | 0.70 | 0.35 | 1.40 | 4.20 |
| **Total** | | | | 7.68 | 3.81 | 15.25 | 45.75 |
| Fixed Operating Costs (million USD/yr) | | | | | | | |
| Labor (SF from V-R) | DFC x MF | 0.0165 | 0.25 | 1.92 | 1.62 | 2.28 | 3.01 |
| Maintenance (3% of A) | | | 0.03 | 4.42 | 2.69 | 7.26 | 16.01 |
| General Overhead (65% of labor + maint) | | | 0.65 | 4.12 | 2.80 | 6.21 | 12.36 |
| Direct Overhead (45% of Labor) | | | 0.45 | 0.86 | 0.73 | 1.03 | 1.35 |
| Insurance (0.5% of TIC) | | | 0.005 | 1.02 | 0.63 | 1.69 | 3.74 |
| **Total** | | | | 12.35 | 8.47 | 18.47 | 36.47 |

## Table 63. Canada cellulosic ethanol (TC) capital and operating costs, ag residue

| Country | Canada | Code = | 7 | | | | |
|---|---|---|---|---|---|---|---|
| **Cost component** | **Units** | **Cost Factor** | **Scale Factor** | **Canada Cellulosic Ethanol (TC)** | **Canada Cellulosic Ethanol (TC)** | **Canada Cellulosic Ethanol (TC)** | **Canada Cellulosic Ethanol (TC)** |
| Year $ | $ | | | 2005 | 2005 | 2005 | 2005 |
| Cost Index (2 = M&S, 3 = CE) | | | 3 | 130.93 | 130.93 | 130.93 | 130.93 |
| Plant Size | dry short ton/day | | | 1,800 | 900 | 3,601 | 10,801 |
| | dry tonne/day | | | 1,632 | 816 | 3,264 | 9,793 |
| Stream Factor | % | | | 95% | 95% | 95% | 95% |
| Feed | Dry short ton/yr | | | 6.242E+05 | 3.121E+05 | .248E+06 | 3.745E+06 |
| | GJ/short ton | | | 1.794E+01 | 1.794E+01 | .794E+01 | 1.794E+01 |
| | GJ/yr | | | 1.119E+07 | 5.598E+06 | 2.239E+07 | 6.717E+07 |
| Feed Cost | USD/dt | | | **$ 10.00** | **10.00** | **10.00** | **$ 10.00** |
| Yield (gal/Dry US Ton) | | | | | | | |
| Ethanol | gal/short ton | | | 80.1 | 80.1 | 80.1 | 80.1 |
| Mixed alcohols | gal/short ton | | | 94.1 | 94.1 | 94.1 | 94.1 |
| Process Efficiency - to ethanol | % HHV | | | 39.6% | 39.6% | 39.6% | 39.6% |
| Process efficiency - overall | % HHV | | | 47.4% | 47.4% | 47.4% | 47.4% |
| Ethanol | gal/yr | | | 5.000E+07 | 2.500E+07 | .000E+08 | 3.000E+08 |
| | Mil Gal/YR | | | **50.00** | **25.00** | **100.00** | **300.00** |
| | tpy | | | 1.657E+05 | 8.288E+04 | 3.315E+05 | 9.945E+05 |
| | GJ/yr | | | 4.436E+06 | 2.218E+06 | 8.873E+06 | 2.662E+07 |
| | gal/Stream day | | | 1.442E+05 | 7.210E+04 | 2.884E+05 | 8.652E+05 |
| | bbl/s stream day | | | 3.433E+03 | .717E+03 | 6.867E+03 | 2.060E+04 |
| Higher Alcohols | | | | | | | |
| | gal/ton | | | 14.0 | 14.0 | 14.0 | 14.0 |
| | gal/yr | | | 8.74E+06 | 4.37E+06 | 1.75E+07 | 5.24E+07 |
| **Capital Cost (million USD)** | | | | | | | |
| Feed Handling & Drying | | | 0.70 | 30.23 | 18.6 | 49.11 | 105.96 |
| Gasification | | | 0.80 | 16.81 | 9.65 | 29.27 | 70.48 |
| Tar Reforming & Quench | | | 0.70 | 50.03 | 30.80 | 81.28 | 175.38 |
| Acid Gas & Sulfur Removal | | | 0.85 | 18.89 | 10.48 | 34.06 | 86.64 |
| Alcohol synthesis - Compression | | | 0.65 | 20.85 | 13.29 | 32.71 | 66.81 |
| Alcohol Synthesis - Other | | | 0.65 | 5.99 | 3.82 | 9.41 | 19.21 |
| Alcohol Separation | | | 0.65 | 9.38 | 5.98 | 14.72 | 30.07 |
| Steam System & Power Generation | | | 0.70 | 21.89 | 13.48 | 35.56 | 76.73 |
| Cooling Water & Other Utilities | | | 0.70 | 4.69 | 2.89 | 7.62 | 16.44 |
| **Direct Fixed Capital (DFC), also called TIC** | | | | 178.76 | 109.00 | 293.74 | 647.72 |
| Engineering | DFC x MF | 0.12 | | 21.45 | 13.08 | 35.25 | 77.73 |
| Construction | DFC x MF | 0.13 | | 23.24 | 14.17 | 38.19 | 84.20 |
| Contractor & Legal | DFC x MF | 0.08 | | 14.30 | 8.72 | 23.50 | 51.82 |
| Process/Project Contingency | DFC x MF | 0.0412 | | 7.37 | 4.49 | 12.10 | 26.69 |
| **Total Plant Cost (TPC)** | | | | 245.12 | 149.46 | 402.78 | 888.15 |
| AFUDC | | | | | | | |
| **Total Plant Investment (TPI)** | | | | 245.12 | 149.46 | 402.78 | 888.15 |
| Land | | | 0.60 | 2.21 | 2.2 | 5.08 | 15.23 |
| Startup | | | | 0.00 | 0.00 | 0.00 | 0.00 |
| **Total Capital Cost (TCC)** | | | | 247.33 | 151.67 | 407.86 | 903.38 |
| **Contingency/TPI** | | | | | | | |
| Working Capital | DFC x MF | 0.05 | | 12.26 | 7.47 | 20.14 | 44.41 |
| **Variable Operating Costs (million USD/yr)** | | | | | | | |
| Feed | | | | 6.24 | 3.12 | 12.48 | 37.45 |
| Utilities | | | | 0.3 | 0.12 | 0.49 | 1.47 |
| Other | | | | 1.0 | 0.50 | 2.00 | 6.00 |
| Catalysts and Chemicals | | | | 0.70 | 0.35 | 1.40 | 4.20 |
| **Total** | | | | 8.24 | 4.09 | 16.37 | 49.12 |
| Fixed Operating Costs (million USD/yr) | | | | | | | |
| Labor (SF from V-R) | DFC x MF | 0.0165 | 0.25 | 3.8 | 3.19 | 4.52 | 5.95 |
| Maintenance (3% of A) | | | 0.03 | 5.36 | 3.27 | 8.81 | 19.43 |
| General Overhead (65% of labor + maint) | | | 0.65 | 5.9 | 4.20 | 8.66 | 16.50 |
| Direct Overhead (45% of Labor) | | | 0.45 | 1.7 | 1.44 | 2.03 | 2.68 |
| Insurance (0.5% of TIC) | | | 0.005 | 1.24 | 0.76 | 2.04 | 4.52 |
| **Total** | | | | 18.06 | 12.86 | 26.07 | 49.07 |

## Table 64. Caribbean cellulosic ethanol (TC) capital and operating costs, bagasse

| **Country** | **Caribbean** | **Code =** | 9 | | | | |
|---|---|---|---|---|---|---|---|
| **Cost component** | **Units** | **Cost Factor** | **Scale Factor** | **Caribbean Cellulosic Ethanol (TC)** | **Caribbean Cellulosic Ethanol (TC)** | **Caribbean Cellulosic Ethanol (TC)** | **Caribbean Cellulosic Ethanol (TC)** |
| Year $ | $ | | | 2005 | 2005 | 2005 | 2005 |
| Cost Index (2 = M&S, 3 = CE) | | | 3 | 130.93 | 130.93 | 130.93 | 130.93 |
| Plant Size | dry short ton/day | | | 1,800 | 900 | 3,601 | 10,801 |
| | dry tonne/day | | | 1,632 | 816 | 3,264 | 9,793 |
| Stream Factor | % | | | 95% | 95% | 95% | 95% |
| Feed | Dry short ton/yr | | | 6.242E+05 | 3.121E+05 | 1.248E+06 | 3.745E+06 |
| | GJ/short ton | | | 1.794E+01 | 1.794E+01 | 1.794E+01 | 1.794E+01 |
| | GJ/yr | | | 1.119E+07 | 5.598E+06 | 2.239E+07 | 6.717E+07 |
| Feed Cost | USD/dt | | | **$ 7.60** | **$ 7.60** | **$ 7.60** | **$ 7.60** |
| Yield (gal/Dry US Ton) | | | | | | | |
| Ethanol | gal/short ton | | | 80.1 | 80.1 | 80.1 | 80.1 |
| Mixed alcohols | gal/short ton | | | 94.1 | 94.1 | 94.1 | 94.1 |
| Process Efficiency - to ethanol | % HHV | | | 39.6% | 39.6% | 39.6% | 39.6% |
| Process efficiency - overall | % HHV | | | 47.4% | 47.4% | 47.4% | 47.4% |
| Ethanol | gal/yr | | | 5.000E+07 | 2.500E+07 | 1.000E+08 | 3.000E+08 |
| | Mil Gal/YR | | | **50.00** | **25.00** | **100.00** | **300.00** |
| | tpy | | | 1.657E+05 | 8.288E+04 | 3.315E+05 | 9.945E+05 |
| | GJ/yr | | | 4.436E+06 | 2.218E+06 | 8.873E+06 | 2.662E+07 |
| | gal/Stream day | | | 1.442E+05 | 7.210E+04 | 2.884E+05 | 8.652E+05 |
| | bbl/s stream day | | | 3.433E+03 | 1.717E+03 | 6.867E+03 | 2.060E+04 |
| Higher Alcohols | | | | | | | |
| | gal/ton | | | 14.0 | 14.0 | 14.0 | 14.0 |
| | gal/yr | | | 8.74E+06 | 4.37E+06 | 1.75E+07 | 5.24E+07 |
| **Capital Cost (million USD)** | | | | | | | |
| Feed Handling & Drying | | | 0.70 | 26.18 | 16.11 | 42.53 | 91.75 |
| Gasification | | | 0.80 | 14.55 | 8.36 | 25.34 | 61.03 |
| Tar Reforming & Quench | | | 0.70 | 43.33 | 26.67 | 70.39 | 151.87 |
| Acid Gas & Sulfur Removal | | | 0.85 | 16.36 | 9.08 | 29.49 | 75.03 |
| Alcohol synthesis - Compression | | | 0.65 | 18.05 | 11.51 | 28.33 | 57.86 |
| Alcohol Synthesis - Other | | | 0.65 | 5.19 | 3.31 | 8.14 | 16.63 |
| Alcohol Separation | | | 0.65 | 8.12 | 5.18 | 12.75 | 26.04 |
| Steam System & Power Generation | | | 0.70 | 18.96 | 11.67 | 30.79 | 66.44 |
| Cooling Water & Other Utilities | | | 0.70 | 4.06 | 2.50 | 6.60 | 14.24 |
| **Direct Fixed Capital (DFC), also called TIC** | | | | 154.80 | 94.39 | 254.37 | 560.89 |
| Engineering | DFC x MF | 0.12 | | 18.58 | 11.33 | 30.52 | 67.31 |
| Construction | DFC x MF | 0.13 | | 20.12 | 12.27 | 33.07 | 72.92 |
| Contractor & Legal | DFC x MF | 0.08 | | 12.38 | 7.55 | 20.35 | 44.87 |
| Process/Project Contingency | DFC x MF | 0.0412 | | 6.38 | 3.89 | 10.48 | 23.11 |
| **Total Plant Cost (TPC)** | | | | 212.26 | 129.42 | 348.79 | 769.10 |
| AFUDC | | | | | | | |
| **Total Plant Investment (TPI)** | | | | 212.26 | 129.42 | 348.79 | 769.10 |
| Land | | | 0.60 | 2.21 | 2.21 | 5.08 | 15.23 |
| Startup | | | | 0.00 | 0.00 | 0.00 | 0.00 |
| **Total Capital Cost (TCC)** | | | | 214.47 | 131.63 | 353.86 | 784.33 |
| **Contingency/TPI** | | | | | | | |
| Working Capital | DFC x MF | 0.05 | | 10.61 | 6.47 | 17.44 | 38.45 |
| **Variable Operating Costs (million USD/yr)** | | | | | | | |
| Feed | | | | 4.74 | 2.37 | 9.49 | 28.46 |
| Utilities | | | | 0.30 | 0.12 | 0.49 | 1.47 |
| Other | | | | 1.00 | 0.50 | 2.00 | 6.00 |
| Catalysts and Chemicals | | | | 0.70 | 0.35 | 1.40 | 4.20 |
| **Total** | | | | 6.74 | 3.34 | 13.38 | 40.13 |
| Fixed Operating Costs (million USD/yr) | | | | | | | |
| Labor (SF from V-R) | DFC x MF | 0.0165 | 0.25 | 2.17 | 1.83 | 2.59 | 3.40 |
| Maintenance (3% of A) | | | 0.03 | 4.64 | 2.83 | 7.63 | 16.83 |
| General Overhead (65% of labor + maint) | | | 0.65 | 4.43 | 3.03 | 6.64 | 13.15 |
| Direct Overhead (45% of Labor) | | | 0.45 | 0.98 | 0.82 | 1.16 | 1.53 |
| Insurance (0.5% of TIC) | | | 0.005 | 1.07 | 0.66 | 1.77 | 3.92 |
| **Total** | | | | 13.30 | 9.17 | 19.79 | 38.83 |

## Table 65. China cellulosic ethanol (TC) capital and operating costs, bagasse

| Country | China | Code = | 3 | | | | |
|---|---|---|---|---|---|---|---|
| **Cost component** | **Units** | **Cost Factor** | **Scale Factor** | **China Cellulosic Ethanol (TC)** | **China Cellulosic Ethanol (TC)** | **China Cellulosic Ethanol (TC)** | **China Cellulosic Ethanol (TC)** |
| Year $ | $ | | | 2005 | 2005 | 2005 | 2005 |
| Cost Index (2 = M&S, 3 = CE) | | | 3 | 130.93 | 130.93 | 130.93 | 130.93 |
| Plant Size | dry short ton/day | | | 1,800 | 900 | 3,601 | 10,801 |
| | dry tonne/day | | | 1,632 | 816 | 3,264 | 9,793 |
| Stream Factor | % | | | 95% | 95% | 95% | 95% |
| Feed | Dry short ton/yr | | | 6.242E+05 | 3.121E+05 | 1.248E+06 | 3.745E+06 |
| | GJ/short ton | | | 1.794E+01 | 1.794E+01 | 1.794E+01 | 1.794E+01 |
| | GJ/yr | | | 1.119E+07 | 5.598E+06 | 2.239E+07 | 6.717E+07 |
| Feed Cost | USD/dt | | | **$ 9.10** | **$ 9.10** | **$ 9.10** | **$ 9.10** |
| Yield (gal/Dry US Ton) | | | | | | | |
| Ethanol | gal/short ton | | | 80.1 | 80.1 | 80.1 | 80.1 |
| Mixed alcohols | gal/short ton | | | 94.1 | 94.1 | 94.1 | 94.1 |
| Process Efficiency - to ethanol | % HHV | | | 39.6% | 39.6% | 39.6% | 39.6% |
| Process efficiency - overall | % HHV | | | 47.4% | 47.4% | 47.4% | 47.4% |
| Ethanol | gal/yr | | | 5.000E+07 | 2.500E+07 | 1.000E+08 | 3.000E+08 |
| | Mil Gal/YR | | | **50.00** | **25.00** | **100.00** | **300.00** |
| | tpy | | | 1.657E+05 | 8.288E+04 | 3.315E+05 | 9.945E+05 |
| | GJ/yr | | | 4.436E+06 | 2.218E+06 | 8.873E+06 | 2.662E+07 |
| | gal/Stream day | | | 1.442E+05 | 7.210E+04 | 2.884E+05 | 8.652E+05 |
| | bbl/s stream day | | | 3.433E+03 | 1.717E+03 | 6.867E+03 | 2.060E+04 |
| Higher Alcohols | | | | | | | |
| | gal/ton | | | 14.0 | 14.0 | 14.0 | 14.0 |
| | gal/yr | | | 8.74E+06 | 4.37E+06 | 1.75E+07 | 5.24E+07 |
| **Capital Cost (million USD)** | | | | | | | |
| Feed Handling & Drying | | | 0.70 | 25.53 | 15.72 | 41.48 | 89.49 |
| Gasification | | | 0.80 | 14.20 | 8.15 | 24.72 | 59.52 |
| Tar Reforming & Quench | | | 0.70 | 42.26 | 26.01 | 68.65 | 148.12 |
| Acid Gas & Sulfur Removal | | | 0.85 | 15.96 | 8.85 | 28.76 | 73.18 |
| Alcohol synthesis - Compression | | | 0.65 | 17.61 | 11.22 | 27.63 | 56.43 |
| Alcohol Synthesis - Other | | | 0.65 | 5.06 | 3.23 | 7.94 | 16.22 |
| Alcohol Separation | | | 0.65 | 7.92 | 5.05 | 12.43 | 25.39 |
| Steam System & Power Generation | | | 0.70 | 18.49 | 11.38 | 30.04 | 64.80 |
| Cooling Water & Other Utilities | | | 0.70 | 3.96 | 2.44 | 6.44 | 13.89 |
| **Direct Fixed Capital (DFC), also called TIC** | | | | 150.98 | 92.06 | 248.09 | 547.05 |
| Engineering | DFC x MF | 0.12 | | 18.12 | 11.05 | 29.77 | 65.65 |
| Construction | DFC x MF | 0.13 | | 19.63 | 11.97 | 32.25 | 71.12 |
| Contractor & Legal | DFC x MF | 0.08 | | 12.08 | 7.36 | 19.85 | 43.76 |
| Process/Project Contingency | DFC x MF | 0.0412 | | 6.22 | 3.79 | 10.22 | 22.54 |
| **Total Plant Cost (TPC)** | | | | 207.03 | 126.23 | 340.18 | 750.12 |
| AFUDC | | | | | | | |
| **Total Plant Investment (TPI)** | | | | 207.03 | 126.23 | 340.18 | 750.12 |
| Land | | | 0.60 | 2.21 | 2.21 | 5.08 | 15.23 |
| Startup | | | | 0.00 | 0.00 | 0.00 | 0.00 |
| **Total Capital Cost (TCC)** | | | | 209.24 | 128.44 | 345.26 | 765.35 |
| **Contingency/TPI** | | | | | | | |
| Working Capital | DFC x MF | 0.05 | | 10.35 | 6.31 | 17.01 | 37.51 |
| **Variable Operating Costs (million USD/yr)** | | | | | | | |
| Feed | | | | 5.68 | 2.84 | 11.36 | 34.08 |
| Utilities | | | | 0.30 | 0.12 | 0.49 | 1.47 |
| Other | | | | 1.00 | 0.50 | 2.00 | 6.00 |
| Catalysts and Chemicals | | | | 0.70 | 0.35 | 1.40 | 4.20 |
| **Total** | | | | 7.68 | 3.81 | 15.25 | 45.75 |
| Fixed Operating Costs (million USD/yr) | | | | | | | |
| Labor (SF from V-R) | DFC x MF | 0.0165 | 0.25 | 2.52 | 2.12 | 2.99 | 3.94 |
| Maintenance (3% of A) | | | 0.03 | 4.53 | 2.76 | 7.44 | 16.41 |
| General Overhead (65% of labor + maint) | | | 0.65 | 4.58 | 3.17 | 6.78 | 13.23 |
| Direct Overhead (45% of Labor) | | | 0.45 | 1.13 | 0.95 | 1.35 | 1.77 |
| Insurance (0.5% of TIC) | | | 0.005 | 1.05 | 0.64 | 1.73 | 3.83 |
| **Total** | | | | 13.81 | 9.64 | 20.29 | 39.18 |

## Table 66. Colombia cellulosic ethanol (TC) capital and operating costs, bagasse

| Country | Colombia | Code = | 6 | | | | |
|---|---|---|---|---|---|---|---|
| Cost component | Units | Cost Factor | Scale Factor | Colombia Cellulosic Ethanol (TC) | Colombia Cellulosic Ethanol (TC) | Colombia Cellulosic Ethanol (TC) | Colombia Cellulosic Ethanol (TC) |
| Year $ | $ | | | 2005 | 2005 | 2005 | 2005 |
| Cost Index (2 = M&S, 3 = CE) | | | 3 | 130.93 | 130.93 | 130.93 | 130.93 |
| Plant Size | dry short ton/day | | | 1,800 | 900 | 3,601 | 10,801 |
| | dry tonne/day | | | 1,632 | 816 | 3,264 | 9,793 |
| Stream Factor | % | | | 95% | 95% | 95% | 95% |
| Feed | Dry short ton/yr | | | 6.242E+05 | 3.121E+05 | 1.248E+06 | 3.745E+06 |
| | GJ/short ton | | | 1.794E+01 | 1.794E+01 | 1.794E+01 | 1.794E+01 |
| | GJ/yr | | | 1.119E+07 | 5.598E+06 | 2.239E+07 | 6.717E+07 |
| Feed Cost | USD/dt | | | **$ 14.50** | **$ 14.50** | **$ 14.50** | **$ 14.50** |
| Yield (gal/Dry US Ton) | | | | | | | |
| Ethanol | gal/short ton | | | 80.1 | 80.1 | 80.1 | 80.1 |
| Mixed alcohols | gal/short ton | | | 94.1 | 94.1 | 94.1 | 94.1 |
| Process Efficiency - to ethanol | % HHV | | | 39.6% | 39.6% | 39.6% | 39.6% |
| Process efficiency - overall | % HHV | | | 47.4% | 47.4% | 47.4% | 47.4% |
| Ethanol | gal/yr | | | 5.000E+07 | 2.500E+07 | 1.000E+08 | 3.000E+08 |
| | Mil Gal/YR | | | **50.00** | **25.00** | **100.00** | **300.00** |
| | tpy | | | 1.657E+05 | 8.288E+04 | 3.315E+05 | 9.945E+05 |
| | GJ/yr | | | 4.436E+06 | 2.218E+06 | 8.873E+06 | 2.662E+07 |
| | gal/Stream day | | | 1.442E+05 | 7.210E+04 | 2.884E+05 | 8.652E+05 |
| | bbl/s stream day | | | 3.433E+03 | 1.717E+03 | 6.867E+03 | 2.060E+04 |
| Higher Alcohols | | | | | | | |
| | gal/ton | | | 14.0 | 14.0 | 14.0 | 14.0 |
| | gal/yr | | | 8.74E+06 | 4.37E+06 | 1.75E+07 | 5.24E+07 |
| | GJ/gal | | | | | | |
| | gal/ethanol eq/yr | | | | | | |
| | gal/ton EtOH eq | | | | | | |
| | BBI EtOH eq/day | | | | | | |
| **Capital Cost (million USD)** | | | | | | | |
| Feed Handling & Drying | | | 0.70 | 23.07 | 14.20 | 37.48 | 80.87 |
| Gasification | | | 0.80 | 12.83 | 7.37 | 22.34 | 53.79 |
| Tar Reforming & Quench | | | 0.70 | 38.19 | 23.51 | 62.04 | 133.85 |
| Acid Gas & Sulfur Removal | | | 0.85 | 14.42 | 8.00 | 25.99 | 66.13 |
| Alcohol synthesis - Compression | | | 0.65 | 15.91 | 10.14 | 24.97 | 50.99 |
| Alcohol Synthesis - Other | | | 0.65 | 4.57 | 2.92 | 7.18 | 14.66 |
| Alcohol Separation | | | 0.65 | 7.16 | 4.56 | 11.24 | 22.95 |
| Steam System & Power Generation | | | 0.70 | 16.71 | 10.28 | 27.14 | 58.56 |
| Cooling Water & Other Utilities | | | 0.70 | 3.58 | 2.20 | 5.82 | 12.55 |
| **Direct Fixed Capital (DFC), also called TIC** | | | | 136.44 | 83.19 | 224.19 | 494.35 |
| Engineering | DFC x MF | 0.12 | | 16.37 | 9.98 | 26.90 | 59.32 |
| Construction | DFC x MF | 0.13 | | 17.74 | 10.81 | 29.14 | 64.27 |
| Contractor & Legal | DFC x MF | 0.08 | | 10.91 | 6.66 | 17.94 | 39.55 |
| Process/Project Contingency | DFC x MF | 0.0412 | | 5.62 | 3.43 | 9.24 | 20.37 |
| **Total Plant Cost (TPC)** | | | | 187.08 | 114.07 | 307.41 | 677.86 |
| AFUDC | | | | | | | |
| **Total Plant Investment (TPI)** | | | | 187.08 | 114.07 | 307.41 | 677.86 |
| Land | | | 0.60 | 2.21 | 2.21 | 5.08 | 15.23 |
| Startup | | | | 0.00 | 0.00 | 0.00 | 0.00 |
| **Total Capital Cost (TCC)** | | | | 189.29 | **116.28** | **312.49** | 693.09 |
| **Contingency/TPI** | | | | | | | |
| Working Capital | DFC x MF | 0.05 | | 9.35 | 5.70 | 15.37 | 33.89 |
| **Variable Operating Costs (million USD/yr)** | | | | | | | |
| Feed | | | | 9.05 | 4.53 | 18.10 | 54.31 |
| Utilities | | | | 0.30 | 0.12 | 0.49 | 1.47 |
| Other | | | | 1.00 | 0.50 | 2.00 | 6.00 |
| Catalysts and Chemicals | | | | 0.70 | 0.35 | 1.40 | 4.20 |
| **Total** | | | | 11.05 | 5.50 | 21.99 | 65.98 |
| Fixed Operating Costs (million USD/yr) | | | | | | | |
| Labor (SF from V-R) | DFC x MF | 0.0165 | 0.25 | 1.44 | 1.21 | 1.71 | 2.26 |
| Maintenance (3% of A) | | | 0.03 | 4.09 | 2.50 | 6.73 | 14.83 |
| General Overhead (65% of labor + maint) | | | 0.65 | 3.60 | 2.41 | 5.49 | 11.11 |
| Direct Overhead (45% of Labor) | | | 0.45 | 0.65 | 0.55 | 0.77 | 1.02 |
| Insurance (0.5% of TIC) | | | 0.005 | 0.95 | 0.58 | 1.56 | 3.47 |
| **Total** | | | | 10.73 | 7.25 | 16.26 | 32.67 |

## Table 67. India cellulosic ethanol (TC) capital and operating costs, bagasse

| Country | India | Code = | 4 | | | | |
|---|---|---|---|---|---|---|---|
| **Cost component** | **Units** | **Cost Factor** | **Scale Factor** | **India Cellulosic Ethanol (TC)** | **India Cellulosic Ethanol (TC)** | **India Cellulosic Ethanol (TC)** | **India Cellulosic Ethanol (TC)** |
| Year $ | $ | | | 2005 | 2005 | 2005 | 2005 |
| Cost Index (2 = M&S, 3 = CE) | | | 3 | 130.93 | 130.93 | 130.93 | 130.93 |
| Plant Size | dry short ton/day | | | 1,800 | 900 | 3,601 | 10,801 |
| | dry tonne/day | | | 1,632 | 816 | 3,264 | 9,793 |
| Stream Factor | % | | | 95% | 95% | 95% | 95% |
| Feed | Dry short ton/yr | | | 6.242E+05 | 3.121E+05 | 1.248E+06 | 3.745E+06 |
| | GJ/short ton | | | 1.794E+01 | 1.794E+01 | 1.794E+01 | 1.794E+01 |
| | GJ/yr | | | 1.119E+07 | 5.598E+06 | 2.239E+07 | 6.717E+07 |
| Feed Cost | USD/dt | | | **$ 7.30** | **$ 7.30** | **$ 7.30** | **7.30** |
| Yield (gal/Dry US Ton) | | | | | | | |
| Ethanol | gal/short ton | | | 80.1 | 80.1 | 80.1 | 80.1 |
| Mixed alcohols | gal/short ton | | | 94.1 | 94.1 | 94.1 | 94.1 |
| Process Efficiency - to ethanol | % HHV | | | 39.6% | 39.6% | 39.6% | 39.6% |
| Process efficiency - overall | % HHV | | | 47.4% | 47.4% | 47.4% | 47.4% |
| Ethanol | gal/yr | | | 5.000E+07 | 2.500E+07 | 1.000E+08 | 3.000E+08 |
| | Mil Gal/YR | | | **50.00** | **25.00** | **100.00** | **300.00** |
| | tpy | | | 1.657E+05 | 8.288E+04 | 3.315E+05 | 9.945E+05 |
| | GJ/yr | | | 4.436E+06 | 2.218E+06 | 8.873E+06 | 2.662E+07 |
| | gal/Stream day | | | 1.442E+05 | 7.210E+04 | 2.884E+05 | 8.652E+05 |
| | bbl/s stream day | | | 3.433E+03 | 1.717E+03 | 6.867E+03 | 2.060E+04 |
| Higher Alcohols | | | | | | | |
| | gal/ton | | | 14.0 | 14.0 | 14.0 | 14.0 |
| | gal/yr | | | 8.74E+06 | 4.37E+06 | 1.75E+07 | 5.24E+07 |
| **Capital Cost (million USD)** | | | | | | | |
| Feed Handling & Drying | | | 0.70 | 24.75 | 15.23 | 40.20 | 86.74 |
| Gasification | | | 0.80 | 13.76 | 7.90 | 23.96 | 57.6 |
| Tar Reforming & Quench | | | 0.70 | 40.96 | 25.21 | 66.54 | 143.5 |
| Acid Gas & Sulfur Removal | | | 0.85 | 15.47 | 8.58 | 27.88 | 70.9 |
| Alcohol synthesis - Compression | | | 0.65 | 17.07 | 10.88 | 26.78 | 54.6 |
| Alcohol Synthesis - Other | | | 0.65 | 4.91 | 3.13 | 7.70 | 15.7 |
| Alcohol Separation | | | 0.65 | 7.68 | 4.89 | 12.05 | 24.6 |
| Steam System & Power Generation | | | 0.70 | 17.92 | 11.03 | 29.11 | 62.8 |
| Cooling Water & Other Utilities | | | 0.70 | 3.84 | 2.36 | 6.24 | 13.4 |
| **Direct Fixed Capital (DFC), also called TIC** | | | | 146.34 | 89.23 | 240.46 | 530.2 |
| Engineering | DFC x MF | 0.12 | | 17.56 | 10.71 | 28.86 | 63.6 |
| Construction | DFC x MF | 0.13 | | 19.02 | 11.60 | 31.26 | 68.9 |
| Contractor & Legal | DFC x MF | 0.08 | | 11.71 | 7.14 | 19.24 | 42.4 |
| Process/Project Contingency | DFC x MF | 0.0412 | | 6.03 | 3.68 | 9.91 | 21.8 |
| **Total Plant Cost (TPC)** | | | | 200.66 | 122.35 | 329.72 | 727.0 |
| AFUDC | | | | | | | |
| **Total Plant Investment (TPI)** | | | | 200.66 | 122.35 | 329.72 | 727.0 |
| Land | | | 0.60 | 2.21 | 2.21 | 5.08 | 15.2 |
| Startup | | | | 0.00 | 0.00 | 0.00 | 0.0 |
| **Total Capital Cost (TCC)** | | | | 202.87 | 124.56 | 334.80 | 742.2 |
| **Contingency/TPI** | | | | | | | |
| Working Capital | DFC x MF | 0.05 | | 10.03 | 6.12 | 16.49 | 36.3 |
| **Variable Operating Costs (million USD/yr)** | | | | | | | |
| Feed | | | | 4.56 | 2.28 | 9.11 | 27.3 |
| Utilities | | | | 0.30 | 0.12 | 0.49 | 1.47 |
| Other | | | | 1.00 | 0.50 | 2.00 | 6.00 |
| Catalysts and Chemicals | | | | 0.70 | 0.35 | 1.40 | 4.20 |
| **Total** | | | | 6.56 | 3.25 | 13.00 | 39.01 |
| Fixed Operating Costs (million USD/yr) | | | | | | | |
| Labor (SF from V-R) | DFC x MF | 0.0165 | 0.25 | 0.85 | 0.71 | 1.01 | 1.33 |
| Maintenance (3% of A) | | | 0.03 | 4.39 | 2.68 | 7.21 | 15.91 |
| General Overhead (65% of labor + maint) | | | 0.65 | 3.40 | 2.20 | 5.34 | 11.20 |
| Direct Overhead (45% of Labor) | | | 0.45 | 0.38 | 0.32 | 0.45 | 0.60 |
| Insurance (0.5% of TIC) | | | 0.005 | 1.01 | 0.62 | 1.67 | 3.71 |
| **Total** | | | | 10.04 | 6.53 | 15.69 | 32.74 |

**Table 68. Mexico cellulosic ethanol (TC) capital and operating costs, bagasse**

| Country | Mexico | Code = | 8 | | | | |
|---|---|---|---|---|---|---|---|
| **Cost component** | **Units** | **Cost Factor** | **Scale Factor** | **Mexico Cellulosic Ethanol (TC)** | **Mexico Cellulosic Ethanol (TC)** | **Mexico Cellulosic Ethanol (TC)** | **Mexico Cellulosic Ethanol (TC)** |
| Year $ | $ | | | 2005 | 2005 | 2005 | 200 |
| Cost Index (2 = M&S, 3 = CE) | | | 3 | 130.93 | 130.93 | 130.93 | 130.93 |
| Plant Size | dry short ton/day | | | 1,800 | 900 | 3,601 | 10,801 |
| | dry tonne/day | | | 1,632 | 816 | 3,264 | 9,793 |
| Stream Factor | % | | | 95% | 95% | 95% | 95% |
| Feed | Dry short ton/yr | | | 6.242E+05 | 3.121E+05 | .248E+06 | 3.745E+06 |
| | GJ/short ton | | | 1.794E+01 | 1.794E+01 | .794E+01 | .794E+01 |
| | GJ/yr | | | 1.119E+07 | 5.596E+06 | .239E+07 | 6.717E+07 |
| Feed Cost | USD/dt | | | **15.40** | **$ 15.40** | **15.40** | **15.40** |
| Yield (gal/Dry US Ton) | | | | | | | |
| Ethanol | gal/short ton | | | 80.1 | 80.1 | 80.1 | 80.1 |
| Mixed alcohols | gal/short ton | | | 94.1 | 94.1 | 94.1 | 94.1 |
| Process Efficiency - to ethanol | % HHV | | | 39.6% | 39.6% | 39.6% | 39.6% |
| Process efficiency - overall | % HHV | | | 47.4% | 47.4% | 47.4% | 47.4% |
| Ethanol | gal/yr | | | 5.000E+07 | 2.500E+07 | .000E+08 | 3.000E+08 |
| | Mil Gal/YR | | | **50.00** | **25.00** | **100.00** | **300.00** |
| | tpy | | | 1.657E+05 | 8.286E+04 | .315E+05 | 9.945E+05 |
| | GJ/yr | | | 4.436E+06 | 2.218E+06 | .873E+06 | 2.662E+07 |
| | gal/Stream day | | | 1.442E+05 | 7.210E+04 | .884E+05 | 8.652E+05 |
| | bbl/s stream day | | | 3.433E+03 | 1.717E+03 | .867E+03 | 2.060E+04 |
| Higher Alcohols | | | | | | | |
| | gal/ton | | | 14.0 | 14.0 | 14.0 | 14.0 |
| | gal/yr | | | 8.74E+06 | 4.37E+06 | 1.75E+07 | 5.24E+07 |
| **Capital Cost (million USD)** | | | | | | | |
| Feed Handling & Drying | | | 0.70 | 18.6 | 1.49 | 32 | 65.41 |
| Gasification | | | 0.80 | 10.3 | 5.96 | 07 | 43.51 |
| Tar Reforming & Quench | | | 0.70 | 30.8 | 9.0 | 18 | 108.27 |
| Acid Gas & Sulfur Removal | | | 0.85 | 11.6 | 6.47 | 02 | 53.49 |
| Alcohol synthesis - Compression | | | 0.65 | 12.8 | 8.20 | 20 | 41.25 |
| Alcohol Synthesis - Other | | | 0.65 | 3.7 | 2.36 | 81 | 11.86 |
| Alcohol Separation | | | 0.65 | 5.7 | 3.69 | 09 | 18.56 |
| Steam System & Power Generation | | | 0.70 | 13.5 | 8.32 | 96 | 47.37 |
| Cooling Water & Other Utilities | | | 0.70 | 2.9 | 1.78 | 70 | 10.15 |
| **Direct Fixed Capital (DFC), also called TIC** | | | | 110.3 | 67.29 | 34 | 399.86 |
| Engineering | DFC x MF | 0.12 | | 13.2 | 8.07 | 76 | 47.98 |
| Construction | DFC x MF | 0.13 | | 14.3 | 8.75 | 57 | 51.98 |
| Contractor & Legal | DFC x MF | 0.08 | | 8.8 | 5.38 | 51 | 31.99 |
| Process/Project Contingency | DFC x MF | 0.0412 | | 4.5 | 2.77 | 47 | 16.47 |
| **Total Plant Cost (TPC)** | | | | 151.3 | 92.26 | 65 | 548.28 |
| AFUDC | | | | | | | |
| **Total Plant Investment (TPI)** | | | | 151.3 | 92.26 | 65 | 548.28 |
| Land | | | 0.60 | 2.2 | 2.2 | 08 | 15.23 |
| Startup | | | | 0.0 | 0.00 | 00 | 0.00 |
| **Total Capital Cost (TCC)** | | | | 153.5 | 94.47 | 72 | 563.51 |
| **Contingency/TPI** | | | | | | | |
| Working Capital | DFC x MF | 0.05 | | 7.6 | 4.6 | 43 | 27.41 |
| **Variable Operating Costs (million USD/yr)** | | | | | | | |
| Feed | | | | 9.61 | 4.81 | 19.23 | 57.68 |
| Utilities | | | | 0.30 | 0.12 | 0.49 | 1.47 |
| Other | | | | 1.00 | 0.50 | 2.00 | 6.00 |
| Catalysts and Chemicals | | | | 0.70 | 0.35 | 1.40 | 4.20 |
| **Total** | | | | 11.61 | 5.78 | 23.12 | 69.35 |
| Fixed Operating Costs (million USD/yr) | | | | | | | |
| Labor (SF from V-R) | DFC x MF | 0.0165 | 0.25 | 0.55 | 0.46 | 0.65 | 0.86 |
| Maintenance (3% of A) | | | 0.03 | 3.31 | 2.02 | 5.44 | 12.00 |
| General Overhead (65% of labor + maint) | | | 0.65 | 2.51 | 1.61 | 3.96 | 8.35 |
| Direct Overhead (45% of Labor) | | | 0.45 | 0.25 | 0.21 | 0.29 | 0.39 |
| Insurance (0.5% of TIC) | | | 0.005 | 0.77 | 0.47 | 1.27 | 2.82 |
| **Total** | | | | 7.38 | 4.77 | 11.61 | 24.41 |

## Table 69. Cellulosic ethanol (TC) summary yields, costs, and plant gate prices, 25 MM GPY, bagasse

| Technology | Units | USA | Argentina | Brazil | Canada | Caribbean Basin | China | Colombia | India | Mexico |
|---|---|---|---|---|---|---|---|---|---|---|
| **Plant Size** | MM gal/yr | 100 | 100 | 100 | 100 | 100 | 100 | 100 | 100 | 100 |
| **Yields** | | | | | | | | | | |
| Primary Product Yield | gal/ton | 80.1 | 80.1 | 80.1 | | 80.1 | 80.1 | 80.1 | 80.1 | 80.1 |
| By Product A | | Mix Alcohol | Mix Alcohol | Mix Alcohol | | Mix Alcohol | Mix Alcohol | Mix Alcohol | Mix Alcohol | Mix Alcohol |
| Unit | | gal/ton | gal/ton | gal/ton | | gal/ton | gal/ton | gal/ton | gal/ton | gal/ton |
| Yield unit | | 14 | 14 | 14 | | 14 | 14 | 14 | 14 | 14 |
| Annual Yield | MM gal/yr | 17.5 | 17.5 | 17.5 | | 17.5 | 17.5 | 17.5 | 17.5 | 17.5 |
| Price unit | | 1.15 | 1.15 | 1.15 | | 1.15 | 1.15 | 1.15 | 1.15 | 1.15 |
| MM USD/yr | | 20.1 | 20.1 | 20.1 | | 20.1 | 20.1 | 20.1 | 20.1 | 20.1 |
| By Product B | | | | | | | | | | |
| Unit | | | | | | | | | | |
| Yield unit | | | | | | | | | | |
| Annual Yield | | | | | | | | | | |
| Price unit | | | | | | | | | | |
| Annual Cost | MM USD/yr | | | | | | | | | |
| **Energy Inputs** | | | | | | | | | | |
| Electricity | | N/A | N/A | N/A | | N/A | N/A | N/A | N/A | N/A |
| Cost | $/kWh | | | | | | | | | |
| Amount | kWh/yr | | | | | | | | | |
| Annual Cost | MM USD/yr | | | | | | | | | |
| Fuel | | N/A | N/A | N/A | | N/A | N/A | N/A | N/A | N/A |
| Type | | | | | | | | | | |
| Unit | | | | | | | | | | |
| Amount | | | | | | | | | | |
| Cost | USD/unit | | | | | | | | | |
| Annual Cost | MM USD/yr | | | | | | | | | |
| **Capital & Operating Costs** | | | | | | | | | | |
| Capital | MMUSD | 314.21 | 337.00 | 320.98 | | 353.86 | 345.26 | 312.49 | 334.8 | 253.72 |
| Operating w/o feed | MM USD/yr | 22.27 | 22.38 | 19.31 | | 23.68 | 24.18 | 20.15 | 19.58 | 15.50 |
| Feed | | | | | | | | | | |
| Unit | | ton | ton | ton | | ton | ton | ton | ton | ton |
| Cost/unit | USD/unit | 35 | 9.1 | 7.6 | | 7.6 | 9.1 | 14.5 | 7.3 | 15.4 |
| Rate | unit/day | 3601 | 3601 | 3601 | | 3601 | 3601 | 3601 | 3601 | 3601 |
| Yearly Cost | MM USD/yr | 43.7 | 11.36 | 9.49 | | 9.49 | 11.36 | 18.1 | 9.11 | 19.23 |
| **Product Plant Gate Cost** | | | | | | | | | | |
| Basis | | | | | | | | | | |
| Value | USD/gal | 0.93 | 0.62 | 0.54 | | 0.62 | 0.65 | 0.63 | 0.57 | 0.54 |

## Table 70.Cellulosic ethanol (TC) summary yields, costs, and plant gate prices, 100 MM GPY, bagasse

| Technology | Units | USA | Argentina | Brazil | Canada | Caribbean Basin | China | Colombia | India | Mexico |
|---|---|---|---|---|---|---|---|---|---|---|
| **Plant Size** | MM gal/yr | 100 | 100 | 100 | 100 | 100 | 100 | 100 | 100 | 100 |
| **Yields** | | | | | | | | | | |
| Primary Product Yield | gal/ton | 80.1 | 80.1 | 80.1 | | 80.1 | 80.1 | 80.1 | 80.1 | 80.1 |
| By Product A | | Mix Alcohol | Mix Alcohol | Mix Alcohol | | Mix Alcohol | Mix Alcohol | Mix Alcohol | Mix Alcohol | Mix Alcohol |
| Unit | | gal/ton | gal/ton | gal/ton | | gal/ton | gal/ton | gal/ton | gal/ton | gal/ton |
| Yield unit | | 14 | 14 | 14 | | 14 | 14 | 14 | 14 | 14 |
| Annual Yield | MM gal/yr | 17.5 | 17.5 | 17.5 | | 17.5 | 17.5 | 17.5 | 17.5 | 17.5 |
| Price unit | | 1.15 | 1.15 | 1.15 | | 1.15 | 1.15 | 1.15 | 1.15 | 1.15 |
| MM USD/yr | | 20.1 | 20.1 | 20.1 | | 20.1 | 20.1 | 20.1 | 20.1 | 20.1 |
| By Product B | | | | | | | | | | |
| Unit | | | | | | | | | | |
| Yield unit | | | | | | | | | | |
| Annual Yield | | | | | | | | | | |
| Price unit | | | | | | | | | | |
| Annual Cost | MM USD/yr | | | | | | | | | |
| **Energy Inputs** | | | | | | | | | | |
| Electricity | | N/A | N/A | N/A | | N/A | N/A | N/A | N/A | N/A |
| Cost | $/kWh | | | | | | | | | |
| Amount | kWh/yr | | | | | | | | | |
| Annual Cost | MM USD/yr | | | | | | | | | |
| Fuel | | N/A | N/A | N/A | | N/A | N/A | N/A | N/A | N/A |
| Type | | | | | | | | | | |
| Unit | | | | | | | | | | |
| Amount | | | | | | | | | | |
| Cost | USD/unit | | | | | | | | | |
| Annual Cost | MM USD/yr | | | | | | | | | |
| **Capital & Operating Costs** | | | | | | | | | | |
| Capital | MMUSD | 314.21 | 337.00 | 320.98 | | 353.86 | 345.26 | 312.49 | 334.8 | 253.72 |
| Operating w/o feed | MM USD/yr | 22.27 | 22.38 | 19.31 | | 23.68 | 24.18 | 20.15 | 19.58 | 15.50 |
| Feed | | | | | | | | | | |
| Unit | | ton | ton | ton | | ton | ton | ton | ton | ton |
| Cost/unit | USD/unit | 35 | 9.1 | 7.6 | | 7.6 | 9.1 | 14.5 | 7.3 | 15.4 |
| Rate | unit/day | 3601 | 3601 | 3601 | | 3601 | 3601 | 3601 | 3601 | 3601 |
| Yearly Cost | MM USD/yr | 43.7 | 11.36 | 9.49 | | 9.49 | 11.36 | 18.1 | 9.11 | 19.23 |
| **Product Plant Gate Cost** | | | | | | | | | | |
| Basis | | | | | | | | | | |
| Value | USD/gal | 0.93 | 0.62 | 0.54 | | 0.62 | 0.65 | 0.63 | 0.57 | 0.54 |

## Table 71. Cellulosic ethanol (TC) summary yields, costs, and plant gate prices, 25 MM GPY, ag residues

| Technology | Units | USA | Argentina | Brazil | Canada | Caribbean Basin | China | Colombia | India | Mexico |
|---|---|---|---|---|---|---|---|---|---|---|
| **Plant Size** | MM gal/yr | 25 | 25 | 25 | 25 | 25 | 25 | 25 | 25 | 25 |
| **Yields** | | | | | | | | | | |
| Primary Product Yield | gal/ton | 80 | 80 | 80 | 80 | 80 | 80 | 80 | 80 | 80 |
| By Product A | | Mix Alcohol | Mix Alcohol | Mix Alcohol | Mix Alcohol | Mix Alcohol | Mix Alcohol | Mix Alcohol | Mix Alcohol | Mix Alcohol |
| Unit | | gal/ton | gal/ton | gal/ton | gal/ton | gal/ton | gal/ton | gal/ton | gal/ton | gal/ton |
| Yield unit | | 14 | 14 | 14 | 14 | 14 | 14 | 14 | 14 | 14 |
| Annual Yield | MM gal/yr | 4.37 | 4.37 | 4.37 | 4.37 | 4.37 | 4.37 | 4.37 | 4.37 | 4.37 |
| Price unit | | 1.15 | 1.15 | 1.15 | 1.15 | 1.15 | 1.15 | 1.15 | 1.15 | 1.15 |
| MM USD/yr | | 5.03 | 5.03 | 5.03 | 5.03 | 5.03 | 5.03 | 5.03 | 5.03 | 5.03 |
| By Product B | | | | | | | | | | |
| Unit | | | | | | | | | | |
| Yield unit | | | | | | | | | | |
| Annual Yield | | | | | | | | | | |
| Price unit | | | | | | | | | | |
| Annual Cost | MM USD/yr | | | | | | | | | |
| **Energy Inputs** | | | | | | | | | | |
| Electricity | | N/A | N/A | N/A | N/A | N/A | N/A | N/A | N/A | N/A |
| Cost | $/kWh | | | | | | | | | |
| Amount | kWh/yr | | | | | | | | | |
| Annual Cost | MM USD/yr | | | | | | | | | |
| Fuel | | N/A | N/A | N/A | N/A | N/A | N/A | N/A | N/A | N/A |
| Type | | | | | | | | | | |
| Unit | | | | | | | | | | |
| Amount | | | | | | | | | | |
| Cost | USD/unit | | | | | | | | | |
| Annual Cost | MM USD/yr | | | | | | | | | |
| **Capital & Operating Costs** | | | | | | | | | | |
| Capital | MMUSD | 116.92 | 125.38 | 119.43 | 151.67 | 131.63 | 128.44 | 116.28 | 124.56 | 94.47 |
| Operating w/o feed | MM USD/yr | 9.7 | 9.44 | 7.51 | 13.84 | 10.14 | 10.62 | 8.22 | 7.51 | 5.74 |
| Feed | | | | | | | | | | |
| Unit | | ton | ton | ton | ton | ton | ton | ton | ton | ton |
| Cost/unit | USD/unit | 35 | 47.2 | 30.8 | 10 | 14.5 | 28.1 | 14.5 | 7.3 | 47.2 |
| Rate | unit/day | 900 | 900 | 900 | 900 | 900 | 900 | 900 | 900 | 900 |
| Yearly Cost | MM USD/yr | 10.92 | 14.73 | 9.61 | 3.12 | 4.53 | 8.77 | 4.53 | 2.28 | 14.73 |
| **Product Plant Gate Cost** | | | | | | | | | | |
| Basis | | | | | | | | | | |
| Value | USD/gal | 1.33 | 1.49 | 1.17 | 1.32 | 1.12 | 1.31 | 1.01 | 0.89 | 1.20 |

## Table 72. Cellulosic ethanol (TC) summary yields, costs, and plant gate prices, 100 MM GPY, ag residues

| Technology | Units | USA | Argentina | Brazil | Canada | Caribbean Basin | China | Colombia | India | Mexico |
|---|---|---|---|---|---|---|---|---|---|---|
| **Plant Size** | MM gal/yr | 100 | 100 | 100 | 100 | 100 | 100 | 100 | 100 | 100 |
| **Yields** | | | | | | | | | | |
| Primary Product Yield | gal/ton | 80.1 | 80.1 | 80.1 | 80.1 | 80.1 | 80.1 | 80.1 | 80.1 | 80.1 |
| By Product A | | Mix Alcohol | Mix Alcohol | Mix Alcohol | Mix Alcohol | Mix Alcohol | Mix Alcohol | Mix Alcohol | Mix Alcohol | Mix Alcohol |
| Unit | | gal/ton | gal/ton | gal/ton | gal/ton | gal/ton | gal/ton | gal/ton | gal/ton | gal/ton |
| Yield unit | | 14 | 14 | 14 | 14 | 14 | 14 | 14 | 14 | 14 |
| Annual Yield | MM gal/yr | 17.5 | 17.5 | 17.5 | 17.5 | 17.5 | 17.5 | 17.5 | 17.5 | 17.5 |
| Price unit | | 1.15 | 1.15 | 1.15 | 1.15 | 1.15 | 1.15 | 1.15 | 1.15 | 1.15 |
| MM USD/yr | | 20.1 | 20.1 | 20.1 | 20.1 | 20.1 | 20.1 | 20.1 | 20.1 | 20.1 |
| By Product B | | | | | | | | | | |
| Unit | | | | | | | | | | |
| Yield unit | | | | | | | | | | |
| Annual Yield | | | | | | | | | | |
| Price unit | | | | | | | | | | |
| Annual Cost | MM USD/yr | | | | | | | | | |
| **Energy Inputs** | | | | | | | | | | |
| Electricity | | N/A | N/A | N/A | N/A | N/A | N/A | N/A | N/A | N/A |
| Cost | $/kWh | | | | | | | | | |
| Amount | kWh/yr | | | | | | | | | |
| Annual Cost | MM USD/yr | | | | | | | | | |
| Fuel | | N/A | N/A | N/A | N/A | N/A | N/A | N/A | N/A | N/A |
| Type | | | | | | | | | | |
| Unit | | | | | | | | | | |
| Amount | | | | | | | | | | |
| Cost | USD/unit | | | | | | | | | |
| Annual Cost | MM USD/yr | | | | | | | | | |
| **Capital & Operating Costs** | | | | | | | | | | |
| Capital | MMUSD | 314.21 | 337.00 | 320.98 | 407.86 | 353.86 | 345.26 | 312.49 | 334.8 | 253.72 |
| Operating w/o feed | MM USD/yr | 22.27 | 22.36 | 19.31 | 29.96 | 23.68 | 24.18 | 20.15 | 19.58 | 15.50 |
| Feed | | | | | | | | | | |
| Unit | | ton | ton | ton | ton | ton | ton | ton | ton | ton |
| Cost/unit | USD/unit | 35 | 47.2 | 30.8 | 10 | 14.5 | 28.1 | 14.5 | 7.3 | 47.2 |
| Rate | unit/day | 3601 | 3601 | 3601 | 3601 | 3601 | 3601 | 3601 | 3601 | 3601 |
| Yearly Cost | MM USD/yr | 43.7 | 58.93 | 38.45 | 12.48 | 18.1 | 35.08 | 18.1 | 9.11 | 58.93 |
| **Product Plant Gate Cost** | | | | | | | | | | |
| Basis | | | | | | | | | | |
| Value | USD/gal | 0.93 | 1.10 | 0.83 | 0.79 | 0.71 | 0.88 | 0.63 | 0.57 | 0.93 |

## Table 73. Cellulosic ethanol (TC) summary yields, costs, and plant gate prices, 25 MM GPY, wood/perennial

| Technology | Units | USA | Argentina | Brazil | Canada | Caribbean Basin | China | Colombia | India | Mexico |
|---|---|---|---|---|---|---|---|---|---|---|
| **Plant Size** | MM gal/yr | 25 | 25 | 25 | 25 | 25 | 25 | 25 | 25 | 25 |
| **Yields** | | | | | | | | | | |
| Primary Product Yield | gal/ton | 80 | 80 | 80 | 80 | 80 | 80 | 80 | 80 | 80 |
| By Product A | | Mix Alcohol | Mix Alcohol | Mix Alcohol | Mix Alcohol | Mix Alcohol | Mix Alcohol | Mix Alcohol | Mix Alcohol | Mix Alcohol |
| Unit | | gal/ton | gal/ton | gal/ton | gal/ton | gal/ton | gal/ton | gal/ton | gal/ton | gal/ton |
| Yield unit | | 14 | 14 | 14 | 14 | 14 | 14 | 14 | 14 | 14 |
| Annual Yield | MM gal/yr | 4.37 | 4.37 | 4.37 | 4.37 | 4.37 | 4.37 | 4.37 | 4.37 | 4.37 |
| Price unit | | 1.15 | 1.15 | 1.15 | 1.15 | 1.15 | 1.15 | 1.15 | 1.15 | 1.15 |
| MM USD/yr | | 5.03 | 5.03 | 5.03 | 5.03 | 5.03 | 5.03 | 5.03 | 5.03 | 5.03 |
| By Product B | | | | | | | | | | |
| Unit | | | | | | | | | | |
| Yield unit | | | | | | | | | | |
| Annual Yield | | | | | | | | | | |
| Price unit | | | | | | | | | | |
| Annual Cost | MM USD/yr | | | | | | | | | |
| **Energy Inputs** | | | | | | | | | | |
| Electricity | | N/A | N/A | N/A | N/A | N/A | N/A | N/A | N/A | N/A |
| Cost | $/kWh | | | | | | | | | |
| Amount | kWh/yr | | | | | | | | | |
| Annual Cost | MM USD/yr | | | | | | | | | |
| Fuel | | N/A | N/A | N/A | N/A | N/A | N/A | N/A | N/A | N/A |
| Type | | | | | | | | | | |
| Unit | | | | | | | | | | |
| Amount | | | | | | | | | | |
| Cost | USD/unit | | | | | | | | | |
| Annual Cost | MM USD/yr | | | | | | | | | |
| **Capital & Operating Costs** | | | | | | | | | | |
| Capital | MMUSD | 116.92 | 125.38 | 119.43 | 151.67 | 131.63 | 128.44 | 116.28 | 124.56 | 94.47 |
| Operating w/o feed | MM USD/yr | 9.7 | 9.44 | 7.51 | 13.84 | 10.14 | 10.62 | 8.22 | 7.51 | 5.74 |
| Feed | | | | | | | | | | |
| Unit | | ton | ton | ton | ton | ton | ton | ton | ton | ton |
| Cost/unit | USD/unit | 35 | 47.2 | 47.2 | 31.8 | 47.2 | 47.2 | 47.2 | 47.2 | 47.2 |
| Rate | unit/day | 900 | 900 | 900 | 900 | 900 | 900 | 900 | 900 | 900 |
| Yearly Cost | MM USD/yr | 10.92 | 14.73 | 14.73 | 9.93 | 14.73 | 14.73 | 14.73 | 14.73 | 14.73 |
| **Product Plant Gate Cost** | | | | | | | | | | |
| Basis | | | | | | | | | | |
| Value | USD/gal | 1.33 | 1.49 | 1.37 | 1.60 | 1.53 | 1.59 | 1.39 | 1.39 | 1.20 |

## Table 74. Cellulosic ethanol (TC) summary yields, costs, and plant gate prices, 100 MM GPY, wood/perennial

| Technology | Units | USA | Argentina | Brazil | Canada | Caribbean Basin | China | Colombia | India | Mexico |
|---|---|---|---|---|---|---|---|---|---|---|
| **Plant Size** | MM gal/yr | 100 | 100 | 100 | 100 | 100 | 100 | 100 | 100 | 100 |
| **Yields** | | | | | | | | | | |
| Primary Product Yield | gal/ton | 80.1 | 80.1 | 80.1 | 80.1 | 80.1 | 80.1 | 80.1 | 80.1 | 80.1 |
| By Product A | | Mix Alcohol | Mix Alcohol | Mix Alcohol | Mix Alcohol | Mix Alcohol | Mix Alcohol | Mix Alcohol | Mix Alcohol | Mix Alcohol |
| Unit | | gal/ton | gal/ton | gal/ton | gal/ton | gal/ton | gal/ton | gal/ton | gal/ton | gal/ton |
| Yield unit | | 14 | 14 | 14 | 14 | 14 | 14 | 14 | 14 | 14 |
| Annual Yield | MM gal/yr | 17.5 | 17.5 | 17.5 | 17.5 | 17.5 | 17.5 | 17.5 | 17.5 | 17.5 |
| Price unit | | 1.15 | 1.15 | 1.15 | 1.15 | 1.15 | 1.15 | 1.15 | 1.15 | 1.15 |
| MM USD/yr | | 20.1 | 20.1 | 20.1 | 20.1 | 20.1 | 20.1 | 20.1 | 20.1 | 20.1 |
| By Product B | | | | | | | | | | |
| Unit | | | | | | | | | | |
| Yield unit | | | | | | | | | | |
| Annual Yield | | | | | | | | | | |
| Price unit | | | | | | | | | | |
| Annual Cost | MM USD/yr | | | | | | | | | |
| **Energy Inputs** | | | | | | | | | | |
| Electricity | | N/A | N/A | N/A | N/A | N/A | N/A | N/A | N/A | N/A |
| Cost | $/kWh | | | | | | | | | |
| Amount | kWh/yr | | | | | | | | | |
| Annual Cost | MM USD/yr | | | | | | | | | |
| Fuel | | N/A | N/A | N/A | N/A | N/A | N/A | N/A | N/A | N/A |
| Type | | | | | | | | | | |
| Unit | | | | | | | | | | |
| Amount | | | | | | | | | | |
| Cost | USD/unit | | | | | | | | | |
| Annual Cost | MM USD/yr | | | | | | | | | |
| **Capital & Operating Costs** | | | | | | | | | | |
| Capital | MMUSD | 314.21 | 337.00 | 320.98 | 407.86 | 353.86 | 345.26 | 312.49 | 334.8 | 253.72 |
| Operating w/o feed | MM USD/yr | 22.27 | 22.36 | 19.31 | 29.96 | 23.68 | 24.18 | 20.15 | 19.58 | 15.50 |
| Feed | | | | | | | | | | |
| Unit | | ton | ton | ton | ton | ton | ton | ton | ton | ton |
| Cost/unit | USD/unit | 35 | 47.2 | 47.2 | 31.8 | 47.2 | 47.2 | 47.2 | 47.2 | 47.2 |
| Rate | unit/day | 3601 | 3601 | 3601 | 3601 | 3601 | 3601 | 3601 | 3601 | 3601 |
| Yearly Cost | MM USD/yr | 43.7 | 58.936 | 58.93 | 39.7 | 58.93 | 58.93 | 58.93 | 58.93 | 58.93 |
| **Product Plant Gate Cost** | | | | | | | | | | |
| Basis | | | | | | | | | | |
| Value | USD/gal | 0.93 | 1.10 | 1.04 | 1.06 | 1.12 | 1.13 | 1.04 | 1.05 | 0.93 |

# 7. Residual Fuel Oil - Pyrolysis

Pyrolysis is one of the renewable processes for producing a liquid fuel from biomass. It offers advantages of a bio-oil that can be easily stored and transported. The primary oil can be used as a boiler fuel, or can be upgraded to chemicals or transportation fuels. Fast pyrolysis has achieved commercial status–the BIOSYN and Dynamotive processes–for production of specialty and chemicals and boiler fuels. Good reviews of pyrolysis are given by Czernik and Bridgwater (2004) and Mohan et al. (2006). This discussion of fast pyrolysis is excerpted from Putsche (2004) and Ringer et al. (2006).

The production of bio-oil from biomass is based on the fast pyrolysis process, which is composed of five major processing areas: feedstock handling and drying, pyrolysis, char combustion, product recovery and steam generation. Figure 24 is a block flow diagram of a fast pyrolysis process.

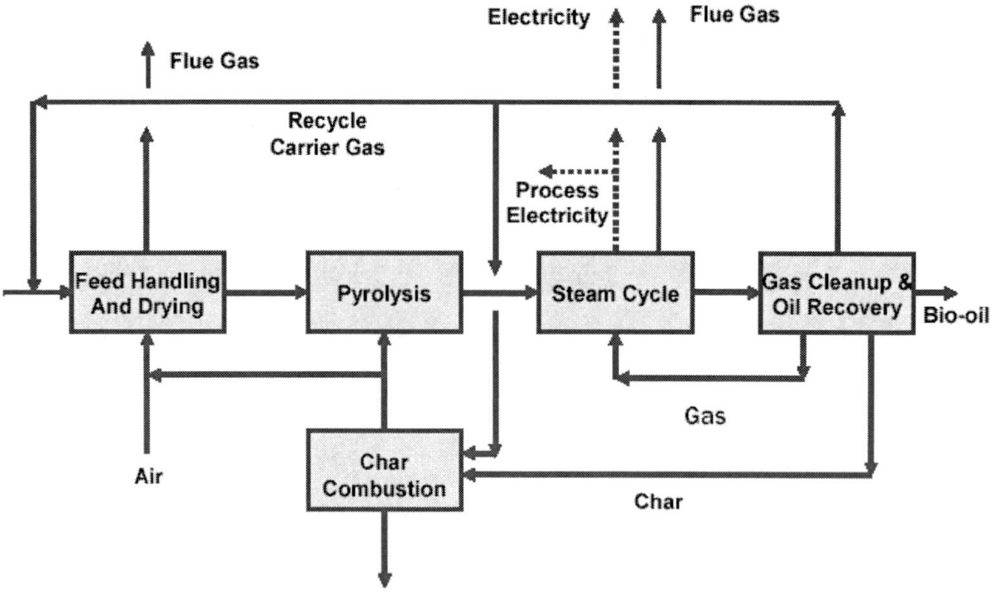


Figure 24. Fast pyrolysis process flow diagram

In the feedstock handling section, the biomass is reduced in size to <1-5mm and dried to 5-10% moisture. It is then sent to pyrolysis where it is heated to 400-500°C in an oxygen-deficient atmosphere to degrade the biomass into a mix of gases, bio-oils, and char. Char is removed using high efficiency cyclones and is combusted to fuel the pyrolysis reaction. To maximize the yield of bio-oils, the reaction is rapidly quenched through heat exchange or direct liquid (e.g., water or recycled bio-oils) injection. The bio-oils are present in the gas stream as aerosols and require scrubbers and/or wet electrostatic precipitators for efficient capture. After cleaning, some of the clean pyrolysis gases are recycled to fluidize the bed and the remaining gases are combusted for process heat. Where feasible, heat is recovered from the pyrolysis gases to generate steam for electricity production.

Capital and operating costs are given in Table 75.

**Table 75. USA base case, fast pyrolysis capital and operating costs**

| Country | USA | Code = | 1 | | | | | |
|---|---|---|---|---|---|---|---|---|
| | | | | USA | USA | USA | USA | USA |
| Cost component | Units | Cost Factor | Scale Factor | Pyrolytic Fuel Oil | Pyrolytic Fuel Oil | Pyrolytic Fuel Oil | Pyrolytic Fuel Oil | Pyrolytic Fuel Oil |
| Year $ | $ | | | 2003 | 2005 | 2005 | 2005 | 2005 |
| Cost Index (2 = M&S, 3 = CE) | | | 3 | 112.42 | 130.93 | 130.93 | 130.93 | 130.93 |
| Plant Size | dry short ton/day | | | 2,205 | 529 | 2,113 | 5,512 | 11,023 |
| | dry tonne/day | | | 2,000 | 480 | 1,917 | 5,000 | 10,000 |
| Stream Factor | % | | | 95% | 95% | 95% | 95% | 95% |
| Feed | dtpy | | | 7.64E+05 | 1.83E+05 | 7.33E+05 | 1.91E+06 | 3.82E+06 |
| | GJ/ton | | | 1.79E+01 | 1.79E+01 | 1.79E+01 | 1.79E+01 | 1.79E+01 |
| | GJ/yr | | | 1.37E+07 | 3.29E+06 | 1.31E+07 | 3.43E+07 | 6.86E+07 |
| Feed Cost | $/dt | | | $ 35.00 | $ 35.00 | $ 35.00 | $ 35.00 | $ 35.00 |
| Process Efficiency (overall) | % | | | 67.5% | 67.5% | 67.5% | 67.5% | 67.5% |
| Process Efficiency (to RFO) | % | | | 67.0% | 67.0% | 67.0% | 67.0% | 67.0% |
| RFO Production | Yield, wt% | | | 70% | 70% | 70% | 70% | 70% |
| | Ton/yr | | | 5.35E+05 | 1.28E+05 | 5.13E+05 | 1.34E+06 | 2.68E+06 |
| | HHV, GJ/ton | | | 17.16 | 17.16 | 17.16 | 17.16 | 17.16 |
| | gal/yr | | | 1.04E+08 | 2.50E+07 | 1.00E+08 | 2.61E+08 | 5.22E+08 |
| | Mil Gal/YR | | | 104.4 | 25.0 | 100.0 | 260.9 | 521.8 |
| | tpy | | | 5.35E+05 | 1.28E+05 | 5.13E+05 | 1.34E+06 | 2.68E+06 |
| | GJ/yr | | | 9.18E+06 | 2.20E+06 | 8.80E+06 | 2.30E+07 | 4.59E+07 |
| | gal/ton | | | 1.37E+02 | 1.37E+02 | 1.37E+02 | 1.37E+02 | 1.37E+02 |
| | gal/Stream day | | | 3.01E+05 | 7.22E+04 | 2.88E+05 | 7.52E+05 | 1.50E+06 |
| | bbl/s day | | | 7.17E+03 | 1.72E+03 | 6.87E+03 | 1.79E+04 | 3.58E+04 |
| | gal EtOH eq/yr | | | 1.03E+08 | 2.48E+07 | 9.92E+07 | 2.59E+08 | 5.17E+08 |
| | gal EtOH eq/ton | | | 1.35E+02 | 1.35E+02 | 1.35E+02 | 1.35E+02 | 1.35E+02 |
| | bbl EtOH eq/day | | | 6.75E+03 | 1.62E+03 | 6.47E+03 | 1.69E+04 | 3.38E+04 |
| Electricity | kWh/y | 2358.2 | kWh/h | 1.96E+07 | 4.71E+06 | 1.88E+07 | 4.91E+07 | 9.81E+07 |
| | GJ eq/yr | | | 6.89E+04 | 1.65E+04 | 6.61E+04 | 1.72E+05 | 3.45E+05 |
| **Capital Cost (million USD)** | | | | | | | | |
| Feed Preparation | | | 0.70 | 12.81 | 5.49 | 14.49 | 28.34 | 46.03 |
| Pyrolysis | | | 0.80 | 9.01 | 3.35 | 10.14 | 21.83 | 38.02 |
| Quench | | | 0.70 | 3.30 | 1.42 | 3.73 | 7.30 | 11.86 |
| Heat Recovery | | | 0.85 | 2.62 | 0.91 | 2.95 | 6.65 | 11.99 |
| Product Recovery and Storage | | | 0.85 | 1.76 | 0.81 | 1.99 | 3.71 | 5.83 |
| Recycle | | | 0.65 | 2.24 | 1.03 | 2.54 | 4.74 | 7.43 |
| Steam and Power Production | | | 0.70 | 7.20 | 3.09 | 8.14 | 15.93 | 25.87 |
| Utilities/auxiliaries | | | 0.70 | 5.85 | 2.51 | 6.61 | 12.94 | 21.01 |
| **Direct Fixed Capital (DFC), also called TIC** | | | | 44.79 | 18.61 | 50.59 | 101.44 | 168.05 |
| Engineering | DFC x MF | 0.12 | | 5.37 | 2.23 | 6.07 | 12.17 | 20.17 |
| Construction | DFC x MF | 0.13 | | 5.82 | 2.42 | 6.58 | 13.19 | 21.85 |
| Contractor & Legal | DFC x MF | 0.08 | | 3.58 | 1.49 | 4.05 | 8.12 | 13.44 |
| Process/Project Contingency | DFC x MF | 0.0412 | | 1.85 | 0.77 | 2.08 | 4.18 | 6.92 |
| **Total Plant Cost (TPC)** | | | | 61.42 | 25.51 | 69.37 | 139.10 | 230.44 |
| AFUDC | | | | | | | | |
| **Total Plant Investment (TPI)** | | | | 61.42 | 25.51 | 69.37 | 139.10 | 230.44 |
| Land | | | 0.60 | 2.21 | 0.94 | 2.15 | 3.83 | 7.66 |
| **Startup** | | | | | | | | |
| **Total Capital Cost (TCC)** | | | | 63.63 | 26.45 | 71.52 | 142.93 | 238.10 |
| **Contingency/TPI** | | | | | | | | |
| Working Capital | DFC x MF | 0.05 | | 3.07 | 1.28 | 3.47 | 6.95 | 11.52 |
| **Variable Operating Costs (million USD/yr)** | | | | | | | | |
| Feed | | | | 26.76 | 6.42 | 25.65 | 66.89 | 133.78 |
| Utilities | | | | 4.00 | 0.96 | 3.83 | 10.00 | 20.00 |
| Other | | | | 1.00 | 0.24 | 0.96 | 2.50 | 5.00 |
| Catalysts and Chemicals | | | | 0.70 | 0.17 | 0.67 | 1.75 | 3.50 |
| **Total** | | | | 32.46 | 7.79 | 31.11 | 81.14 | 162.28 |
| Fixed Operating Costs (million USD/yr) | | | | | | | | |
| Labor (SF from V-R) | DFC x MF | 0.05 | 0.25 | 2.24 | 1.83 | 2.58 | 3.28 | 3.90 |
| Maintenance (3% of A) | | | 0.03 | 1.34 | 0.56 | 1.52 | 3.04 | 5.04 |
| General Overhead (65% of labor + maint) | | | 0.65 | 2.33 | 1.55 | 2.66 | 4.11 | 5.81 |
| Direct Overhead (45% of Labor) | | | 0.45 | 1.01 | 0.82 | 1.16 | 1.48 | 1.76 |
| Insurance (0.5% of TIC) | | | 0.005 | 0.32 | 0.13 | 0.36 | 0.71 | 1.19 |
| **Total** | | | | 7.24 | 4.89 | 8.28 | 12.62 | 17.70 |

Figure 25 presents a summary of the plant gate price breakdown for the USA base case. The feedstock costs by country for bagasse, agricultural residues, and wood/perennial feeds used are given in Table 19. Plant gate prices are given in Table 76.

Table 77 throughTable 84 give the capital and operating cost details for individual countries. Table 85 throughTable 90 give the summary operating cost details for the three types of cellulosic feeds for the individual countries.

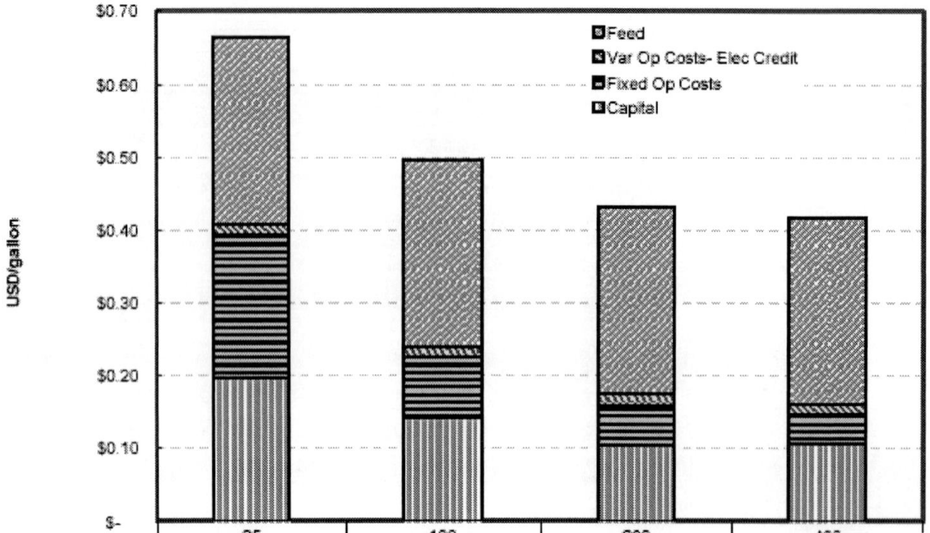


| | 25 | 100 | 200 | 400 |
|---|---|---|---|---|
| Feed | $0.26 | $0.26 | $0.26 | $0.26 |
| Var Op Costs- Elec Credit | $0.01 | $0.01 | $0.01 | $0.01 |
| Fixed Op Costs | $0.20 | $0.08 | $0.06 | $0.04 |
| Capital | $0.20 | $0.14 | $0.10 | $0.11 |

Plant Size (mil gal/yr)

Figure 25. Residual fuel oil price as a function of plant size

**Table 76. Pyrolytic fuel oil plant gate prices**

| Plant Size, MM GPY | 25 | 100 | 25 | 100 | 25 | 100 |
|---|---|---|---|---|---|---|
| Feed | Bagasse | | Ag Residues | | Wood /Perennials | |
| Country | Fuel Oil Plant Gate Cost (USD/gal fuel oil) | | | | | |
| USA | 0.66 | 0.50 | 0.66 | 0.50 | 0.66 | 0.50 |
| Argentina | 0.46 | 0.30 | 0.74 | 0.58 | 0.74 | 0.58 |
| Brazil | 0.37 | 0.26 | 0.54 | 0.43 | 0.66 | 0.55 |
| Canada | -- | -- | 0.63 | 0.38 | 0.79 | 0.54 |
| Caribbean Basin | 0.47 | 0.30 | 0.52 | 0.35 | 0.76 | 0.59 |
| China | 0.50 | 0.32 | 0.64 | 0.46 | 0.78 | 0.60 |
| Colombia | 0.45 | 0.32 | 0.45 | 0.32 | 0.69 | 0.56 |
| India | 0.36 | 0.26 | 0.36 | 0.26 | 0.66 | 0.55 |
| Mexico | 0.36 | 0.29 | 0.60 | 0.52 | 0.60 | 0.52 |

## Table 77. Argentina residual fuel oil capital and operating costs, bagasse

| Country | Argentina | Code = | 5 | | | | |
|---|---|---|---|---|---|---|---|
| | | | | Argentina | Argentina | Argentina | Argentina |
| **Cost component** | **Units** | **Cost Factor** | **Scale Factor** | **Pyrolytic Fuel Oil** | **Pyrolytic Fuel Oil** | **Pyrolytic Fuel Oil** | **Pyrolytic Fuel Oil** |
| Year $ | $ | | | 2005 | 2005 | 2005 | 2005 |
| Cost Index (2 = M&S, 3 = CE) | | | 3 | 130.93 | 130.93 | 130.93 | 130.93 |
| Plant Size | dry short ton/day | | | 529 | 2,113 | 4,225 | 8,449 |
| | dry tonne/day | | | 480 | 1,917 | 3,833 | 7,665 |
| Stream Factor | % | | | 95% | 95% | 95% | 95% |
| Feed | dtpy | | | 1.83E+05 | 7.33E+05 | 1.47E+06 | 2.93E+06 |
| | GJ/ton | | | 1.79E+01 | 1.79E+01 | 1.79E+01 | 1.79E+01 |
| | GJ/yr | | | 3.29E+06 | 1.31E+07 | 2.63E+07 | 5.25E+07 |
| Feed Cost | $/dt | | | $ 9.10 | $ 9.10 | $ 9.10 | 9.10 |
| Process Efficiency (overall) | % | | | 67.5% | 67.5% | 67.5% | 67.5% |
| Process Efficiency (to RFO) | % | | | 67.0% | 67.0% | 67.0% | 67.0% |
| RFO Production | Yield, wt% | | | 70% | 70% | 70% | 70% |
| | Ton/yr | | | 1.28E+05 | 5.13E+05 | 1.03E+06 | 2.05E+06 |
| | HHV, GJ/ton | | | 17.16 | 17.16 | 17.16 | 17.16 |
| | gal/yr | | | 2.50E+07 | 1.00E+08 | 2.00E+08 | 4.00E+08 |
| | Mil Gal/YR | | | 25.0 | 100.0 | 200.0 | 400.0 |
| | tpy | | | 1.28E+05 | 5.13E+05 | 1.03E+06 | 2.05E+06 |
| | GJ/yr | | | 2.20E+06 | 8.80E+06 | 1.76E+07 | 3.52E+07 |
| | gal/ton | | | 1.37E+02 | 1.37E+02 | 1.37E+02 | 1.37E+02 |
| | gal/Stream day | | | 7.22E+04 | 2.88E+05 | 5.77E+05 | 1.15E+06 |
| | bbl/s day | | | 1.72E+03 | 6.87E+03 | 1.37E+04 | 2.75E+04 |
| | gal EtOH eq/yr | | | 2.48E+07 | 9.92E+07 | 1.98E+08 | 3.97E+08 |
| | gal EtOH eq/ton | | | 1.35E+02 | 1.35E+02 | 1.35E+02 | 1.35E+02 |
| | bbl EtOH eq/day | | | 1.62E+03 | 6.47E+03 | 1.29E+04 | 2.59E+04 |
| Electricity | kWh/y | 2358.2 | kWh/h | 4.71E+06 | 1.88E+07 | 3.76E+07 | 7.52E+07 |
| | GJ eq/yr | | | 1.65E+04 | 6.61E+04 | 1.32E+05 | 2.64E+05 |
| **Capital Cost (million USD)** | | | | | | | |
| Feed Preparation | | | 0.70 | 5.90 | 15.55 | 25.26 | 41.03 |
| Pyrolysis | | | 0.80 | 3.60 | 10.89 | 18.95 | 33.00 |
| Quench | | | 0.70 | 1.52 | 4.01 | 6.51 | 10.57 |
| Heat Recovery | | | 0.85 | 0.97 | 3.16 | 5.70 | 10.27 |
| Product Recovery and Storage | | | 0.65 | 0.87 | 2.14 | 3.36 | 5.26 |
| Recycle | | | 0.65 | 1.11 | 2.73 | 4.28 | 6.71 |
| Steam and Power Production | | | 0.70 | 3.32 | 8.74 | 14.20 | 23.06 |
| Utilities/auxiliaries | | | 0.70 | 2.69 | 7.10 | 11.53 | 18.73 |
| **Direct Fixed Capital (DFC), also called TIC** | | | | 19.98 | 54.32 | 89.79 | 148.65 |
| Engineering | DFC x MF | 0.12 | | 2.40 | 6.52 | 10.77 | 17.84 |
| Construction | DFC x MF | 0.13 | | 2.60 | 7.06 | 11.67 | 19.32 |
| Contractor & Legal | DFC x MF | 0.08 | | 1.60 | 4.35 | 7.18 | 11.89 |
| Process/Project Contingency | DFC x MF | 0.0412 | | 0.82 | 2.24 | 3.70 | 6.12 |
| **Total Plant Cost (TPC)** | | | | 27.40 | 74.48 | 123.12 | 203.83 |
| AFUDC | | | | | | | |
| **Total Plant Investment (TPI)** | | | | 27.40 | 74.48 | 123.12 | 203.83 |
| Land | | | 0.60 | 0.94 | 2.15 | 3.27 | 6.53 |
| **Startup** | | | | | | | |
| **Total Capital Cost (TCC)** | | | | 28.33 | 76.64 | 126.38 | 210.36 |
| **Contingency/TPI** | | | | | | | |
| Working Capital | DFC x MF | 0.05 | | 1.37 | 3.72 | 6.16 | 10.19 |
| **Variable Operating Costs (million USD/yr)** | | | | | | | |
| Feed | | | | 1.67 | 6.67 | 13.33 | 26.66 |
| Utilities | | | | 0.96 | 3.83 | 7.67 | 15.33 |
| Other | | | | 0.24 | 0.96 | 1.92 | 3.83 |
| Catalysts and Chemicals | | | | 0.17 | 0.67 | 1.34 | 2.68 |
| **Total** | | | | 3.04 | 12.13 | 24.26 | 48.51 |
| Fixed Operating Costs (million USD/yr) | | | | | | | |
| Labor (SF from V-R) | DFC x MF | 0.05 | 0.25 | 1.55 | 2.19 | 2.60 | 3.10 |
| Maintenance (3% of A) | | | 0.03 | 0.60 | 1.63 | 2.69 | 4.46 |
| General Overhead (65% of labor + maint) | | | 0.65 | 1.40 | 2.48 | 3.44 | 4.91 |
| Direct Overhead (45% of Labor) | | | 0.45 | 0.70 | 0.99 | 1.17 | 1.39 |
| Insurance (0.5% of TIC) | | | 0.005 | 0.14 | 0.38 | 0.63 | 1.05 |
| **Total** | | | | 4.38 | 7.67 | 10.55 | 14.91 |

## Table 78. Brazil residual fuel oil capital and operating costs, bagasse

| Country | Brazil | Code = | 2 | | | | |
|---|---|---|---|---|---|---|---|
| | | | | Brazil | Brazil | Brazil | Brazil |
| Cost component | Units | Cost Factor | Scale Factor | Pyrolytic Fuel Oil | Pyrolytic Fuel Oil | Pyrolytic Fuel Oil | Pyrolytic Fuel Oil |
| Year $ | $ | | | 2005 | 2005 | 2005 | 2005 |
| Cost Index (2 = M&S, 3 = CE) | | | 3 | 130.93 | 130.93 | 130.93 | 130.93 |
| Plant Size | dry short ton/day | | | 529 | 2,113 | 4,225 | 8,449 |
| | dry tonne/day | | | 480 | 1,917 | 3,833 | 7,665 |
| Stream Factor | % | | | 95% | 95% | 95% | 95% |
| Feed | dtpy | | | 1.83E+05 | 7.33E+05 | 1.47E+06 | 2.93E+06 |
| | GJ/ton | | | 1.79E+01 | 1.79E+01 | 1.79E+01 | 1.79E+01 |
| | GJ/yr | | | 3.29E+06 | 1.31E+07 | 2.63E+07 | 5.25E+07 |
| Feed Cost | $/dt | | | 7.60 | $ 7.60 | 7.60 | 7.60 |
| Process Efficiency (overall) | % | | | 67.5% | 67.5% | 67.5% | 67.5% |
| Process Efficiency (to RFO) | % | | | 67.0% | 67.0% | 67.0% | 67.0% |
| RFO Production | Yield, wt% | | | 70% | 70% | 70% | 70% |
| | Ton/yr | | | 1.28E+05 | 5.13E+05 | 1.03E+06 | 2.05E+06 |
| | HHV, GJ/ton | | | 17.16 | 17.16 | 17.16 | 17.16 |
| | gal/yr | | | 2.50E+07 | 1.00E+08 | 2.00E+08 | 4.00E+08 |
| | Mil Gal/YR | | | 25.0 | 100.0 | 200.0 | 400.0 |
| | tpy | | | 1.28E+05 | 5.13E+05 | 1.03E+06 | 2.05E+06 |
| | GJ/yr | | | 2.20E+06 | 8.80E+06 | 1.76E+07 | 3.52E+07 |
| | gal/ton | | | 1.37E+02 | 1.37E+02 | 1.37E+02 | 1.37E+02 |
| | gal/Stream day | | | 7.22E+04 | 2.88E+05 | 5.77E+05 | 1.15E+06 |
| | bbl/s day | | | 1.72E+03 | 6.87E+03 | 1.37E+04 | 2.75E+04 |
| | gal EtOH eq/yr | | | 2.48E+07 | 9.92E+07 | 1.98E+08 | 3.97E+08 |
| | gal EtOH eq/ton | | | 1.35E+02 | 1.35E+02 | 1.35E+02 | 1.35E+02 |
| | bbl EtOH eq/day | | | 1.62E+03 | 6.47E+03 | 1.29E+04 | 2.59E+04 |
| Electricity | kWh/y | 2358.2 | kWh/h | 4.71E+06 | 1.88E+07 | 3.76E+07 | 7.52E+07 |
| | GJ eq/yr | | | 1.65E+04 | 6.61E+04 | 1.32E+05 | 2.64E+05 |
| **Capital Cost (million USD)** | | | | | | | |
| Feed Preparation | | | 0.70 | 5.62 | 14.80 | 24.04 | 39.05 |
| Pyrolysis | | | 0.80 | 3.42 | 10.36 | 18.04 | 31.40 |
| Quench | | | 0.70 | 1.45 | 3.81 | 6.19 | 10.06 |
| Heat Recovery | | | 0.85 | 0.93 | 3.01 | 5.42 | 9.78 |
| Product Recovery and Storage | | | 0.65 | 0.83 | 2.04 | 3.19 | 5.01 |
| Recycle | | | 0.65 | 1.06 | 2.60 | 4.07 | 6.39 |
| Steam and Power Production | | | 0.70 | 3.16 | 8.32 | 13.51 | 21.95 |
| Utilities/auxiliaries | | | 0.70 | 2.56 | 6.76 | 10.97 | 17.83 |
| **Direct Fixed Capital (DFC), also called TIC** | | | | 19.01 | 51.70 | 85.45 | 141.47 |
| Engineering | DFC x MF | 0.12 | | 2.28 | 6.20 | 10.25 | 16.98 |
| Construction | DFC x MF | 0.13 | | 2.47 | 6.72 | 11.11 | 18.39 |
| Contractor & Legal | DFC x MF | 0.08 | | 1.52 | 4.14 | 6.84 | 11.32 |
| Process/Project Contingency | DFC x MF | 0.0412 | | 0.78 | 2.13 | 3.52 | 5.83 |
| **Total Plant Cost (TPC)** | | | | 26.07 | 70.89 | 117.18 | 193.99 |
| AFUDC | | | | | | | |
| **Total Plant Investment (TPI)** | | | | 26.07 | 70.89 | 117.18 | 193.99 |
| Land | | | 0.60 | 0.94 | 2.15 | 3.27 | 6.53 |
| **Startup** | | | | | | | |
| **Total Capital Cost (TCC)** | | | | 27.01 | 73.04 | 120.44 | 200.52 |
| **Contingency/TPI** | | | | | | | |
| Working Capital | DFC x MF | 0.05 | | 1.30 | 3.54 | 5.86 | 9.70 |
| **Variable Operating Costs (million USD/yr)** | | | | | | | |
| Feed | | | | 1.39 | 5.57 | 11.13 | 22.27 |
| Utilities | | | | 0.96 | 3.83 | 7.67 | 15.33 |
| Other | | | | 0.24 | 0.96 | 1.92 | 3.83 |
| Catalysts and Chemicals | | | | 0.17 | 0.67 | 1.34 | 2.68 |
| **Total** | | | | 2.76 | 11.03 | 22.06 | 44.11 |
| Fixed Operating Costs (million USD/yr) | | | | | | | |
| Labor (SF from V-R) | DFC x MF | 0.05 | 0.25 | 0.78 | 1.10 | 1.31 | 1.56 |
| Maintenance (3% of A) | | | 0.03 | 0.57 | 1.55 | 2.56 | 4.24 |
| General Overhead (65% of labor + maint) | | | 0.65 | 0.88 | 1.72 | 2.52 | 3.77 |
| Direct Overhead (45% of Labor) | | | 0.45 | 0.35 | 0.49 | 0.59 | 0.70 |
| Insurance (0.5% of TIC) | | | 0.005 | 0.14 | 0.37 | 0.60 | 1.00 |
| **Total** | | | | 2.71 | 5.23 | 7.58 | 11.27 |

## Table 79. Canada residual fuel oil capital and operating costs, ag residues

| Country | Canada | Code = | 7 | | | | |
|---|---|---|---|---|---|---|---|
| | | | | Canada | Canada | Canada | Canada |
| Cost component | Units | Cost Factor | Scale Factor | Pyrolytic Fuel Oil | Pyrolytic Fuel Oil | Pyrolytic Fuel Oil | Pyrolytic Fuel Oil |
| Year $ | $ | | | 2005 | 2005 | 2005 | 2005 |
| Cost Index (2 = M&S, 3 = CE) | | | 3 | 130.93 | 130.93 | 130.93 | 130.93 |
| Plant Size | dry short ton/day | | | 529 | 2,113 | 4,225 | 8,449 |
| | dry tonne/day | | | 480 | 1,917 | 3,833 | 7,665 |
| Stream Factor | % | | | 95% | 95% | 95% | 95% |
| Feed | dtpy | | | 1.83E+05 | 7.33E+05 | 1.47E+06 | 2.93E+06 |
| | GJ/ton | | | 1.79E+01 | 1.79E+01 | 1.79E+01 | 1.79E+01 |
| | GJ/yr | | | 3.29E+06 | 1.31E+07 | 2.63E+07 | 5.25E+07 |
| Feed Cost | $/dt | | | 10.00 | 10.00 | $ 10.00 | 10.00 |
| Process Efficiency (overall) | % | | | 67.5% | 67.5% | 67.5% | 67.5% |
| Process Efficiency (to RFO) | % | | | 67.0% | 67.0% | 67.0% | 67.0% |
| RFO Production | Yield, wt% | | | 70% | 70% | 70% | 70% |
| | Ton/yr | | | 1.28E+05 | 5.13E+05 | 1.03E+06 | 2.05E+06 |
| | HHV, GJ/ton | | | 17.16 | 17.16 | 17.16 | 17.16 |
| | gal/yr | | | 2.50E+07 | 1.00E+08 | 2.00E+08 | 4.00E+08 |
| | Mil Gal/YR | | | 25.0 | 100.0 | 200.0 | 400.0 |
| | tpy | | | 1.28E+05 | 5.13E+05 | 1.03E+06 | 2.05E+06 |
| | GJ/yr | | | 2.20E+06 | 8.80E+06 | 1.76E+07 | 3.52E+07 |
| | gal/ton | | | 1.37E+02 | 1.37E+02 | 1.37E+02 | 1.37E+02 |
| | gal/Stream day | | | 7.22E+04 | 2.88E+05 | 5.77E+05 | 1.15E+06 |
| | bbl/s day | | | 1.72E+03 | 6.87E+03 | 1.37E+04 | 2.75E+04 |
| | gal EtOH eq/yr | | | 2.48E+07 | 9.92E+07 | 1.98E+08 | 3.97E+08 |
| | gal EtOH eq/ton | | | 1.35E+02 | 1.35E+02 | 1.35E+02 | 1.35E+02 |
| | bbl EtOH eq/day | | | 1.62E+03 | 6.47E+03 | 1.29E+04 | 2.59E+04 |
| Electricity | kWh/y | 2358.2 | kWh/h | 4.71E+06 | 1.88E+07 | 3.76E+07 | 7.52E+07 |
| | GJ eq/yr | | | 1.65E+04 | 6.61E+04 | 1.32E+05 | 2.64E+05 |
| **Capital Cost (million USD)** | | | | | | | |
| Feed Preparation | | | 0.70 | 7.16 | 18.87 | 30.65 | 49.79 |
| Pyrolysis | | | 0.80 | 4.36 | 13.21 | 23.00 | 40.04 |
| Quench | | | 0.70 | 1.84 | 4.86 | 7.90 | 12.83 |
| Heat Recovery | | | 0.85 | 1.18 | 3.84 | 6.92 | 12.47 |
| Product Recovery and Storage | | | 0.65 | 1.06 | 2.60 | 4.07 | 6.39 |
| Recycle | | | 0.65 | 1.35 | 3.31 | 5.19 | 8.15 |
| Steam and Power Production | | | 0.70 | 4.02 | 10.61 | 17.23 | 27.99 |
| Utilities/auxiliaries | | | 0.70 | 3.27 | 8.61 | 13.99 | 22.73 |
| **Direct Fixed Capital (DFC), also called TIC** | | | | 24.24 | 65.92 | 108.96 | 180.38 |
| Engineering | DFC x MF | 0.12 | | 2.91 | 7.91 | 13.07 | 21.65 |
| Construction | DFC x MF | 0.13 | | 3.15 | 8.57 | 14.16 | 23.45 |
| Contractor & Legal | DFC x MF | 0.08 | | 1.94 | 5.27 | 8.72 | 14.43 |
| Process/Project Contingency | DFC x MF | 0.0412 | | 1.00 | 2.72 | 4.49 | 7.43 |
| **Total Plant Cost (TPC)** | | | | 33.24 | 90.38 | 149.40 | 247.34 |
| AFUDC | | | | | | | |
| **Total Plant Investment (TPI)** | | | | 33.24 | 90.38 | 149.40 | 247.34 |
| Land | | | 0.60 | 0.94 | 2.15 | 3.27 | 6.53 |
| **Startup** | | | | | | | |
| **Total Capital Cost (TCC)** | | | | 34.18 | 92.54 | 152.67 | 253.87 |
| **Contingency/TPI** | | | | | | | |
| Working Capital | DFC x MF | 0.05 | | 1.66 | 4.52 | 7.47 | 12.37 |
| **Variable Operating Costs (million USD/yr)** | | | | | | | |
| Feed | | | | 1.83 | 7.33 | 14.65 | 29.30 |
| Utilities | | | | 0.96 | 3.83 | 7.67 | 15.33 |
| Other | | | | 0.24 | 0.96 | 1.92 | 3.83 |
| Catalysts and Chemicals | | | | 0.17 | 0.67 | 1.34 | 2.68 |
| **Total** | | | | 3.20 | 12.79 | 25.57 | 51.14 |
| Fixed Operating Costs (million USD/yr) | | | | | | | |
| Labor (SF from V-R) | DFC x MF | 0.05 | 0.25 | 3.06 | 4.33 | 5.15 | 6.12 |
| Maintenance (3% of A) | | | 0.03 | 0.73 | 1.98 | 3.27 | 5.41 |
| General Overhead (65% of labor + maint) | | | 0.65 | 2.46 | 4.10 | 5.47 | 7.50 |
| Direct Overhead (45% of Labor) | | | 0.45 | 1.38 | 1.95 | 2.32 | 2.76 |
| Insurance (0.5% of TIC) | | | 0.005 | 0.17 | 0.46 | 0.76 | 1.27 |
| **Total** | | | | 7.80 | 12.82 | 16.97 | 23.06 |

# Table 80. Caribbean basin residual fuel oil capital and operating costs, bagasse

**Country** **Caribbean** **Code =** 9

| Cost component | Units | Cost Factor | Scale Factor | Caribbean Pyrolytic Fuel Oil | Caribbean Pyrolytic Fuel Oil | Caribbean Pyrolytic Fuel Oil | Caribbean Pyrolytic Fuel Oil |
|---|---|---|---|---|---|---|---|
| Year $ | $ | | | 2005 | 2005 | 2005 | 2005 |
| Cost Index (2 = M&S, 3 = CE) | | | 3 | 130.93 | 130.93 | 130.93 | 130.93 |
| Plant Size | dry short ton/day | | | 529 | 2,113 | 4,225 | 8,449 |
| | dry tonne/day | | | 480 | 1,917 | 3,833 | 7,665 |
| Stream Factor | % | | | 95% | 95% | 95% | 95% |
| Feed | dtpy | | | 1.83E+05 | 7.33E+05 | 1.47E+06 | 2.93E+06 |
| | GJ/ton | | | 1.79E+01 | 1.79E+01 | 1.79E+01 | 1.79E+01 |
| | GJ/yr | | | 3.29E+06 | 1.31E+07 | 2.63E+07 | 5.25E+07 |
| Feed Cost | $/dt | | | $ 7.60 | $ 7.60 | $ 7.60 | $ 7.60 |
| Process Efficiency (overall) | % | | | 67.5% | 67.5% | 67.5% | 67.5% |
| Process Efficiency (to RFO) | % | | | 67.0% | 67.0% | 67.0% | 67.0% |
| RFO Production | Yield, wt% | | | 70% | 70% | 70% | 70% |
| | Ton/yr | | | 1.28E+05 | 5.13E+05 | 1.03E+06 | 2.05E+06 |
| | HHV, GJ/ton | | | 17.16 | 17.16 | 17.16 | 17.16 |
| | gal/yr | | | 2.50E+07 | 1.00E+08 | 2.00E+08 | 4.00E+08 |
| | Mil Gal/YR | | | 25.0 | 100.0 | 200.0 | 400.0 |
| | tpy | | | 1.28E+05 | 5.13E+05 | 1.03E+06 | 2.05E+06 |
| | GJ/yr | | | 2.20E+06 | 8.80E+06 | 1.76E+07 | 3.52E+07 |
| | gal/ton | | | 1.37E+02 | 1.37E+02 | 1.37E+02 | 1.37E+02 |
| | gal/Stream day | | | 7.22E+04 | 2.88E+05 | 5.77E+05 | 1.15E+06 |
| | bbl/s day | | | 1.72E+03 | 6.87E+03 | 1.37E+04 | 2.75E+04 |
| | gal EtOH eq/yr | | | 2.48E+07 | 9.92E+07 | 1.98E+08 | 3.97E+08 |
| | gal EtOH eq/ton | | | 1.35E+02 | 1.35E+02 | 1.35E+02 | 1.35E+02 |
| | bbl EtOH eq/day | | | 1.62E+03 | 6.47E+03 | 1.29E+04 | 2.59E+04 |
| Electricity | kWh/y | 2358.2 | kWh/h | 4.71E+06 | 1.88E+07 | 3.76E+07 | 7.52E+07 |
| | GJ eq/yr | | | 1.65E+04 | 6.61E+04 | 1.32E+05 | 2.64E+05 |
| **Capital Cost (million USD)** | | | | | | | |
| Feed Preparation | | | 0.70 | 6.20 | 16.34 | 26.54 | 43.12 |
| Pyrolysis | | | 0.80 | 3.78 | 11.44 | 19.92 | 34.67 |
| Quench | | | 0.70 | 1.60 | 4.21 | 6.84 | 11.11 |
| Heat Recovery | | | 0.85 | 1.02 | 3.32 | 5.99 | 10.79 |
| Product Recovery and Storage | | | 0.65 | 0.91 | 2.25 | 3.53 | 5.53 |
| Recycle | | | 0.65 | 1.17 | 2.87 | 4.50 | 7.06 |
| Steam and Power Production | | | 0.70 | 3.48 | 9.19 | 14.92 | 24.24 |
| Utilities/auxiliaries | | | 0.70 | 2.83 | 7.46 | 12.12 | 19.68 |
| **Direct Fixed Capital (DFC), also called TIC** | | | | 20.99 | 57.08 | 94.35 | 156.20 |
| Engineering | DFC x MF | 0.12 | | 2.52 | 6.85 | 11.32 | 18.74 |
| Construction | DFC x MF | 0.13 | | 2.73 | 7.42 | 12.27 | 20.31 |
| Contractor & Legal | DFC x MF | 0.08 | | 1.68 | 4.57 | 7.55 | 12.50 |
| Process/Project Contingency | DFC x MF | 0.0412 | | 0.86 | 2.35 | 3.89 | 6.44 |
| **Total Plant Cost (TPC)** | | | | 28.79 | 78.27 | 129.37 | 214.18 |
| AFUDC | | | | | | | |
| **Total Plant Investment (TPI)** | | | | 28.79 | 78.27 | 129.37 | 214.18 |
| Land | | | 0.60 | 0.94 | 2.15 | 3.27 | 6.53 |
| **Startup** | | | | | | | |
| **Total Capital Cost (TCC)** | | | | 29.73 | 80.42 | 132.64 | 220.71 |
| **Contingency/TPI** | | | | | | | |
| Working Capital | DFC x MF | 0.05 | | 1.44 | 3.91 | 6.47 | 10.71 |
| **Variable Operating Costs (million USD/yr)** | | | | | | | |
| Feed | | | | 1.39 | 5.57 | 11.13 | 22.27 |
| Utilities | | | | 0.96 | 3.83 | 7.67 | 15.33 |
| Other | | | | 0.24 | 0.96 | 1.92 | 3.83 |
| Catalysts and Chemicals | | | | 0.17 | 0.67 | 1.34 | 2.68 |
| **Total** | | | | 2.76 | 11.03 | 22.06 | 44.11 |
| Fixed Operating Costs (million USD/yr) | | | | | | | |
| Labor (SF from V-R) | DFC x MF | 0.05 | 0.25 | 1.75 | 2.48 | 2.95 | 3.50 |
| Maintenance (3% of A) | | | 0.03 | 0.63 | 1.71 | 2.83 | 4.69 |
| General Overhead (65% of labor + maint) | | | 0.65 | 1.55 | 2.72 | 3.76 | 5.32 |
| Direct Overhead (45% of Labor) | | | 0.45 | 0.79 | 1.12 | 1.33 | 1.58 |
| Insurance (0.5% of TIC) | | | 0.005 | 0.15 | 0.40 | 0.66 | 1.10 |
| **Total** | | | | 4.87 | 8.43 | 11.52 | 16.20 |

## Table 81. China residual fuel oil capital and operating costs, bagasse

| Country | China | Code = | 3 | | | | |
|---|---|---|---|---|---|---|---|
| | | | | China | China | China | China |
| Cost component | Units | Cost Factor | Scale Factor | Pyrolytic Fuel Oil | Pyrolytic Fuel Oil | Pyrolytic Fuel Oil | Pyrolytic Fuel Oil |
| Year $ | $ | | | 2005 | 2005 | 2005 | 2005 |
| Cost Index (2 = M&S, 3 = CE) | | | 3 | 130.93 | 130.93 | 130.93 | 130.93 |
| Plant Size | dry short ton/day | | | 529 | 2,113 | 4,225 | 8,449 |
| | dry tonne/day | | | 480 | 1,917 | 3,833 | 7,665 |
| Stream Factor | % | | | 95% | 95% | 95% | 95% |
| Feed | dtpy | | | 1.83E+05 | 7.33E+05 | 1.47E+06 | 2.93E+06 |
| | GJ/ton | | | 1.79E+01 | 1.79E+01 | 1.79E+01 | 1.79E+01 |
| | GJ/yr | | | 3.29E+06 | 1.31E+07 | 2.63E+07 | 5.25E+07 |
| Feed Cost | $/dt | | | 9.10 | $ 9.10 | $ 9.10 | $ 9.10 |
| Process Efficiency (overall) | % | | | 67.5% | 67.5% | 67.5% | 67.5% |
| Process Efficiency (to RFO) | % | | | 67.0% | 67.0% | 67.0% | 67.0% |
| RFO Production | Yield, wt% | | | 70% | 70% | 70% | 70% |
| | Ton/yr | | | 1.28E+05 | 5.13E+05 | 1.03E+06 | 2.05E+06 |
| | HHV, GJ/ton | | | 17.16 | 17.16 | 17.16 | 17.16 |
| | gal/yr | | | 2.50E+07 | 1.00E+08 | 2.00E+08 | 4.00E+08 |
| | Mil Gal/YR | | | 25.0 | 100.0 | 200.0 | 400.0 |
| | tpy | | | 1.28E+05 | 5.13E+05 | 1.03E+06 | 2.05E+06 |
| | GJ/yr | | | 2.20E+06 | 8.80E+06 | 1.76E+07 | 3.52E+07 |
| | gal/ton | | | 1.37E+02 | 1.37E+02 | 1.37E+02 | 1.37E+02 |
| | gal/Stream day | | | 7.22E+04 | 2.88E+05 | 5.77E+05 | 1.15E+06 |
| | bbl/s day | | | 1.72E+03 | 6.87E+03 | 1.37E+04 | 2.75E+04 |
| | gal EtOH eq/yr | | | 2.48E+07 | 9.92E+07 | 1.98E+08 | 3.97E+08 |
| | gal EtOH eq/ton | | | 1.35E+02 | 1.35E+02 | 1.35E+02 | 1.35E+02 |
| | bbl EtOH eq/day | | | 1.62E+03 | 6.47E+03 | 1.29E+04 | 2.59E+04 |
| Electricity | kWh/y | 2358.2 | kWh/h | 4.71E+06 | 1.88E+07 | 3.76E+07 | 7.52E+07 |
| | GJ eq/yr | | | 1.65E+04 | 6.61E+04 | 1.32E+05 | 2.64E+05 |
| **Capital Cost (million USD)** | | | | | | | |
| Feed Preparation | | | 0.70 | 6.05 | 15.94 | 25.89 | 42.05 |
| Pyrolysis | | | 0.80 | 3.69 | 11.16 | 19.43 | 33.82 |
| Quench | | | 0.70 | 1.56 | 4.11 | 6.67 | 10.84 |
| Heat Recovery | | | 0.85 | 1.00 | 3.24 | 5.84 | 10.53 |
| Product Recovery and Storage | | | 0.65 | 0.89 | 2.19 | 3.44 | 5.40 |
| Recycle | | | 0.65 | 1.14 | 2.80 | 4.39 | 6.88 |
| Steam and Power Production | | | 0.70 | 3.40 | 8.96 | 14.55 | 23.64 |
| Utilities/auxiliaries | | | 0.70 | 2.76 | 7.28 | 11.82 | 19.20 |
| **Direct Fixed Capital (DFC), also called TIC** | | | | 20.48 | 55.67 | 92.02 | 152.35 |
| Engineering | DFC x MF | 0.12 | | 2.46 | 6.68 | 11.04 | 18.28 |
| Construction | DFC x MF | 0.13 | | 2.66 | 7.24 | 11.96 | 19.81 |
| Contractor & Legal | DFC x MF | 0.08 | | 1.64 | 4.45 | 7.36 | 12.19 |
| Process/Project Contingency | DFC x MF | 0.0412 | | 0.84 | 2.29 | 3.79 | 6.28 |
| **Total Plant Cost (TPC)** | | | | 28.08 | 76.34 | 126.18 | 208.90 |
| AFUDC | | | | | | | |
| **Total Plant Investment (TPI)** | | | | 28.08 | 76.34 | 126.18 | 208.90 |
| Land | | | 0.60 | 0.94 | 2.15 | 3.27 | 6.53 |
| **Startup** | | | | | | | |
| **Total Capital Cost (TCC)** | | | | 29.02 | 78.49 | 129.45 | 215.43 |
| **Contingency/TPI** | | | | | | | |
| Working Capital | DFC x MF | 0.05 | | 1.40 | 3.82 | 6.31 | 10.44 |
| **Variable Operating Costs (million USD/yr)** | | | | | | | |
| Feed | | | | 1.67 | 6.67 | 13.33 | 26.66 |
| Utilities | | | | 0.96 | 3.83 | 7.67 | 15.33 |
| Other | | | | 0.24 | 0.96 | 1.92 | 3.83 |
| Catalysts and Chemicals | | | | 0.17 | 0.67 | 1.34 | 2.68 |
| **Total** | | | | 3.04 | 12.13 | 24.26 | 48.51 |
| Fixed Operating Costs (million USD/yr) | | | | | | | |
| Labor (SF from V-R) | DFC x MF | 0.05 | 0.25 | 2.03 | 2.87 | 3.41 | 4.06 |
| Maintenance (3% of A) | | | 0.03 | 0.61 | 1.67 | 2.76 | 4.57 |
| General Overhead (65% of labor + maint) | | | 0.65 | 1.72 | 2.95 | 4.01 | 5.61 |
| Direct Overhead (45% of Labor) | | | 0.45 | 0.91 | 1.29 | 1.54 | 1.83 |
| Insurance (0.5% of TIC) | | | 0.005 | 0.15 | 0.39 | 0.65 | 1.08 |
| **Total** | | | | 5.42 | 9.17 | 12.37 | 17.14 |

## Table 82. Colombia residual fuel oil capital and operating costs, bagasse

| Country | Colombia | Code = | 6 | | | | |
|---|---|---|---|---|---|---|---|
| | | | | Colombia | Colombia | Colombia | Colombia |
| Cost component | Units | Cost Factor | Scale Factor | Pyrolytic Fuel Oil | Pyrolytic Fuel Oil | Pyrolytic Fuel Oil | Pyrolytic Fuel Oil |
| Year $ | $ | | | 2005 | 2005 | 2005 | 2005 |
| Cost Index (2 = M&S, 3 = CE) | | | 3 | 130.93 | 130.93 | 130.93 | 130.93 |
| Plant Size | dry short ton/day | | | 529 | 2,113 | 4,225 | 8,449 |
| | dry tonne/day | | | 480 | 1,917 | 3,833 | 7,665 |
| Stream Factor | % | | | 95% | 95% | 95% | 95% |
| Feed | dtpy | | | 1.83E+05 | 7.33E+05 | 1.47E+06 | 2.93E+06 |
| | GJ/ton | | | 1.79E+01 | 1.79E+01 | 1.79E+01 | 1.79E+01 |
| | GJ/yr | | | 3.29E+06 | 1.31E+07 | 2.63E+07 | 5.25E+07 |
| Feed Cost | $/dt | | | $ 14.50 | $ 14.50 | $ 14.50 | $ 14.50 |
| Process Efficiency (overall) | % | | | 67.5% | 67.5% | 67.5% | 67.5% |
| Process Efficiency (to RFO) | % | | | 67.0% | 67.0% | 67.0% | 67.0% |
| RFO Production | Yield, wt% | | | 70% | 70% | 70% | 70% |
| | Ton/yr | | | 1.28E+05 | 5.13E+05 | 1.03E+06 | 2.05E+06 |
| | HHV, GJ/ton | | | 17.16 | 17.16 | 17.16 | 17.16 |
| | gal/yr | | | 2.50E+07 | 1.00E+08 | 2.00E+08 | 4.00E+08 |
| | Mil Gal/YR | | | 25.0 | 100.0 | 200.0 | 400.0 |
| | tpy | | | 1.28E+05 | 5.13E+05 | 1.03E+06 | 2.05E+06 |
| | GJ/yr | | | 2.20E+06 | 8.80E+06 | 1.76E+07 | 3.52E+07 |
| | gal/ton | | | 1.37E+02 | 1.37E+02 | 1.37E+02 | 1.37E+02 |
| | gal/Stream day | | | 7.22E+04 | 2.88E+05 | 5.77E+05 | 1.15E+06 |
| | bbl/s day | | | 1.72E+03 | 6.87E+03 | 1.37E+04 | 2.75E+04 |
| | gal EtOH eq/yr | | | 2.48E+07 | 9.92E+07 | 1.98E+08 | 3.97E+08 |
| | gal EtOH eq/ton | | | 1.35E+02 | 1.35E+02 | 1.35E+02 | 1.35E+02 |
| | bbl EtOH eq/day | | | 1.62E+03 | 6.47E+03 | 1.29E+04 | 2.59E+04 |
| Electricity | kWh/y | 2358.2 | kWh/h | 4.71E+06 | 1.88E+07 | 3.76E+07 | 7.52E+07 |
| | GJ eq/yr | | | 1.65E+04 | 6.61E+04 | 1.32E+05 | 2.64E+05 |
| **Capital Cost (million USD)** | | | | | | | |
| Feed Preparation | | | 0.70 | 5.46 | 14.4 | 23.40 | 38.00 |
| Pyrolysis | | | 0.80 | 3.33 | 10.0 | 17.55 | 30.56 |
| Quench | | | 0.70 | 1.41 | 3.7 | 6.03 | 9.79 |
| Heat Recovery | | | 0.85 | 0.90 | 2.9 | 5.28 | 9.51 |
| Product Recovery and Storage | | | 0.65 | 0.81 | 1.9 | 3.11 | 4.88 |
| Recycle | | | 0.65 | 1.03 | 2.5 | 3.96 | 6.22 |
| Steam and Power Production | | | 0.70 | 3.07 | 8.1 | 13.15 | 21.36 |
| Utilities/auxiliaries | | | 0.70 | 2.49 | 6.5 | 10.68 | 17.35 |
| **Direct Fixed Capital (DFC), also called TIC** | | | | 18.50 | 50.3 | 83.16 | 137.67 |
| Engineering | DFC x MF | 0.12 | | 2.22 | 6.0 | 9.98 | 16.52 |
| Construction | DFC x MF | 0.13 | | 2.41 | 6.5 | 10.81 | 17.90 |
| Contractor & Legal | DFC x MF | 0.08 | | 1.48 | 4.0 | 6.65 | 11.01 |
| Process/Project Contingency | DFC x MF | 0.0412 | | 0.76 | 2.0 | 3.43 | 5.67 |
| **Total Plant Cost (TPC)** | | | | 25.37 | 68.9 | 114.03 | 188.78 |
| AFUDC | | | | | | | |
| **Total Plant Investment (TPI)** | | | | 25.37 | 68.98 | 114.03 | 188.78 |
| Land | | | 0.60 | 0.94 | 2.15 | 3.27 | 6.53 |
| **Startup** | | | | | | | |
| **Total Capital Cost (TCC)** | | | | 26.31 | 71.14 | 117.29 | 195.30 |
| **Contingency/TPI** | | | | | | | |
| Working Capital | DFC x MF | 0.05 | | 1.27 | 3.45 | 5.70 | 9.44 |
| **Variable Operating Costs (million USD/yr)** | | | | | | | |
| Feed | | | | 2.66 | 10.62 | 21.24 | 42.48 |
| Utilities | | | | 0.96 | 3.83 | 7.67 | 15.33 |
| Other | | | | 0.24 | 0.96 | 1.92 | 3.83 |
| Catalysts and Chemicals | | | | 0.17 | 0.67 | 1.34 | 2.68 |
| **Total** | | | | 4.03 | 16.09 | 32.17 | 64.33 |
| Fixed Operating Costs (million USD/yr) | | | | | | | |
| Labor (SF from V-R) | DFC x MF | 0.05 | 0.25 | 1.16 | 1.64 | 1.95 | 2.32 |
| Maintenance (3% of A) | | | 0.03 | 0.56 | 1.51 | 2.49 | 4.13 |
| General Overhead (65% of labor + maint) | | | 0.65 | 1.12 | 2.05 | 2.89 | 4.20 |
| Direct Overhead (45% of Labor) | | | 0.45 | 0.52 | 0.74 | 0.88 | 1.05 |
| Insurance (0.5% of TIC) | | | 0.005 | 0.13 | 0.36 | 0.59 | 0.98 |
| **Total** | | | | 3.49 | 6.30 | 8.81 | 12.67 |

# Table 83. India residual fuel oil capital and operating costs, bagasse

| Country | India | Code = | 4 | | | | |
|---|---|---|---|---|---|---|---|
| | | | | India | India | India | India |
| Cost component | Units | Cost Factor | Scale Factor | Pyrolytic Fuel Oil | Pyrolytic Fuel Oil | Pyrolytic Fuel Oil | Pyrolytic Fuel Oil |
| Year $ | $ | | | 2005 | 2005 | 2005 | 2005 |
| Cost Index (2 = M&S, 3 = CE) | | | 3 | 130.93 | 130.93 | 130.93 | 130.93 |
| Plant Size | dry short ton/day | | | 529 | 2,113 | 4,225 | 8,449 |
| | dry tonne/day | | | 480 | 1,917 | 3,833 | 7,665 |
| Stream Factor | % | | | 95% | 95% | 95% | 95% |
| Feed | dtpy | | | 1.83E+05 | 7.33E+05 | 1.47E+06 | 2.93E+06 |
| | GJ/ton | | | 1.79E+01 | 1.79E+01 | 1.79E+01 | 1.79E+01 |
| | GJ/yr | | | 3.29E+06 | 1.31E+07 | 2.63E+07 | 5.25E+07 |
| Feed Cost | $/dt | | | 7.30 | $ 7.30 | $ 7.30 | $ 7.30 |
| Process Efficiency (overall) | % | | | 67.5% | 67.5% | 67.5% | 67.5% |
| Process Efficiency (to RFO) | % | | | 67.0% | 67.0% | 67.0% | 67.0% |
| RFO Production | Yield, wt% | | | 70% | 70% | 70% | 70% |
| | Ton/yr | | | 1.28E+05 | 5.13E+05 | 1.03E+06 | 2.05E+06 |
| | HHV, GJ/ton | | | 17.16 | 17.16 | 17.16 | 17.16 |
| | gal/yr | | | 2.50E+07 | 1.00E+08 | 2.00E+08 | 4.00E+08 |
| | Mil Gal/YR | | | 25.0 | 100.0 | 200.0 | 400.0 |
| | tpy | | | 1.28E+05 | 5.13E+05 | 1.03E+06 | 2.05E+06 |
| | GJ/yr | | | 2.20E+06 | 8.80E+06 | 1.76E+07 | 3.52E+07 |
| | gal/ton | | | 1.37E+02 | 1.37E+02 | 1.37E+02 | 1.37E+02 |
| | gal/Stream day | | | 7.22E+04 | 2.88E+05 | 5.77E+05 | 1.15E+06 |
| | bbl/s day | | | 1.72E+03 | 6.87E+03 | 1.37E+04 | 2.75E+04 |
| | gal EtOH eq/yr | | | 2.48E+07 | 9.92E+07 | 1.98E+08 | 3.97E+08 |
| | gal EtOH eq/ton | | | 1.35E+02 | 1.35E+02 | 1.35E+02 | 1.35E+02 |
| | bbl EtOH eq/day | | | 1.62E+03 | 6.47E+03 | 1.29E+04 | 2.59E+04 |
| Electricity | kWh/y | 2358.2 | kWh/h | 4.71E+06 | 1.88E+07 | 3.76E+07 | 7.52E+07 |
| | GJ eq/yr | | | 1.65E+04 | 6.61E+04 | 1.32E+05 | 2.64E+05 |
| **Capital Cost (million USD)** | | | | | | | |
| Feed Preparation | | | 0.70 | 5.86 | 15.45 | 25.09 | 40.76 |
| Pyrolysis | | | 0.80 | 3.57 | 10.82 | 18.83 | 32.78 |
| Quench | | | 0.70 | 1.5 | 3.98 | 6.47 | 10.50 |
| Heat Recovery | | | 0.85 | 0.97 | 3.14 | 5.66 | 10.20 |
| Product Recovery and Storage | | | 0.65 | 0.86 | 2.12 | 3.33 | 5.23 |
| Recycle | | | 0.65 | 1.10 | 2.71 | 4.25 | 6.67 |
| Steam and Power Production | | | 0.70 | 3.29 | 8.68 | 14.10 | 22.91 |
| Utilities/auxiliaries | | | 0.70 | 2.68 | 7.05 | 11.45 | 18.61 |
| **Direct Fixed Capital (DFC), also called TIC** | | | | 19.85 | 53.96 | 89.19 | 147.66 |
| Engineering | DFC x MF | 0.12 | | 2.38 | 6.48 | 10.70 | 17.72 |
| Construction | DFC x MF | 0.13 | | 2.58 | 7.01 | 11.60 | 19.20 |
| Contractor & Legal | DFC x MF | 0.08 | | 1.59 | 4.32 | 7.14 | 11.81 |
| Process/Project Contingency | DFC x MF | 0.0412 | | 0.82 | 2.22 | 3.67 | 6.08 |
| **Total Plant Cost (TPC)** | | | | 27.21 | 73.99 | 122.30 | 202.48 |
| AFUDC | | | | | | | |
| **Total Plant Investment (TPI)** | | | | 27.21 | 73.99 | 122.30 | 202.48 |
| Land | | | 0.60 | 0.94 | 2.15 | 3.27 | 6.53 |
| **Startup** | | | | | | | |
| **Total Capital Cost (TCC)** | | | | 28.15 | 76.14 | 125.57 | 209.01 |
| **Contingency/TPI** | | | | | | | |
| Working Capital | DFC x MF | 0.05 | | 1.36 | 3.70 | 6.12 | 10.12 |
| **Variable Operating Costs (million USD/yr)** | | | | | | | |
| Feed | | | | 1.34 | 5.35 | 10.69 | 21.39 |
| Utilities | | | | 0.96 | 3.83 | 7.67 | 15.33 |
| Other | | | | 0.24 | 0.96 | 1.92 | 3.83 |
| Catalysts and Chemicals | | | | 0.17 | 0.67 | 1.34 | 2.68 |
| **Total** | | | | 2.71 | 10.81 | 21.62 | 43.23 |
| Fixed Operating Costs (million USD/yr) | | | | | | | |
| Labor (SF from V-R) | DFC x MF | 0.05 | 0.25 | 0.68 | 0.97 | 1.15 | 1.37 |
| Maintenance (3% of A) | | | 0.03 | 0.60 | 1.62 | 2.68 | 4.43 |
| General Overhead (65% of labor + maint) | | | 0.65 | 0.83 | 1.68 | 2.49 | 3.77 |
| Direct Overhead (45% of Labor) | | | 0.45 | 0.31 | 0.43 | 0.52 | 0.61 |
| Insurance (0.5% of TIC) | | | 0.005 | 0.14 | 0.38 | 0.63 | 1.05 |
| **Total** | | | | 2.56 | 5.08 | 7.45 | 11.22 |

## Table 84. Mexico residual fuel oil capital and operating costs, bagasse

| Country | Mexico | Code = | 8 | | | | |
|---|---|---|---|---|---|---|---|
| | | | | Mexico | Mexico | Mexico | Mexico |
| **Cost component** | **Units** | **Cost Factor** | **Scale Factor** | **Pyrolytic Fuel Oil** | **Pyrolytic Fuel Oil** | **Pyrolytic Fuel Oil** | **Pyrolytic Fuel Oil** |
| Year $ | $ | | | 2005 | 2005 | 2005 | 2005 |
| Cost Index (2 = M&S, 3 = CE) | | | 3 | 130.93 | 130.93 | 130.93 | 130.93 |
| Plant Size | dry short ton/day | | | 529 | 2,113 | 4,225 | 8,449 |
| | dry tonne/day | | | 480 | 1,917 | 3,833 | 7,665 |
| | | | | | | | |
| Stream Factor | % | | | 95% | 95% | 95% | 95% |
| Feed | dtpy | | | 1.83E+05 | 7.33E+05 | 1.47E+06 | 2.93E+06 |
| | GJ/ton | | | 1.79E+01 | 1.79E+01 | 1.79E+01 | 1.79E+01 |
| | GJ/yr | | | 3.29E+06 | 1.31E+07 | 2.63E+07 | 5.25E+07 |
| Feed Cost | $/dt | | | 15.40 | $ 15.40 | $ 15.40 | $ 15.40 |
| Process Efficiency (overall) | % | | | 67.5% | 67.5% | 67.5% | 67.5% |
| Process Efficiency (to RFO) | % | | | 67.0% | 67.0% | 67.0% | 67.0% |
| RFO Production | Yield, wt% | | | 70% | 70% | 70% | 70% |
| | Ton/yr | | | 1.28E+05 | 5.13E+05 | 1.03E+06 | 2.05E+06 |
| | HHV, GJ/ton | | | 17.16 | 17.16 | 17.16 | 17.16 |
| | gal/yr | | | 2.50E+07 | 1.00E+08 | 2.00E+08 | 4.00E+08 |
| | Mil Gal/YR | | | 25.0 | 100.0 | 200.0 | 400.0 |
| | tpy | | | 1.28E+05 | 5.13E+05 | 1.03E+06 | 2.05E+06 |
| | GJ/yr | | | 2.20E+06 | 8.80E+06 | 1.76E+07 | 3.52E+07 |
| | gal/ton | | | 1.37E+02 | 1.37E+02 | 1.37E+02 | 1.37E+02 |
| | gal/Stream day | | | 7.22E+04 | 2.88E+05 | 5.77E+05 | 1.15E+06 |
| | bbl/s day | | | 1.72E+03 | 6.87E+03 | 1.37E+04 | 2.75E+04 |
| | gal EtOH eq/yr | | | 2.48E+07 | 9.92E+07 | 1.98E+08 | 3.97E+08 |
| | gal EtOH eq/ton | | | 1.35E+02 | 1.35E+02 | 1.35E+02 | 1.35E+02 |
| | bbl EtOH eq/day | | | 1.62E+03 | 6.47E+03 | 1.29E+04 | 2.59E+04 |
| Electricity | kWh/y | 2358.2 | kWh/ | 4.71E+06 | 1.88E+07 | 3.76E+07 | 7.52E+07 |
| | GJ eq/yr | | | 1.65E+04 | 6.61E+04 | 1.32E+05 | 2.64E+05 |
| | | | | | | | |
| **Capital Cost (million USD)** | | | | | | | |
| Feed Preparation | | | 0.70 | 4.42 | 11.65 | 18.92 | 30.74 |
| Pyrolysis | | | 0.80 | 2.69 | 8.16 | 14.20 | 24.72 |
| Quench | | | 0.70 | 1.14 | 3.00 | 4.88 | 7.92 |
| Heat Recovery | | | 0.85 | 0.73 | 2.37 | 4.27 | 7.70 |
| Product Recovery and Storage | | | 0.65 | 0.65 | 1.60 | 2.51 | 3.94 |
| Recycle | | | 0.65 | 0.83 | 2.04 | 3.21 | 5.03 |
| Steam and Power Production | | | 0.70 | 2.48 | 6.55 | 10.64 | 17.28 |
| Utilities/auxiliaries | | | 0.70 | 2.02 | 5.32 | 8.64 | 14.03 |
| **Direct Fixed Capital (DFC), also called TIC** | | | | 14.97 | 40.69 | 67.26 | 111.35 |
| Engineering | DFC x MF | 0.12 | | 1.80 | 4.88 | 8.07 | 13.36 |
| Construction | DFC x MF | 0.13 | | 1.95 | 5.29 | 8.74 | 14.48 |
| Contractor & Legal | DFC x MF | 0.08 | | 1.20 | 3.26 | 5.38 | 8.91 |
| Process/Project Contingency | DFC x MF | 0.0412 | | 0.62 | 1.68 | 2.77 | 4.59 |
| **Total Plant Cost (TPC)** | | | | 20.52 | 55.80 | 92.23 | 152.69 |
| AFUDC | | | | | | | |
| **Total Plant Investment (TPI)** | | | | 20.52 | 55.80 | 92.23 | 152.69 |
| Land | | | 0.60 | 0.94 | 2.15 | 3.27 | 6.53 |
| **Startup** | | | | | | | |
| **Total Capital Cost (TCC)** | | | | 21.46 | 57.95 | 95.49 | 159.22 |
| **Contingency/TPI** | | | | | | | |
| Working Capital | DFC x MF | 0.05 | | 1.03 | 2.79 | 4.61 | 7.63 |
| **Variable Operating Costs (million USD/yr)** | | | | | | | |
| Feed | | | | 2.83 | 11.28 | 22.56 | 45.12 |
| Utilities | | | | 0.96 | 3.83 | 7.67 | 15.33 |
| Other | | | | 0.24 | 0.96 | 1.92 | 3.83 |
| Catalysts and Chemicals | | | | 0.17 | 0.67 | 1.34 | 2.68 |
| **Total** | | | | 4.19 | 16.75 | 33.49 | 66.96 |
| Fixed Operating Costs (million USD/yr) | | | | | | | |
| Labor (SF from V-R) | DFC x MF | 0.05 | 0.25 | 0.44 | 0.62 | 0.74 | 0.88 |
| Maintenance (3% of A) | | | 0.03 | 0.45 | 1.22 | 2.02 | 3.34 |
| General Overhead (65% of labor + maint) | | | 0.65 | 0.58 | 1.20 | 1.79 | 2.74 |
| Direct Overhead (45% of Labor) | | | 0.45 | 0.20 | 0.28 | 0.33 | 0.40 |
| Insurance (0.5% of TIC) | | | 0.005 | 0.11 | 0.29 | 0.48 | 0.80 |
| **Total** | | | | 1.77 | 3.61 | 5.36 | 8.16 |

### Table 85. Pyrolysis residual fuel oil summary yields, costs, and plant gate prices, 25 MM GPY, bagasse

| Technology | Units | USA | Argentina | Brazil | Canada | Caribbean Basin | China | Colombia | India | Mexico |
|---|---|---|---|---|---|---|---|---|---|---|
| **Plant Size** | MM gal/yr | 25 | 25 | 25 | 25 | 25 | 25 | 25 | 25 | 25 |
| **Yields** | | | | | | | | | | |
| Primary Product Yield | gal/ton | 136.6 | 136.6 | 136.6 | | 136.6 | 136.6 | 136.6 | 136.6 | 136.6 |
| By Product A | | | | | | | | | | |
| Unit | | Electricity | Electricity | Electricity | | Electricity | Electricity | Electricity | Electricity | Electricity |
| Yield unit | | kWh | kWh | kWh | | kWh | kWh | kWh | kWh | kWh |
| Annual Yield | | 4.71E+06 | 4.71E+06 | 4.71E+06 | | 4.71E+06 | 4.71E+06 | 4.71E+06 | 4.71E+06 | 4.71E+06 |
| Price unit | | 0.04 | 0.04 | 0.04 | | 0.04 | 0.04 | 0.04 | 0.04 | 0.04 |
| MM USD/yr | | 0.187 | 0.187 | 0.187 | | 0.187 | 0.187 | 0.187 | 0.187 | 0.187 |
| By Product B | | | | | | | | | | |
| Unit | | | | | | | | | | |
| Yield unit | | | | | | | | | | |
| Annual Yield | | | | | | | | | | |
| Price unit | | | | | | | | | | |
| Annual Cost | MM USD/yr | | | | | | | | | |
| **Energy Inputs** | | | | | | | | | | |
| Electricity | | N/A | N/A | N/A | | N/A | N/A | N/A | N/A | N/A |
| Cost | $/kWh | | | | | | | | | |
| Amount | kWh/yr | | | | | | | | | |
| Annual Cost | MM USD/yr | | | | | | | | | |
| Fuel | | N/A | N/A | N/A | | N/A | N/A | N/A | N/A | N/A |
| Type | | | | | | | | | | |
| Unit | | | | | | | | | | |
| Amount | | | | | | | | | | |
| Cost | USD/unit | | | | | | | | | |
| Annual Cost | MM USD/yr | | | | | | | | | |
| **Capital & Operating Costs** | | | | | | | | | | |
| Capital | MMUSD | 26.45 | 28.33 | 27.01 | | 29.73 | 29.02 | 26.31 | 28.15 | 21.46 |
| Operating w/o feed | MM USD/yr | 6.26 | 5.75 | 4.08 | | 6.24 | 6.79 | 4.86 | 3.93 | 3.14 |
| Feed | | | | | | | | | | |
| Unit | | ton | ton | ton | | ton | ton | ton | ton | ton |
| Cost/unit | USD/unit | 35 | 9.1 | 7.6 | | 7.6 | 9.1 | 14.5 | 7.3 | 15.4 |
| Rate | unit/day | 529 | 529 | 529 | | 529 | 529 | 529 | 529 | 529 |
| Yearly Cost | MM USD/yr | 6.42 | 1.67 | 1.39 | | 1.39 | 1.67 | 529 | 1.34 | 2.83 |
| **Product Plant Gate Cost** | | | | | | | | | | |
| Basis | | | | | | | | | | |
| Value | USD/gal | 0.66 | 0.46 | 0.46 | | 0.47 | 0.5 | 0.45 | 0.36 | 0.36 |

### Table 86. Pyrolysis residual fuel oil summary yields, costs, and plant gate prices, 100 MM GPY, bagasse

| Technology | Units | USA | Argentina | Brazil | Canada | Caribbean Basin | China | Colombia | India | Mexico |
|---|---|---|---|---|---|---|---|---|---|---|
| **Plant Size** | MM gal/yr | 100 | 100 | 100 | 100 | 100 | 100 | 100 | 100 | 100 |
| **Yields** | | | | | | | | | | |
| Primary Product Yield | gal/ton | 136.6 | 136.6 | 136.6 | 136.6 | 136.6 | 136.6 | 136.6 | 136.6 | 136.6 |
| By Product A | | Electricity | Electricity | Electricity | Electricity | Electricity | Electricity | Electricity | Electricity | Electricity |
| Unit | | | | | | | | | | |
| Yield unit | | kwh | kwh | kwh | kwh | kwh | kwh | kwh | kwh | kwh |
| Annual Yield | | 1.88E+07 | 1.88E+07 | 1.88E+07 | 1.88E+07 | 1.88E+07 | 1.88E+07 | 1.88E+07 | 1.88E+07 | 1.88E+07 |
| Price unit | | 0.04 | 0.04 | 0.04 | 0.04 | 0.04 | 0.04 | 0.04 | 0.04 | 0.04 |
| MM USD/yr | | 0.746 | 0.746 | 0.746 | 0.746 | 0.746 | 0.746 | 0.746 | 0.746 | 0.746 |
| By Product B | | | | | | | | | | |
| Unit | | | | | | | | | | |
| Yield unit | | | | | | | | | | |
| Annual Yield | | | | | | | | | | |
| Price unit | | | | | | | | | | |
| Annual Cost | MM USD/yr | | | | | | | | | |
| **Energy Inputs** | | | | | | | | | | |
| Electricity | | N/A | N/A | N/A | N/A | N/A | N/A | N/A | N/A | N/A |
| Cost | $/kWh | | | | | | | | | |
| Amount | kWh/yr | | | | | | | | | |
| Annual Cost | MM USD/yr | | | | | | | | | |
| Fuel | | N/A | N/A | N/A | N/A | N/A | N/A | N/A | N/A | N/A |
| Type | | | | | | | | | | |
| Unit | | | | | | | | | | |
| Amount | | | | | | | | | | |
| Cost | USD/unit | | | | | | | | | |
| Annual Cost | MM USD/yr | | | | | | | | | |
| **Capital & Operating Costs** | | | | | | | | | | |
| Capital | MMUSD | 71.52 | 76.64 | 73.04 | | 80.42 | 78.49 | 71.14 | 76.14 | 57.95 |
| Operating w/o feed | MM USD/yr | 26.8 | 13.13 | 10.7 | | 13.9 | 14.64 | 11.76 | 10.54 | 9.08 |
| Feed | | | | | | | | | | |
| Unit | | ton | ton | ton | | ton | ton | ton | ton | ton |
| Cost/unit | USD/unit | 35 | 9.1 | 7.6 | | 7.6 | 9.1 | 14.5 | 7.3 | 15.4 |
| Rate | unit/day | 2113 | 2113 | 2113 | | 2113 | 2113 | 2113 | 2113 | 2113 |
| Yearly Cost | MM USD/yr | 26.65 | 6.67 | 5.57 | | 5.57 | 6.67 | 10.62 | 5.35 | 11.28 |
| **Product Plant Gate Cost** | | | | | | | | | | |
| Basis | | | | | | | | | | |
| Value | USD/gal | 0.50 | 0.3 | 0.26 | | 0.3 | 0.32 | 0.32 | 0.26 | 0.29 |

## Table 87. Pyrolysis residual fuel oil summary yields, costs, and plant gate prices, 25 MM GPY, ag residues

| Technology | Units | USA | Argentina | Brazil | Canada | Caribbean Basin | China | Colombia | India | Mexico |
|---|---|---|---|---|---|---|---|---|---|---|
| **Plant Size** | MM gal/yr | 25 | 25 | 25 | 25 | 25 | 25 | 25 | 25 | 25 |
| **Yields** | | | | | | | | | | |
| Primary Product Yield | gal/ton | 136.6 | 136.6 | 136.6 | 136.6 | 136.6 | 136.6 | 136.6 | 136.6 | 136.6 |
| By Product A | | | | | | | | | | |
| Unit | | Electricity | Electricity | Electricity | Electricity | Electricity | Electricity | Electricity | Electricity | Electricity |
| Yield unit | | kWh | kWh | kWh | kWh | kWh | kWh | kWh | kWh | kWh |
| Annual Yield | | 4.71E+06 | 4.71E+06 | 4.71E+06 | 4.71E+06 | 4.71E+06 | 4.71E+06 | 4.71E+06 | 4.71E+06 | 4.71E+06 |
| Price unit | | 0.04 | 0.04 | 0.04 | 0.04 | 0.04 | 0.04 | 0.04 | 0.04 | 0.04 |
| MM USD/yr | | 0.187 | 0.187 | 0.187 | 0.187 | 0.187 | 0.187 | 0.187 | 0.187 | 0.187 |
| By Product B | | | | | | | | | | |
| Unit | | | | | | | | | | |
| Yield unit | | | | | | | | | | |
| Annual Yield | | | | | | | | | | |
| Price unit | | | | | | | | | | |
| Annual Cost | MM USD/yr | | | | | | | | | |
| **Energy Inputs** | | | | | | | | | | |
| Electricity | | N/A | N/A | N/A | N/A | N/A | N/A | N/A | N/A | N/A |
| Cost | $/kWh | | | | | | | | | |
| Amount | kWh/yr | | | | | | | | | |
| Annual Cost | MM USD/yr | | | | | | | | | |
| Fuel | | N/a | N/a | N/a | N/a | N/a | N/a | N/a | N/a | N/a |
| Type | | | | | | | | | | |
| Unit | | | | | | | | | | |
| Amount | | | | | | | | | | |
| Cost | USD/unit | | | | | | | | | |
| Annual Cost | MM USD/yr | | | | | | | | | |
| **Capital & Operating Costs** | | | | | | | | | | |
| Capital | MMUSD | 26.45 | 28.33 | 27.01 | 34.18 | 29.73 | 29.02 | 26.31 | 28.15 | 221.46 |
| Operating w/o feed | MM USD/yr | 6.26 | 5.75 | 4.08 | 9.17 | 6.24 | 6.79 | 4.86 | 3.93 | 3.14 |
| Feed | | | | | | | | | | |
| Unit | | ton | ton | ton | | ton | ton | ton | ton | ton |
| Cost/unit | USD/unit | 35 | 47.2 | 30.8 | 10 | 14.5 | 28.1 | 14.5 | 7.3 | 47.2 |
| Rate | unit/day | 529 | 529 | 529 | 529 | 529 | 529 | 529 | 529 | 529 |
| Yearly Cost | MM USD/yr | 6.42 | 8.66 | 5.65 | 1.83 | 2.66 | 5.16 | 2.66 | 1.34 | 8.66 |
| **Product Plant Gate Cost** | | | | | | | | | | |
| Basis | | | | | | | | | | |
| Value | USD/gal | 0.66 | 0.74 | 0.54 | 0.63 | 0.52 | 0.64 | 0.45 | 0.36 | 0.6 |

## Table 88. Pyrolysis residual fuel oil summary yields, costs, and plant gate prices, 100 MM GPY, ag residues

| Technology | Units | USA | Argentina | Brazil | Canada | Caribbean Basin | China | Colombia | India | Mexico |
|---|---|---|---|---|---|---|---|---|---|---|
| **Plant Size** | MM gal/yr | 100 | 100 | 100 | 100 | 100 | 100 | 100 | 100 | 100 |
| **Yields** | | | | | | | | | | |
| Primary Product Yield | gal/ton | 136.6 | 136.6 | 136.6 | 136.6 | 136.6 | 136.6 | 136.6 | 136.6 | 136.6 |
| By Product A | | Electricity | Electricity | Electricity | Electricity | Electricity | Electricity | Electricity | Electricity | Electricity |
| Unit | | | | | | | | | | |
| Yield unit | | kwh | kwh | kwh | kwh | kwh | kwh | kwh | kwh | kwh |
| Annual Yield | | 1.88E+07 | 1.88E+07 | 1.88E+07 | 1.88E+07 | 1.88E+07 | 1.88E+07 | 1.88E+07 | 1.88E+07 | 1.88E+07 |
| Price unit | | 0.04 | 0.04 | 0.04 | 0.04 | 0.04 | 0.04 | 0.04 | 0.04 | 0.04 |
| MM USD/yr | | 0.746 | 0.746 | 0.746 | 0.746 | 0.746 | 0.746 | 0.746 | 0.746 | 0.746 |
| By Product B | | | | | | | | | | |
| Unit | | | | | | | | | | |
| Yield unit | | | | | | | | | | |
| Annual Yield | | | | | | | | | | |
| Price unit | | | | | | | | | | |
| Annual Cost | MM USD/yr | | | | | | | | | |
| **Energy Inputs** | | | | | | | | | | |
| Electricity | | N/A | N/A | N/A | N/A | N/A | N/A | N/A | N/A | N/A |
| Cost | $/kWh | | | | | | | | | |
| Amount | kWh/yr | | | | | | | | | |
| Annual Cost | MM USD/yr | | | | | | | | | |
| Fuel | | N/A | N/A | N/A | N/A | N/A | N/A | N/A | N/A | N/A |
| Type | | | | | | | | | | |
| Unit | | | | | | | | | | |
| Amount | | | | | | | | | | |
| Cost | USD/unit | | | | | | | | | |
| Annual Cost | MM USD/yr | | | | | | | | | |
| **Capital & Operating Costs** | | | | | | | | | | |
| Capital | MMUSD | 71.52 | 76.64 | 73.04 | 92.54 | 81.42 | 78.49 | 71.14 | 76.14 | 57.95 |
| Operating w/o feed | MM USD/yr | 26.8 | 13.13 | 10.7 | 18.28 | 13.9 | 14.64 | 11.76 | 10.54 | 9.08 |
| Feed | | | | | | | | | | |
| Unit | | ton | ton | ton | ton | ton | ton | ton | ton | ton |
| Cost/unit | USD/unit | 35 | 47.2 | 30.8 | 10 | 14.5 | 28.1 | 14.5 | 7.3 | 47.2 |
| Rate | unit/day | 2113 | 2113 | 2113 | 2113 | 2113 | 2113 | 2113 | 2113 | 2113 |
| Yearly Cost | MM USD/yr | 26.65 | 34.58 | 22.57 | 7.33 | 10.62 | 20.59 | 10.62 | 5.35 | 34.58 |
| **Product Plant Gate Cost** | | | | | | | | | | |
| Basis | | | | | | | | | | |
| Value | USD/gal | 0.50 | 0.58 | 0.43 | 0.38 | 0.35 | 0.46 | 0.32 | 0.26 | 0.52 |

## Table 89. Pyrolysis residual fuel oil summary yields, costs, and plant gate prices, 25 MM GPY, wood/perennials

| Technology | Units | USA | Argentina | Brazil | Canada | Caribbean Basin | China | Colombia | India | Mexico |
|---|---|---|---|---|---|---|---|---|---|---|
| **Plant Size** | MM gal/yr | 25 | 25 | 25 | 25 | 25 | 25 | 25 | 25 | 25 |
| **Yields** | | | | | | | | | | |
| Primary Product Yield | gal/ton | 136.6 | | | | | | | | |
| By Product A | | | | | | | | | | |
| Unit | | Electricity | Electricity | Electricity | Electricity | Electricity | Electricity | Electricity | Electricity | Electricity |
| Yield unit | | kWh | kWh | kWh | kWh | kWh | kWh | kWh | kWh | kWh |
| Annual Yield | | 4.71E+06 | 4.71E+06 | 4.71E+06 | 4.71E+06 | 4.71E+06 | 4.71E+06 | 4.71E+06 | 4.71E+06 | 4.71E+06 |
| Price unit | | 0.04 | 0.04 | 0.04 | 0.04 | 0.04 | 0.04 | 0.04 | 0.04 | 0.04 |
| MM USD/yr | | 0.187 | 0.187 | 0.187 | 0.187 | 0.187 | 0.187 | 0.187 | 0.187 | 0.187 |
| By Product B | | | | | | | | | | |
| Unit | | | | | | | | | | |
| Yield unit | | | | | | | | | | |
| Annual Yield | | | | | | | | | | |
| Price unit | | | | | | | | | | |
| Annual Cost | MM USD/yr | | | | | | | | | |
| **Energy Inputs** | | | | | | | | | | |
| Electricity | | N/A | N/A | N/A | N/A | N/A | N/A | N/A | N/A | N/A |
| Cost | $/kWh | | | | | | | | | |
| Amount | kWh/yr | | | | | | | | | |
| Annual Cost | MM USD/yr | | | | | | | | | |
| Fuel | | N/a | N/a | N/a | N/a | N/a | N/a | N/a | N/a | N/a |
| Type | | | | | | | | | | |
| Unit | | | | | | | | | | |
| Amount | | | | | | | | | | |
| Cost | USD/unit | | | | | | | | | |
| Annual Cost | MM USD/yr | | | | | | | | | |
| **Capital & Operating Costs** | | | | | | | | | | |
| Capital | MMUSD | 26.45 | 28.33 | 27.01 | 34.18 | 29.73 | 29.02 | 26.31 | 28.15 | 21.46 |
| Operating w/o feed | MM USD/yr | 6.26 | 5.75 | 4.08 | 9.17 | 6.24 | 6.79 | 4.86 | 3.93 | 3.14 |
| Feed | | | | | | | | | | |
| Unit | | ton | ton | ton | ton | ton | ton | ton | ton | ton |
| Cost/unit | USD/unit | 35 | 47.2 | 47.2 | 31.8 | 47.2 | 47.2 | 47.2 | 47.2 | 47.2 |
| Rate | unit/day | 529 | 529 | 529 | 529 | 529 | 529 | 529 | 529 | 529 |
| Yearly Cost | MM USD/yr | 6.42 | 8.66 | 8.66 | 5.83 | 8.66 | 8.66 | 8.66 | 8.66 | 8.66 |
| **Product Plant Gate Cost** | | | | | | | | | | |
| Basis | | | | | | | | | | |
| Value | USD/gal | 0.66 | 0.74 | 0.66 | 0.79 | 0.76 | 0.78 | 0.69 | 0.66 | 0.60 |

## Table 90. Pyrolysis residual fuel oil summary yields, costs, and plant gate prices, 100 MM GPY, wood/perennials

| Technology | Units | USA | Argentina | Brazil | Canada | Caribbean Basin | China | Colombia | India | Mexico |
|---|---|---|---|---|---|---|---|---|---|---|
| **Plant Size** | MM gal/yr | 100 | 100 | 100 | 100 | 100 | 100 | 100 | 100 | 100 |
| **Yields** | | | | | | | | | | |
| Primary Product Yield | gal/ton | 136.6 | 136.6 | 136.6 | 136.6 | 136.6 | 136.6 | 136.6 | 136.6 | 136.6 |
| By Product A | | Electricity | Electricity | Electricity | Electricity | Electricity | Electricity | Electricity | Electricity | Electricity |
| Unit | | | | | | | | | | |
| Yield unit | | kwh | kwh | kwh | kwh | kwh | kwh | kwh | kwh | kwh |
| Annual Yield | | 1.88E+07 | 1.88E+07 | 1.88E+07 | 1.88E+07 | 1.88E+07 | 1.88E+07 | 1.88E+07 | 1.88E+07 | 1.88E+07 |
| Price unit | | 0.04 | 0.04 | 0.04 | 0.04 | 0.04 | 0.04 | 0.04 | 0.04 | 0.04 |
| MM USD/yr | | 0.746 | 0.746 | 0.746 | 0.746 | 0.746 | 0.746 | 0.746 | 0.746 | 0.746 |
| By Product B | | | | | | | | | | |
| Unit | | | | | | | | | | |
| Yield unit | | | | | | | | | | |
| Annual Yield | | | | | | | | | | |
| Price unit | | | | | | | | | | |
| Annual Cost | MM USD/yr | | | | | | | | | |
| **Energy Inputs** | | | | | | | | | | |
| Electricity | | N/A | N/A | N/A | N/A | N/A | N/A | N/A | N/A | N/A |
| Cost | $/kWh | | | | | | | | | |
| Amount | kWh/yr | | | | | | | | | |
| Annual Cost | MM USD/yr | | | | | | | | | |
| Fuel | | N/A | N/A | N/A | N/A | N/A | N/A | N/A | N/A | N/A |
| Type | | | | | | | | | | |
| Unit | | | | | | | | | | |
| Amount | | | | | | | | | | |
| Cost | USD/unit | | | | | | | | | |
| Annual Cost | MM USD/yr | | | | | | | | | |
| **Capital & Operating Costs** | | | | | | | | | | |
| Capital | MMUSD | 71.52 | 76.64 | 73.04 | 92.54 | 80.42 | 78.49 | 71.14 | 73.99 | 57.95 |
| Operating w/o feed | MM USD/yr | 26.8 | 13.13 | 10.7 | 18.28 | 13.9 | 14.64 | 11.76 | 10.54 | 9.08 |
| Feed | | | | | | | | | | |
| Unit | | ton | ton | ton | ton | ton | ton | ton | ton | ton |
| Cost/unit | USD/unit | 35 | 47.2 | 47.2 | 31.8 | 47.2 | 47.2 | 47.2 | 47.2 | 47.2 |
| Rate | unit/day | 2113 | 2113 | 2113 | 2113 | 2113 | 2113 | 2113 | 2113 | 2113 |
| Yearly Cost | MM USD/yr | 26.65 | 34.58 | 34.58 | 23.3 | 34.58 | 34.58 | 34.58 | 34.58 | 34.58 |
| **Product Plant Gate Cost** | | | | | | | | | | |
| Basis | | | | | | | | | | |
| Value | USD/gal | 0.50 | 0.58 | 0.55 | 0.54 | 0.59 | 0.60 | 0.56 | 0.55 | 0.52 |

# 8. BIODIESEL

The biodiesel process used for the technoeconomic analysis is based on the model plant presented by Haas et al. (2006) for conversion of soy oil to biodiesel. The facility, Figure 26, contains three processing areas: (1) a transesterification unit to convert soy oil to fatty acid methyl esters (FAME) and crude glycerin, (2) a biodiesel purification unit, and (3) and a glycerin purification unit. For information purposes, the flow diagram also shows a dilute acid esterification unit for recycled greases.

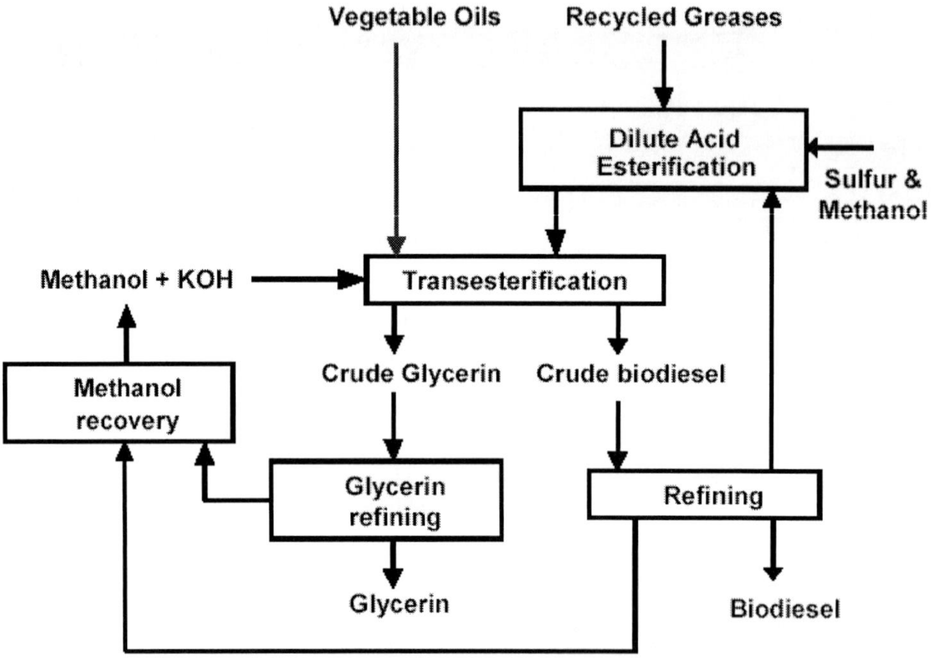


Figure 26. Biodiesel process flow diagram

**Table 91. Historic crush values**

| Year | soybeans $/bu | Meal-48% $/ton A | Oil Cents/lb | Crush $ | Beans $/ton B | Oil $/ton C | Meal/Beans A/B | Oil/Beans C/B |
|---|---|---|---|---|---|---|---|---|
| 2000 | 4.5 | 173.60 | 14.15 | 0.87 | 152.09 | 283.00 | 1.141 | 1.861 |
| 2001 | 4.4 | 167.70 | 16.46 | 1.15 | 146.73 | 329.20 | 1.143 | 2.244 |
| 2002 | 5.5 | 181.60 | 22.04 | 0.93 | 185.26 | 440.80 | 0.980 | 2.379 |
| 2003 | 4.3 | 256.05 | 29.97 | 4.64 | 145.39 | 599.40 | 1.761 | 4.123 |
| 2004 | 5.7 | 182.90 | 23.01 | 0.85 | 192.29 | 460.20 | 0.951 | 2.393 |
| 2005 | 5.6 | 172.50 | 22.00 | 0.65 | 187.60 | 440.00 | 0.919 | 2.345 |

**Table 92. Historic soybean oil to soybean price ratio based on crush**

| Year | C/B |
|---|---|
| 2000 | 1.8607 |
| 2001 | 2.2435 |
| 2002 | 2.3790 |
| 2004 | 2.3932 |
| 2005 | 2.3454 |
| AVE | **2.2444** |

The model is based on soy oil cost, but feedstock supply curves developed by ORNL are based on soybeans. To estimate the soy oil cost used as input in the cash flow model, an average soybean crush spread value was used. Soybean crush spread (Chicago Board of Trade 2006) is the dollar value quoted as the difference between the combined sales values of the products (soybean oil and soybean meal) and the cost of the raw soybeans. Crush spread is also known as "crush" and "GPM" (gross processing margin). When a bushel of soybeans weighing 60 pounds is crushed, the typical result is 11 pounds of soybean oil, 44 pounds of 48% protein soybean meal, 4 pounds of hulls, and 1 pound of waste. If the hulls are retained the result is 48 pounds of 44% soybean meal. The following equation is used to convert prices into cents per bushel.

GPM = [(price of soybean meal ($/ton) x 0.0222 +price of soybean oil (¢/lb) x 11/100] – price of soybeans ($/bu)

Historic crush values (USDA 2006) are used to estimate the ratio of oil to beans price, see Table 91 and Table 92. In the calculation of the average value to use in this study the year 2003 was considered an outlier.

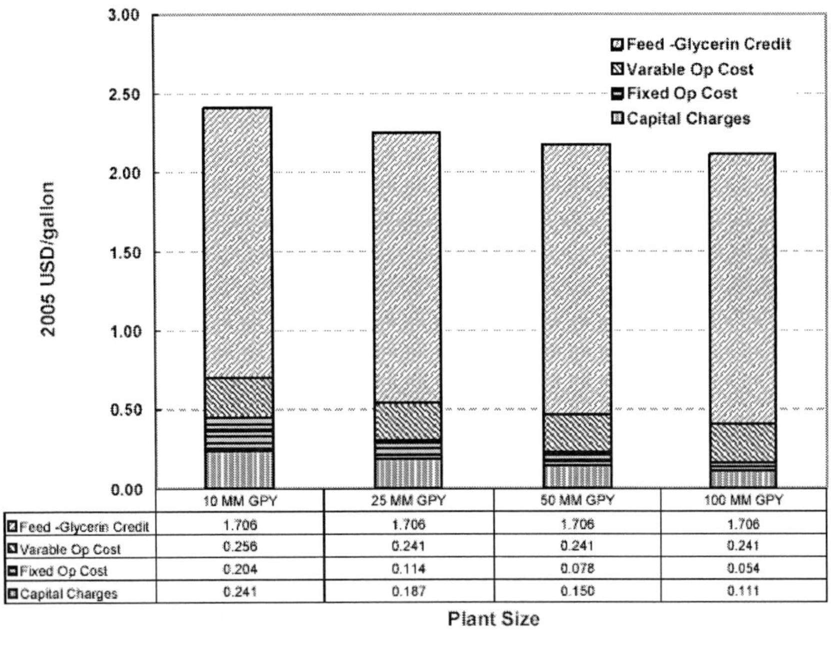


Figure 27. USA base case biodiesel plant gate prices

The capital and operating costs for the USA base case are given in Table 93. Figure 27. USA base case biodiesel plant gate prices, shows the breakdown of plant gate price as a function of plant size. Table 94 through Table 97 give capital and operating costs for selected countries. Table 98 andTable 99 give summary yields, capital costs, operating costs, and plant gate prices for the selected countries for 25 MM GPY and 100 MM GPY plants, respectively.

**Table 93. USA biodiesel capital and operating costs**

| Country | USA | Code = | 1 | | | | |
|---|---|---|---|---|---|---|---|
| | | | | USA | USA | USA | USA |
| Cost component | Units | Cost Factor | Scale Factor | Biodiesel | Biodiesel | Biodiesel | Biodiesel |
| Year $ | $ | | | 2005 | 2005 | 2005 | 2005 |
| Cost Index (2 = M&S, 3 = CE) | | | 3 | 130.93 | 130.93 | 130.93 | 130.93 |
| Plant Size | dry short ton/day | | | 118 | 295 | 589 | 1,178 |
| | dry tonne/day | | | 107 | 267 | 534 | 1,068 |
| Stream Factor | % | | | 95% | 95% | 95% | 95% |
| Feed | Dry short ton/yr | | | 3.704E+04 | 9.259E+04 | 1.852E+05 | 3.704E+05 |
| | GJ/short ton | | | 3.592E+01 | 3.592E+01 | 3.592E+01 | 3.592E+01 |
| | GJ/yr | | | 1.331E+06 | 3.326E+06 | 6.653E+06 | 1.331E+07 |
| Soy Oil Cost | USD/dt | | | **$ 480.00** | **$ 480.00** | **$ 480.00** | **$ 480.00** |
| | USD/gal | | | $ 1.84 | $ 1.84 | $ 1.84 | $ 1.84 |
| Biodiesel | gal/short ton | | | 270.0 | 270.0 | 270.0 | 270.0 |
| | GJ/yr | | | 1.366E+06 | 3.416E+06 | 6.831E+06 | 1.366E+07 |
| | gal/Stream day | | | 3.181E+04 | 7.952E+04 | 1.590E+05 | 3.181E+05 |
| | bbl/s stream day | | | 7.574E+02 | 1.893E+03 | 3.787E+03 | 7.574E+03 |
| | gal/yr | | | 1.000E+07 | 2.500E+07 | 5.000E+07 | 1.000E+08 |
| | Mil Gal/YR | | | **10** | **25** | **50** | **100** |
| Glycerin | ton/day | | | 9.44 | 23.59 | 47.18 | 94.36 |
| | ton/ton feed | | | 0.080 | 0.080 | 0.080 | 0.080 |
| | gal/yr | | | 5.65E+05 | 1.41E+06 | 2.82E+06 | 5.65E+06 |
| Process Efficiency - to biodiesel | % HHV | | | 102.7% | 102.7% | 102.7% | 102.7% |
| Process efficiency - overall | % HHV | | | | | | |
| **Capital Cost (million USD)** | | | | | | | |
| Caustic Refining | | | 0.70 | 3.00 | 6.75 | 10.97 | 17.82 |
| Feed Handling | | | 0.70 | 0.15 | 0.29 | 0.47 | 0.76 |
| Esterification | | | 0.70 | 0.42 | 0.81 | 1.31 | 2.13 |
| Separation | | | 0.70 | 2.12 | 4.02 | 6.53 | 10.62 |
| Biodiesel Post-Treatment | | | 0.70 | 1.61 | 3.06 | 4.96 | 8.06 |
| Glycerin Purification & MeOH Recovery | | | 0.70 | 1.00 | 1.90 | 3.08 | 5.01 |
| Storage Facilities | | | 0.70 | 1.44 | 2.73 | 4.43 | 7.20 |
| **Direct Fixed Capital (DFC), also called TIC** | | | | 9.74 | 19.55 | 31.76 | 51.60 |
| Engineering | DFC x MF | 0.12 | | 1.17 | 2.35 | 3.81 | 6.19 |
| Construction | DFC x MF | 0.13 | | 1.27 | 2.54 | 4.13 | 6.71 |
| Contractor & Legal | DFC x MF | 0.08 | | 0.78 | 1.56 | 2.54 | 4.13 |
| Process/Project Contingency | DFC x MF | 0.0412 | | 0.40 | 0.81 | 1.31 | 2.13 |
| **Total Plant Cost (TPC)** | | | | 13.36 | 26.81 | 43.55 | 70.75 |
| AFUDC | | | | | | | |
| **Total Plant Investment (TPI)** | | | | 13.36 | 26.81 | 43.55 | 70.75 |
| Land | | | 0.60 | 2.21 | 3.83 | 5.80 | 11.61 |
| Startup | | | | | | | |
| **Total Capital Cost (TCC)** | | | | 15.57 | 30.64 | 49.36 | 82.36 |
| **Contingency/TPI** | | | | | | | |
| Working Capital | DFC x MF | 0.05 | | 0.67 | 1.34 | 2.18 | 3.54 |
| **Variable Operating Costs (million USD/yr)** | | | | | | | |
| Feed | | | | 17.78 | 44.44 | 88.89 | 177.78 |
| Utilities | | | | 0.62 | 1.16 | 2.32 | 4.64 |
| Methanol | | | | 0.766 | 1.91 | 3.83 | 7.66 |
| Catalysts and Chemicals | | | | 1.18 | 2.94 | 5.89 | 11.78 |
| **Total** | | | | 20.34 | 50.46 | 100.93 | 201.85 |
| Fixed Operating Costs (million USD/yr) | | | | | | | |
| Labor (SF from V-R) | DFC x MF | 0.072 | 0.25 | 0.71 | 0.82 | 0.98 | 1.17 |
| Maintenance (3% of A) | | | 0.03 | 0.29 | 0.59 | 0.95 | 1.55 |
| General Overhead (65% of labor + maint) | | | 0.65 | 0.65 | 0.92 | 1.26 | 1.76 |
| Direct Overhead (45% of Labor) | | | 0.45 | 0.32 | 0.37 | 0.44 | 0.52 |
| Insurance (0.5% of TIC) | | | 0.005 | 0.08 | 0.15 | 0.25 | 0.41 |
| **Total** | | | | 2.04 | 2.85 | 3.88 | 5.42 |

## Table 94. Argentina biodiesel capital and operating costs

| Country | Argentina | Code = | 5 | | | | |
|---|---|---|---|---|---|---|---|
| | | | | Argentina | Argentina | Argentina | Argentina |
| Cost component | Units | Cost Factor | Scale Factor | Biodiesel | Biodiesel | Biodiesel | Biodiesel |
| Year $ | $ | | | 2005 | 2005 | 2005 | 2005 |
| Cost Index (2 = M&S, 3 = CE) | | | 3 | 130.93 | 130.93 | 130.93 | 130.93 |
| Plant Size | dry short ton/day | | | 118 | 295 | 589 | 1,178 |
| | dry tonne/day | | | 107 | 267 | 534 | 1,068 |
| Stream Factor | % | | | 95% | 95% | 95% | 95% |
| Feed | Dry short ton/yr | | | 3.704E+04 | 9.259E+04 | 1.852E+05 | 3.704E+05 |
| | GJ/short ton | | | 3.592E+01 | 3.592E+01 | 3.592E+01 | 3.592E+01 |
| | GJ/yr | | | 1.331E+06 | 3.326E+06 | 6.653E+06 | 1.331E+07 |
| Soy Oil Cost | USD/dt | | | **$ 157.24** | **157.24** | **$ 157.24** | **$ 157.24** |
| | USD/gal | | | $ 0.60 | 0.60 | $ 0.60 | $ 0.60 |
| Biodiesel | gal/short ton | | | 270.0 | 270.0 | 270.0 | 270.0 |
| | GJ/yr | | | 1.366E+06 | 3.416E+06 | 6.831E+06 | 1.366E+07 |
| | gal/Stream day | | | 3.181E+04 | 7.952E+04 | 1.590E+05 | 3.181E+05 |
| | bbl/s stream day | | | 7.574E+02 | 1.893E+03 | 3.787E+03 | 7.574E+03 |
| | gal/yr | | | 1.000E+07 | 2.500E+07 | 5.000E+07 | 1.000E+08 |
| | Mil Gal/YR | | | **10** | **25** | **50** | **100** |
| Glycerin | ton/day | | | 9.44 | 23.59 | 47.18 | 94.36 |
| | ton/ton feed | | | 0.080 | 0.080 | 0.080 | 0.080 |
| | gal/yr | | | 5.65E+05 | 1.41E+06 | 2.82E+06 | 5.65E+06 |
| soy beans | ton/day | | | 6.45E+02 | 1.60E+03 | 3.20E+03 | 6.40E+03 |
| | bu/day | | | 2.16E+04 | 5.37E+04 | 1.07E+05 | 2.14E+05 |
| | bu/yr | | | 6.734E+06 | 1.683E+07 | 3.367E+07 | 6.734E+07 |
| | MM$/yr | | | 12.19 | 30.47 | 60.94 | 121.88 |
| Process Efficiency - to biodiesel | % HHV | | | 102.7% | 102.7% | 102.7% | 102.7% |
| Process efficiency - overall | % HHV | | | | | | |
| **Capital Cost (million USD)** | | | | | | | |
| Caustic Refining | | | 0.70 | 3.23 | 7.25 | 11.78 | 19.14 |
| Feed Handling | | | 0.70 | 0.16 | 0.31 | 0.50 | 0.81 |
| Esterification | | | 0.70 | 0.46 | 0.87 | 1.41 | 2.28 |
| Separation | | | 0.70 | 2.27 | 4.32 | 7.02 | 11.40 |
| Biodiesel Post-Treatment | | | 0.70 | 1.73 | 3.28 | 5.33 | 8.66 |
| Glycerin Purification & MeOH Recovery | | | 0.70 | 1.07 | 2.04 | 3.31 | 5.38 |
| Storage Facilities | | | 0.70 | 1.54 | 2.93 | 4.76 | 7.73 |
| **Direct Fixed Capital (DFC), also called TIC** | | | | 10.46 | 20.99 | 34.10 | 55.40 |
| Engineering | DFC x MF | 0.12 | | 1.26 | 2.52 | 4.09 | 6.65 |
| Construction | DFC x MF | 0.13 | | 1.36 | 2.73 | 4.43 | 7.20 |
| Contractor & Legal | DFC x MF | 0.08 | | 0.84 | 1.68 | 2.73 | 4.43 |
| Process/Project Contingency | DFC x MF | 0.0412 | | 0.43 | 0.86 | 1.41 | 2.28 |
| **Total Plant Cost (TPC)** | | | | 14.35 | 28.79 | 46.76 | 75.97 |
| AFUDC | | | | | | | |
| **Total Plant Investment (TPI)** | | | | 14.35 | 28.79 | 46.76 | 75.97 |
| Land | | | 0.60 | 2.21 | 3.83 | 5.80 | 11.61 |
| Startup | | | | | | | |
| **Total Capital Cost (TCC)** | | | | 16.56 | 32.62 | 52.57 | 87.58 |
| **Contingency/TPI** | | | | | | | |
| Working Capital | DFC x MF | 0.05 | | 0.72 | 1.44 | 2.34 | 3.80 |
| **Variable Operating Costs (million USD/yr)** | | | | | | | |
| Feed | | | | 5.82 | 14.56 | 29.12 | 58.24 |
| Utilities | | | | 0.62 | 1.16 | 2.32 | 4.64 |
| Methanol | | | | 0.766 | 1.91 | 3.83 | 7.66 |
| Catalysts and Chemicals | | | | 1.18 | 2.94 | 5.89 | 11.78 |
| **Total** | | | | 8.39 | 20.58 | 41.15 | 82.31 |
| Fixed Operating Costs (million USD/yr) | | | | | | | |
| Labor (SF from V-R) | DFC x MF | 0.072 | 0.25 | 0.60 | 0.70 | 0.83 | 0.99 |
| Maintenance (3% of A) | | | 0.03 | 0.31 | 0.63 | 1.02 | 1.66 |
| General Overhead (65% of labor + maint) | | | 0.65 | 0.59 | 0.86 | 1.21 | 1.72 |
| Direct Overhead (45% of Labor) | | | 0.45 | 0.27 | 0.32 | 0.37 | 0.45 |
| Insurance (0.5% of TIC) | | | 0.005 | 0.08 | 0.16 | 0.26 | 0.44 |
| **Total** | | | | 1.86 | 2.67 | 3.70 | 5.26 |

**Table 95. Brazil biodiesel capital and operating costs**

| Country | Brazil | Code = | 2 | | | | |
|---|---|---|---|---|---|---|---|
| | | | | Brazil | Brazil | Brazil | Brazil |
| Cost component | Units | Cost Factor | Scale Factor | Biodiesel | Biodiesel | Biodiesel | Biodiesel |
| Year $ | $ | | | 2005 | 2005 | 2005 | 2005 |
| Cost Index (2 = M&S, 3 = CE) | | | 3 | 130.93 | 130.93 | 130.93 | 130.93 |
| Plant Size | dry short ton/day | | | 118 | 295 | 589 | 1,178 |
| | dry tonne/day | | | 107 | 267 | 534 | 1,068 |
| Stream Factor | % | | | 95% | 95% | 95% | 95% |
| Feed | Dry short ton/yr | | | 3.704E+04 | 9.259E+04 | 1.852E+05 | 3.704E+05 |
| | GJ/short ton | | | 3.592E+01 | 3.592E+01 | 3.592E+01 | 3.592E+01 |
| | GJ/yr | | | 1.331E+06 | 3.326E+06 | 6.653E+06 | 1.331E+07 |
| Soy Oil Cost | USD/dt | | | **$ 453.47** | **453.47** | **$ 453.47** | **$ 453.47** |
| | USD/gal | | | $ 1.74 | 1.74 | $ 1.74 | $ 1.74 |
| Biodiesel | gal/short ton | | | 270.0 | 270.0 | 270.0 | 270.0 |
| | GJ/yr | | | 1.366E+06 | 3.416E+06 | 6.831E+06 | 1.366E+07 |
| | gal/Stream day | | | 3.181E+04 | 7.952E+04 | 1.590E+05 | 3.181E+05 |
| | bbl/s stream day | | | 7.574E+02 | 1.893E+03 | 3.787E+03 | 7.574E+03 |
| | gal/yr | | | 1.000E+07 | 2.500E+07 | 5.000E+07 | 1.000E+08 |
| | Mil Gal/YR | | | **10** | **25** | **50** | **100** |
| Glycerin | ton/day | | | 9.44 | 23.59 | 47.18 | 94.36 |
| | ton/ton feed | | | 0.080 | 0.080 | 0.080 | 0.080 |
| | gal/yr | | | 5.65E+05 | 1.41E+06 | 2.82E+06 | 5.65E+06 |
| soy beans | ton/day | | | 6.45E+02 | 1.60E+03 | 3.20E+03 | 6.40E+03 |
| | bu/day | | | 2.16E+04 | 5.37E+04 | 1.07E+05 | 2.14E+05 |
| | bu/yr | | | 6.734E+06 | 1.683E+07 | 3.367E+07 | 6.734E+07 |
| | MM$/yr | | | 35.15 | 87.88 | 175.75 | 351.50 |
| Process Efficiency - to biodiesel | % HHV | | | 102.7% | 102.7% | 102.7% | 102.7% |
| Process efficiency - overall | % HHV | | | | | | |
| **Capital Cost (million USD)** | | | | | | | |
| Caustic Refining | | | 0.70 | 3.07 | 6.90 | 11.21 | 18.21 |
| Feed Handling | | | 0.70 | 0.15 | 0.29 | 0.48 | 0.77 |
| Esterification | | | 0.70 | 0.43 | 0.82 | 1.34 | 2.17 |
| Separation | | | 0.70 | 2.16 | 4.11 | 6.68 | 10.85 |
| Biodiesel Post-Treatment | | | 0.70 | 1.64 | 3.12 | 5.07 | 8.24 |
| Glycerin Purification & MeOH Recovery | | | 0.70 | 1.02 | 1.94 | 3.15 | 5.12 |
| Storage Facilities | | | 0.70 | 1.47 | 2.79 | 4.53 | 7.36 |
| **Direct Fixed Capital (DFC), also called TIC** | | | | 9.96 | 19.98 | 32.46 | 52.73 |
| Engineering | DFC x MF | 0.12 | | 1.19 | 2.40 | 3.89 | 6.33 |
| Construction | DFC x MF | 0.13 | | 1.29 | 2.60 | 4.22 | 6.85 |
| Contractor & Legal | DFC x MF | 0.08 | | 0.80 | 1.60 | 2.60 | 4.22 |
| Process/Project Contingency | DFC x MF | 0.0412 | | 0.41 | 0.82 | 1.34 | 2.17 |
| **Total Plant Cost (TPC)** | | | | 13.65 | 27.40 | 44.50 | 72.30 |
| AFUDC | | | | | | | |
| **Total Plant Investment (TPI)** | | | | 13.65 | 27.40 | 44.50 | 72.30 |
| Land | | | 0.60 | 2.21 | 3.83 | 5.80 | 11.61 |
| Startup | | | | | | | |
| **Total Capital Cost (TCC)** | | | | 15.86 | 31.23 | 50.31 | 83.91 |
| **Contingency/TPI** | | | | | | | |
| Working Capital | DFC x MF | 0.05 | | 0.68 | 1.37 | 2.23 | 3.61 |
| **Variable Operating Costs (million USD/yr)** | | | | | | | |
| Feed | | | | 16.80 | 41.99 | 83.98 | 167.95 |
| Utilities | | | | 0.62 | 1.16 | 2.32 | 4.64 |
| Methanol | | | | 0.766 | 1.91 | 3.83 | 7.66 |
| Catalysts and Chemicals | | | | 1.18 | 2.94 | 5.89 | 11.78 |
| **Total** | | | | 19.36 | 48.01 | 96.01 | 192.02 |
| Fixed Operating Costs (million USD/yr) | | | | | | | |
| Labor (SF from V-R) | DFC x MF | 0.072 | 0.25 | 0.30 | 0.35 | 0.42 | 0.50 |
| Maintenance (3% of A) | | | 0.03 | 0.30 | 0.60 | 0.97 | 1.58 |
| General Overhead (65% of labor + maint) | | | 0.65 | 0.39 | 0.62 | 0.90 | 1.35 |
| Direct Overhead (45% of Labor) | | | 0.45 | 0.14 | 0.16 | 0.19 | 0.22 |
| Insurance (0.5% of TIC) | | | 0.005 | 0.08 | 0.16 | 0.25 | 0.42 |
| **Total** | | | | 1.20 | 1.88 | 2.74 | 4.07 |

## Table 96. China biodiesel capital and operating costs

**Country** **Brazil** **Code =** 2

| Cost component | Units | Cost Factor | Scale Factor | Brazil Biodiesel | Brazil Biodiesel | Brazil Biodiesel | Brazil Biodiesel |
|---|---|---|---|---|---|---|---|
| Year $ | $ | | | 2005 | 2005 | 2005 | 2005 |
| Cost Index (2 = M&S, 3 = CE) | | | 3 | 130.93 | 130.93 | 130.93 | 130.93 |
| Plant Size | dry short ton/day | | | 118 | 295 | 589 | 1,178 |
| | dry tonne/day | | | 107 | 267 | 534 | 1,068 |
| Stream Factor | % | | | 95% | 95% | 95% | 95% |
| Feed | Dry short ton/yr | | | 3.704E+04 | 9.259E+04 | 1.852E+05 | 3.704E+05 |
| | GJ/short ton | | | 3.592E+01 | 3.592E+01 | 3.592E+01 | 3.592E+01 |
| | GJ/yr | | | 1.331E+06 | 3.326E+06 | 6.653E+06 | 1.331E+07 |
| Soy Oil Cost | USD/dt | | | **$ 453.47** | **$ 453.47** | **$ 453.47** | **$ 453.47** |
| | USD/gal | | | $ 1.74 | $ 1.74 | $ 1.74 | $ 1.74 |
| Biodiesel | gal/short ton | | | 270.0 | 270.0 | 270.0 | 270.0 |
| | GJ/yr | | | 1.366E+06 | 3.416E+06 | 6.831E+06 | 1.366E+07 |
| | gal/Stream day | | | 3.181E+04 | 7.952E+04 | 1.590E+05 | 3.181E+05 |
| | bbl/s stream day | | | 7.574E+02 | 1.893E+03 | 3.787E+03 | 7.574E+03 |
| | gal/yr | | | 1.000E+07 | 2.500E+07 | 5.000E+07 | 1.000E+08 |
| | Mil Gal/YR | | | **10** | **25** | **50** | **100** |
| Glycerin | ton/day | | | 9.44 | 23.59 | 47.18 | 94.36 |
| | ton/ton feed | | | 0.080 | 0.080 | 0.080 | 0.080 |
| | gal/yr | | | 5.65E+05 | 1.41E+06 | 2.82E+06 | 5.65E+06 |
| soy beans | ton/day | | | 6.45E+02 | 1.60E+03 | 3.20E+03 | 6.40E+03 |
| | bu/day | | | 2.16E+04 | 5.37E+04 | 1.07E+05 | 2.14E+05 |
| | bu/yr | | | 6.734E+06 | 1.683E+07 | 3.367E+07 | 6.734E+07 |
| | MM$/yr | | | 35.15 | 87.88 | 175.75 | 351.50 |
| Process Efficiency - to biodiesel | % HHV | | | 102.7% | 102.7% | 102.7% | 102.7% |
| Process efficiency - overall | % HHV | | | | | | |
| **Capital Cost (million USD)** | | | | | | | |
| Caustic Refining | | | 0.70 | 3.07 | 6.90 | 11.21 | 18.21 |
| Feed Handling | | | 0.70 | 0.15 | 0.29 | 0.48 | 0.77 |
| Esterification | | | 0.70 | 0.43 | 0.82 | 1.34 | 2.17 |
| Separation | | | 0.70 | 2.16 | 4.11 | 6.68 | 10.85 |
| Biodiesel Post-Treatment | | | 0.70 | 1.64 | 3.12 | 5.07 | 8.24 |
| Glycerin Purification & MeOH Recovery | | | 0.70 | 1.02 | 1.94 | 3.15 | 5.12 |
| Storage Facilities | | | 0.70 | 1.47 | 2.79 | 4.53 | 7.36 |
| **Direct Fixed Capital (DFC), also called TIC** | | | | 9.96 | 19.98 | 32.46 | 52.73 |
| Engineering | DFC x MF | 0.12 | | 1.19 | 2.40 | 3.89 | 6.33 |
| Construction | DFC x MF | 0.13 | | 1.29 | 2.60 | 4.22 | 6.85 |
| Contractor & Legal | DFC x MF | 0.08 | | 0.80 | 1.60 | 2.60 | 4.22 |
| Process/Project Contingency | DFC x MF | 0.0412 | | 0.41 | 0.82 | 1.34 | 2.17 |
| **Total Plant Cost (TPC)** | | | | 13.65 | 27.40 | 44.50 | 72.30 |
| AFUDC | | | | | | | |
| **Total Plant Investment (TPI)** | | | | 13.65 | 27.40 | 44.50 | 72.30 |
| Land | | | 0.60 | 2.21 | 3.83 | 5.80 | 11.61 |
| Startup | | | | | | | |
| **Total Capital Cost (TCC)** | | | | 15.86 | 31.23 | 50.31 | 83.91 |
| **Contingency/TPI** | | | | | | | |
| Working Capital | DFC x MF | 0.05 | | 0.68 | 1.37 | 2.23 | 3.61 |
| **Variable Operating Costs (million USD/yr)** | | | | | | | |
| Feed | | | | 16.80 | 41.99 | 83.98 | 167.95 |
| Utilities | | | | 0.62 | 1.16 | 2.32 | 4.64 |
| Methanol | | | | 0.766 | 1.91 | 3.83 | 7.66 |
| Catalysts and Chemicals | | | | 1.18 | 2.94 | 5.89 | 11.78 |
| **Total** | | | | 19.36 | 48.01 | 96.01 | 192.02 |
| Fixed Operating Costs (million USD/yr) | | | | | | | |
| Labor (SF from V-R) | DFC x MF | 0.072 | 0.25 | 0.30 | 0.35 | 0.42 | 0.50 |
| Maintenance (3% of A) | | | 0.03 | 0.30 | 0.60 | 0.97 | 1.58 |
| General Overhead (65% of labor + maint) | | | 0.65 | 0.39 | 0.62 | 0.90 | 1.35 |
| Direct Overhead (45% of Labor) | | | 0.45 | 0.14 | 0.16 | 0.19 | 0.22 |
| Insurance (0.5% of TIC) | | | 0.005 | 0.08 | 0.16 | 0.25 | 0.42 |
| **Total** | | | | 1.20 | 1.88 | 2.74 | 4.07 |

## Table 97. Colombia biodiesel capital and operating costs

| Country | Colombia | Code = | 6 | | | | | |
|---|---|---|---|---|---|---|---|---|
| | | | | USA | Colombia | Colombia | Colombia | Colombia |
| **Cost component** | **Units** | **Cost Factor** | **Scale Factor** | **Biodiesel** | **Biodiesel** | **Biodiesel** | **Biodiesel** | **Biodiesel** |
| Year $ | $ | | | 2002 | 2005 | 2005 | 2005 | 2005 |
| Cost Index (2 = M&S, 3 = CE) | | | 3 | 110.63 | 130.93 | 130.93 | 130.93 | 130.93 |
| Plant Size | dry short ton/day | | | 118 | 118 | 295 | 589 | 1,178 |
| | dry tonne/day | | | 107 | 107 | 267 | 534 | 1,068 |
| Stream Factor | % | | | 95% | 95% | 95% | 95% | 95% |
| Feed | Dry short ton/yr | | | 3.704E+04 | 3.704E+04 | 9.259E+04 | 1.852E+05 | 3.704E+05 |
| | GJ/short ton | | | 3.592E+01 | 3.592E+01 | 3.592E+01 | 3.592E+01 | 3.592E+01 |
| | GJ/yr | | | 1.331E+06 | 1.331E+06 | 3.326E+06 | 6.653E+06 | 1.331E+07 |
| Soy Oil Cost | USD/dt | | | **$ 155.50** | **$ 155.50** | **$ 155.50** | **$ 155.50** | **$ 155.50** |
| | USD/gal | | | $ 0.60 | $ 0.60 | $ 0.60 | $ 0.60 | $ 0.60 |
| Biodiesel | gal/short ton | | | 270.0 | 270.0 | 270.0 | 270.0 | 270.0 |
| | GJ/yr | | | 1.366E+06 | 1.366E+06 | 3.416E+06 | 6.831E+06 | 1.366E+07 |
| | gal/Stream day | | | 3.181E+04 | 3.181E+04 | 7.952E+04 | 1.590E+05 | 3.181E+05 |
| | bbl/s stream day | | | 7.574E+02 | 7.574E+02 | 1.893E+03 | 3.787E+03 | 7.574E+03 |
| | gal/yr | | | 1.000E+07 | 1.000E+07 | 2.500E+07 | 5.000E+07 | 1.000E+08 |
| | Mil Gal/YR | | | **10** | **10** | **25** | **50** | **100** |
| Glycerin | ton/day | | | 12.61 | 12.61 | 31.53 | 63.06 | 126.12 |
| | ton/ton feed | | | 0.107 | 0.107 | 0.107 | 0.107 | 0.107 |
| | gal/yr | | | 7.55E+05 | 7.55E+05 | 1.89E+06 | 3.77E+06 | 7.55E+06 |
| soy beans | ton/day | | | 6.45E+02 | 6.45E+02 | 1.60E+03 | 3.20E+03 | 6.40E+03 |
| | bu/day | | | 2.16E+04 | 2.16E+04 | 5.37E+04 | 1.07E+05 | 2.14E+05 |
| | bu/yr | | | 6.734E+06 | 6.734E+06 | 1.683E+07 | 3.367E+07 | 6.734E+07 |
| | MM$/yr | | | 12.05 | 12.05 | 30.13 | 60.27 | 120.53 |
| Process Efficiency - to biodiesel | % HHV | | | 102.7% | 102.7% | 102.7% | 102.7% | 102.7% |
| Process efficiency - overall | % HHV | | | | | | | |
| **Capital Cost (million USD)** | | | | | | | | |
| Caustic Refining | | | 0.70 | 2.539 | 2.99 | 6.72 | 10.91 | 17.72 |
| Feed Handling | | | 0.70 | 0.128 | 0.15 | 0.29 | 0.46 | 0.75 |
| Esterification | | | 0.70 | 0.359 | 0.42 | 0.80 | 1.30 | 2.11 |
| Separation | | | 0.70 | 1.790 | 2.11 | 4.00 | 6.50 | 10.56 |
| Biodiesel Post-Treatment | | | 0.70 | 1.359 | 1.60 | 3.04 | 4.94 | 8.02 |
| Glycerin Purification & MeOH Recovery | | | 0.70 | 0.844 | 0.99 | 1.89 | 3.07 | 4.98 |
| Storage Facilities | | | 0.70 | 1.214 | 1.43 | 2.71 | 4.41 | 7.16 |
| **Direct Fixed Capital (DFC), also called TIC** | | | | 8.23 | 9.69 | 19.44 | 31.58 | 51.31 |
| Engineering | DFC x MF | 0.12 | | 0.99 | 1.16 | 2.33 | 3.79 | 6.16 |
| Construction | DFC x MF | 0.13 | | 1.07 | 1.26 | 2.53 | 4.11 | 6.67 |
| Contractor & Legal | DFC x MF | 0.08 | | 0.66 | 0.78 | 1.56 | 2.53 | 4.10 |
| Process/Project Contingency | DFC x MF | 0.0412 | | 0.34 | 0.40 | 0.80 | 1.30 | 2.11 |
| **Total Plant Cost (TPC)** | | | | 11.29 | 13.29 | 26.66 | 43.31 | 70.35 |
| AFUDC | | | | | | | | |
| **Total Plant Investment (TPI)** | | | | 11.29 | 13.29 | 26.66 | 43.31 | 70.35 |
| Land | | | 0.60 | 2.21 | 2.21 | 3.83 | 5.80 | 11.61 |
| Startup | | | | | | | | |
| **Total Capital Cost (TCC)** | | | | 13.50 | 15.50 | 30.49 | 49.11 | 81.96 |
| **Contingency/TPI** | | | | | | | | |
| Working Capital | DFC x MF | 0.05 | | 0.56 | 0.66 | 1.33 | 2.17 | 3.52 |
| **Variable Operating Costs (million USD/yr)** | | | | | | | | |
| Feed | | | | 5.76 | 5.76 | 14.40 | 28.80 | 57.59 |
| Utilities | | | | 0.70 | 0.83 | 2.07 | 4.14 | 8.28 |
| Methanol | | | | 0.87 | 1.024 | 2.56 | 5.12 | 10.24 |
| Catalysts and Chemicals | | | | 1.33 | 1.57 | 3.94 | 7.87 | 15.74 |
| **Total** | | | | 8.65 | 9.19 | 22.96 | 45.93 | 91.86 |
| Fixed Operating Costs (million USD/yr) | | | | | | | | |
| Labor (SF from V-R) | DFC x MF | 0.072 | 0.25 | 0.596 | 0.45 | 0.56 | 0.67 | 0.80 |
| Maintenance (3% of A) | | | 0.03 | 0.25 | 0.29 | 0.58 | 0.95 | 1.54 |
| General Overhead (65% of labor + maint) | | | 0.65 | 0.55 | 0.48 | 0.75 | 1.05 | 1.52 |
| Direct Overhead (45% of Labor) | | | 0.45 | 0.27 | 0.20 | 0.25 | 0.30 | 0.36 |
| Insurance (0.5% of TIC) | | | 0.005 | 0.07 | 0.08 | 0.15 | 0.25 | 0.41 |
| **Total** | | | | 1.73 | 1.50 | 2.30 | 3.22 | 4.63 |

**Table 98. Summaries biodiesel summary yields, costs, and plant gate costs, 25 MM GPY**

| Technology | Units | USA | Argentina | Brazil | Canada | Caribbean Basin | China | Colombia | India | Mexico |
|---|---|---|---|---|---|---|---|---|---|---|
| **Plant Size** | MM gal/yr | 25 | 25 | 25 | | | 25 | | | |
| **Yields** | | | | | | | | | | |
| Primary Product Yield | gal/ton soy oil | 270 | 270 | 270 | | | 270 | | | |
| | gal/ton soybeans | 50 | 50 | 50 | | | 50 | | | |
| | gal/bu soybeans | 1.11 | 1.11 | 1.11 | | | 1.11 | | | |
| By Product A | | | | | | | | | | |
| Unit | | Glycerin | Glycerin | Glycerin | | | Glycerin | | | |
| Yield unit | | | | | | | | | | |
| Annual Yield | MM gpy | 1.89 | 1.89 | 1.89 | | | 1.89 | | | |
| Price unit | USD/gal | 1.27 | 1.27 | 1.27 | | | 1.27 | | | |
| MM USD/yr | | 2.4 | 2.4 | 2.4 | | | 2.4 | | | |
| By Product B | | | | | | | | | | |
| Unit | | meal | meal | meal | | | meal | | | |
| Yield unit | ton/ton beans | 0.7370 | 0.7370 | 0.7370 | | | 0.7370 | | | |
| Annual Yield | tons | 3.92E+05 | 3.92E+05 | 3.92E+05 | | | 3.92E+05 | | | |
| Price unit | USD/ton | 171.55 | 131.42 | 159.91 | | | 292.55 | | | |
| Annual Cost | MM USD/yr | 67.19 | 51.47 | 62.63 | | | 114.58 | | | |
| **Energy Inputs** | | | | | | | | | | |
| Electricity | | | | | | | | | | |
| Cost | $/kWh | 0.04 | 0.04 | 0.04 | | | 0.04 | | | |
| Amount | kWh/yr | 2.50E+06 | 2.50E+06 | 2.50E+06 | | | 2.50E+06 | | | |
| Annual Cost | MM USD/yr | 0.10 | 0.10 | 0.10 | | | 0.10 | | | |
| Fuel | | | | | | | | | | |
| Type | | Methanol | Methanol | Methanol | | | Methanol | | | |
| Unit | | gal | gal | gal | | | gal | | | |
| Amount | Mmgpy | 1.14 | 1.14 | 1.14 | | | 1.14 | | | |
| Cost | USD/unit | 0.90 | 0.90 | 0.90 | | | 0.90 | | | |
| Annual Cost | MM USD/yr | 1.024 | 1.024 | 1.024 | | | 1.024 | | | |
| **Capital & Operating Costs** | | | | | | | | | | |
| Capital | MMUSD | 30.64 | 32.62 | 31.23 | | | 33.33 | | | |
| Operating w/o feed | MM USD/yr | 11.55 | 11.35 | 10.5 | | | 11.87 | | | |
| Feed | | | | | | | | | | |
| Ton Soybean/year | ton/yr | 5.31E+05 | 5.31E+05 | 5.31E+05 | | | 5.31E+05 | | | |
| Cost soybeans | USD/bu | 5.60 | 4.29 | 5.22 | | | 9.55 | | | |
| Ton Beans/ton oil | | 5.427 | 5.427 | 5.427 | | | 5.427 | | | |
| $ton oil/$ ton beans | | 2.345 | 2.345 | 2.345 | | | 2.345 | | | |
| Unit | ton oil | 9.79E+04 | 9.79E+04 | 9.79E+04 | | | 9.79E+04 | | | |
| Cost/unit | USD/lb | 0.24 | 0.184 | 0.224 | | | 0.409 | | | |
| Rate | unit/day | 357 | 357 | 357 | | | 357 | | | |
| Beans yearly cost | MM USD/yr | 99.20 | 75.99 | 92.47 | | | 169.17 | | | |
| **Product Plant Gate Cost** | | | | | | | | | | |
| Basis | | USTL | 10 yr SL | 10 yr SL | | | 9 yr SL | | | |
| Value | USD/gal | 2.23 | 1.85 | 2.11 | | | 3.56 | | | |

**Table 99. Summaries biodiesel summary yields, costs, and plant gate costs, 100 MM GPY**

| Technology | Units | USA | Argentina | Brazil | Canada | Caribbean Basin | China | Colombia | India | Mexico |
|---|---|---|---|---|---|---|---|---|---|---|
| **Plant Size** | MM gal/yr | 100 | 100 | 100 | | | 100 | | | |
| **Yields** | | | | | | | | | | |
| Primary Product Yield | gal/ton soy oil | 270 | 270 | 270 | | | 270 | | | |
| | gal/ton soybeans | 50 | 50 | 50 | | | 50 | | | |
| | gal/bu soybeans | 1.11 | 1.11 | 1.11 | | | 1.11 | | | |
| By Product A | | | | | | | | | | |
| Unit | | glycerin | glycerin | glycerin | | | glycerin | | | |
| Yield unit | | | | | | | | | | |
| Annual Yield | MM gpy | 7.55 | 7.55 | 7.55 | | | 7.55 | | | |
| Price unit | USD/gal | 1.27 | 1.27 | 1.27 | | | 1.27 | | | |
| MM USD/yr | | 9.59 | 9.59 | 9.59 | | | 9.59 | | | |
| By Product B | | | | | | | | | | |
| Unit | | meal | meal | meal | | | meal | | | |
| Yield unit | ton/ton beans | 0.7370 | 0.7370 | 0.7370 | | | 0.7370 | | | |
| Annual Yield | tons | 1.57E+06 | 1.57E+06 | 1.57E+06 | | | 1.57E+06 | | | |
| Price unit | USD/ton | 171.55 | 131.42 | 159.91 | | | 292.55 | | | |
| Annual Cost | MM USD/yr | 268.77 | 205.89 | 250.53 | | | 458.34 | | | |
| **Energy Inputs** | | | | | | | | | | |
| Electricity | | | | | | | | | | |
| Cost | $/kWh | 0.04 | 0.04 | 0.04 | | | 0.04 | | | |
| Amount | kWh/yr | 1.00E+07 | 1.00E+07 | 1.00E+07 | | | 1.00E+07 | | | |
| Annual Cost | MM USD/yr | 0.40 | 0.40 | 0.40 | | | 0.40 | | | |
| Fuel | | | | | | | | | | |
| Type | | Methanol | Methanol | Methanol | | | Methanol | | | |
| Unit | | gal | gal | gal | | | gal | | | |
| Amount | MM gpy | 4.56 | 4.56 | 4.56 | | | 4.56 | | | |
| Cost | USD/unit | 0.90 | 0.90 | 0.90 | | | 0.90 | | | |
| Annual Cost | MM USD/yr | 4.10 | 4.10 | 4.10 | | | 4.10 | | | |
| **Capital & Operating Costs** | | | | | | | | | | |
| Capital | MMUSD | 82.36 | 87.58 | 83.91 | | | 89.46 | | | |
| Operating w/o feed | MM USD/yr | 39.86 | 39.68 | 38.42 | | | 40.45 | | | |
| Feed | | | | | | | | | | |
| Ton Soybean/year | ton/yr | 2.13E+06 | 2.13E+06 | 2.13E+06 | | | 2.13E+06 | | | |
| Cost soybeans | USD/bu | 5.60 | 4.29 | 5.22 | | | 9.55 | | | |
| Ton Beans/ton oil | | 5.427 | 5.427 | 5.427 | | | 5.427 | | | |
| $ton oil/$ ton beans | | 2.345 | 2.345 | 2.345 | | | 2.345 | | | |
| Unit | ton | 3.92E+05 | 3.92E+05 | 3.92E+05 | | | 3.92E+05 | | | |
| Cost/unit | USD/lb | 0.24 | 0.184 | 0.224 | | | 0.409 | | | |
| Rate | unit/day | 1428 | 1428 | 1428 | | | 1428 | | | |
| Beans Yearly Cost | MM USD/yr | 396.81 | 303.98 | 369.88 | | | 676.70 | | | |
| **Product Plant Gate Cost** | | | | | | | | | | |
| Basis | | USTL | 10 yr SL | 10 yr SL | | | 9 yr SL | | | |
| Value | USD/gal | 2.11 | 1.72 | 2.00 | | | 3.42 | | | |

# 9. RENEWABLE DIESEL

Renewable diesel is produced by hydrotreating oil or grease containing fatty acids, e.g., soy oil, canola oil, palm oil, or yellow grease. There are a number of commercial hydrotreating processes. The process chosen is based on the UOP Unionfining and MQD Unionfining process (Hydrocarbon Processing 2004). In the Uniofining process oxygen is removed (deoxygenation) from the bio-oil by decarboxylation (rejecting oxygen as carbon dioxide) and hydrogenation (removing oxygen as water), producing an ultra-low sulfur diesel product. Figure 28 gives the process flow diagram for the hydrotreater. Feed and a hydrogen-rich gas are contacted with a regenerable hydrotreating catalyst in vessel one (1) at typical hydrotreating conditions, e.g., temperature of 350-400 °C, and 500-2000 psia, then reactor products and unreacted hydrogen are cooled and separated (2). The liquid product is stripped to remove light components and hydrogen sulfide (3), or fractionated to recover multiple liquid products. Table 100 gives a comparison of the material balances for production of renewable diesel and biodiesel (Arena et al. 2006). Table 101 presents a comparison of the properties of renewable diesel and biodiesel (Arena et al. 2006).

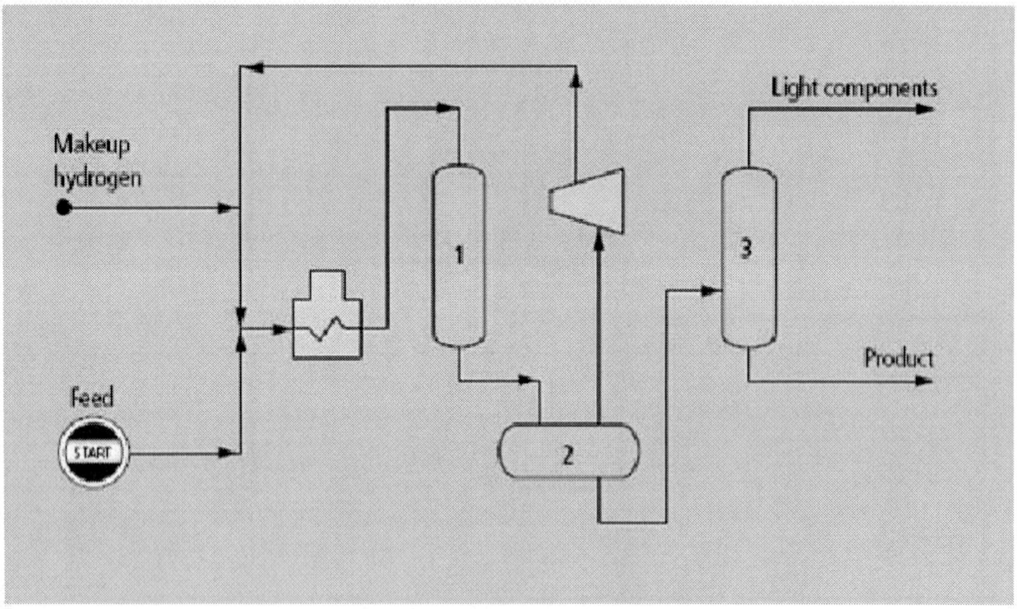


Figure 28. UOP LLC hydrotreating process flow diagram (Hydrocarbon Processing 2004)

Table 102 gives the capital and operating costs for the USA base case. In this analysis hydrogen is assumed to be purchased across the fence, so capital and operating costs associated with hydrogen production are not included. Figure 29 presents the USA case plant gate price breakdown for 0.24 USD/lb soy oil. Table 103 through Table 105give comparable costs for other countries. Table 106 and Table 107 give summary yields and cost information for 25 MM GPY and 100 MM GPY plants.

**Table 100. Renewable diesel and biodiesel yields**

| Feed | Biodiesel | Green diesel |
|---|---|---|
| % Oil or Grease | 100 | 100 |
| % H2 | | 1.5-3.8 |
| % methanol | 8.7 | |
| Products | | |
| % water,C02 | | 12-16 |
| % Lt HC | | 2-5 |
| % diesel | 96 | 83-86 |
| % glycerol | 12 | |
| Operating cost $/gal | .05 | .025 |

**Table 101. Renewable diesel and biodiesel properties**

| | Biodiesel (FAME) | Green diesel |
|---|---|---|
| %O | 11 | 0 |
| Density g/ml | .883 | .78 |
| Sulfur content | <10ppm | <10ppm |
| Heating Value (lower) MJ/kg | 38 | 44 |
| % change in NOx emission | +10 | 0 to -10 |
| Cloud Point C | -5 | -5 to -30 |
| Distillation 10-90% pt | 340-355 | 265-320 |
| Cetane | 50 | 80-90 |

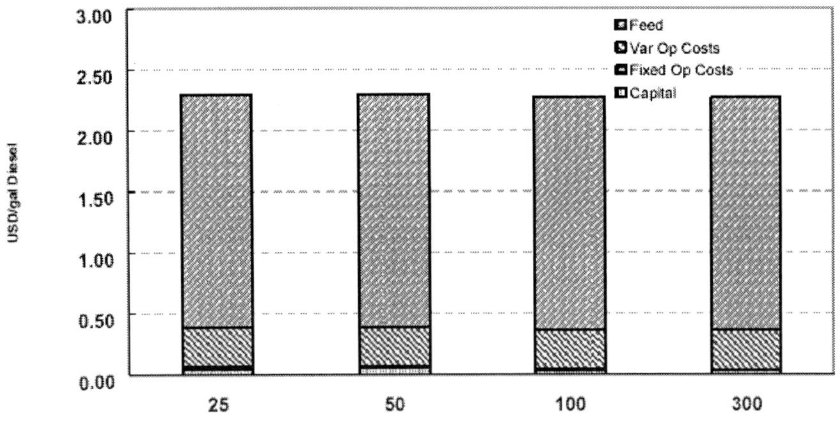


| | 25 | 50 | 100 | 300 |
|---|---|---|---|---|
| Feed | 1.91 | 1.91 | 1.91 | 1.91 |
| Var Op Costs | 0.32 | 0.32 | 0.32 | 0.32 |
| Fixed Op Costs | 0.02 | 0.01 | 0.01 | 0.01 |
| Capital | $0.05 | $0.05 | $0.03 | $0.03 |

Plant Size (Mil gal/yr)

Figure 29. USA plant gate price breakdown, soy oil at $0.24/gal

## Table 102. USA base case renewable diesel capital and operating costs

| Country | USA | Code = | 1 | | | | |
|---|---|---|---|---|---|---|---|
| **Cost component** | **Units** | **Cost Factor** | **Scale Factor** | **USA Renewable Diesel** | **USA Renewable Diesel** | **USA Renewable Diesel** | **USA Renewable Diesel** |
| Year $ | $ | | | 2005 | 2005 | 2005 | 2005 |
| Cost Index (2 = M&S, 3 = CE) | | | 3 | 130.93 | 130.93 | 130.93 | 130.93 |
| Plant Size | dry short ton/day | | | 282 | 565 | 847 | 1,130 |
| | dry tonne/day | | | 256 | 512 | 769 | 1,025 |
| Stream Factor | % | | | 95% | 95% | 95% | 95% |
| Feed | Dry short ton/yr | | | 9.792E+04 | 1.958E+05 | 2.938E+05 | 3.917E+05 |
| | GJ/short ton | | | 3.592E+01 | 3.592E+01 | 3.592E+01 | 3.592E+01 |
| | GJ/yr | | | 3.518E+06 | 7.035E+06 | 1.055E+07 | 1.407E+07 |
| Feed Cost - Soy oil | USD/dt | | | **$ 486.48** | **$ 486.48** | **$ 486.48** | **$ 486.48** |
| | Bbl/SD | | | 1,753 | 3,505 | 5,258 | 7,010 |
| Renewable Diesel | gal/short ton soybean oil | | | 255.3 | 255.3 | 255.3 | 255.3 |
| | gal/ton soybeans | | | 47.0 | 47.0 | 47.0 | 47.0 |
| Process Efficiency - to diesel | % HHV | | | 92.3% | 92.3% | 92.3% | 92.3% |
| Process Effciency - with hydrogen | | | | | | | |
| Diesel | gal/yr | | | 2.500E+07 | 5.000E+07 | 7.500E+07 | 1.000E+08 |
| | Mil Gal/YR | | | **25.00** | **50.00** | **75.00** | **100.00** |
| | tpy | | | 8.127E+04 | 1.625E+05 | 2.438E+05 | 3.251E+05 |
| | GJ/yr | | | 3.248E+06 | 6.495E+06 | 9.743E+06 | 1.299E+07 |
| | gal/Stream day | | | 7.210E+04 | 1.442E+05 | 2.163E+05 | 2.884E+05 |
| | bbl/s stream day | | | 1,717 | 3,433 | 5,150 | 6,866 |
| Electricity requirement | kWh/yr | | | 6.077E+05 | 1.215E+06 | 1.823E+06 | 2.431E+06 |
| **Capital Cost (million USD)** | | | | | | | |
| Process | | 0.9 | 0.70 | 3.73 | 6.07 | 8.06 | 9.85 |
| | | | 0.80 | | | | |
| TIC from Hydrocarbon Processing | 2367 | $/BPSD | 0.70 | | | | |
| | | | 0.85 | | | | |
| | | | 0.65 | | | | |
| | | | 0.65 | | | | |
| | | | 0.65 | | | | |
| Utiliies - 10% assumed | | 0.1 | 0.70 | 0.41 | 0.67 | 0.90 | 1.09 |
| | | | 0.70 | | | | |
| **Direct Fixed Capital (DFC), also called TIC** | | | | 4.15 | 6.74 | 8.95 | 10.95 |
| Engineering | DFC x MF | 0.12 | | 0.50 | 0.81 | 1.07 | 1.31 |
| Construction | DFC x MF | 0.13 | | 0.54 | 0.88 | 1.16 | 1.42 |
| Contractor & Legal | DFC x MF | 0.08 | | 0.33 | 0.54 | 0.72 | 0.88 |
| Process/Project Contingency | DFC x MF | 0.0412 | | 0.17 | 0.28 | 0.37 | 0.45 |
| **Total Plant Cost (TPC)** | | | | 5.69 | 9.24 | 12.27 | 15.01 |
| AFUDC | | | | | | | |
| **Total Plant Investment (TPI)** | | | | 5.69 | 9.24 | 12.27 | 15.01 |
| Land | | | 0.60 | 2.21 | 2.21 | 2.82 | 3.76 |
| Startup | | | | 0.00 | 0.00 | 0.00 | 0.00 |
| **Total Capital Cost (TCC)** | | | | 7.90 | 11.45 | 15.09 | 18.77 |
| **Contingency/TPI** | | | | | | | |
| Working Capital | DFC x MF | 0.05 | | 0.28 | 0.46 | 0.61 | 0.75 |
| **Variable Operating Costs (million USD/yr)** | | | | | | | |
| Feed | | | | 47.64 | 95.27 | 142.91 | 190.54 |
| Utilities (electricity) | | 0.04 | | 0.02 | 0.05 | 0.07 | 0.097 |
| Other | | | | 1.30 | 2.60 | 3.90 | 5.20 |
| Hydrogen | | | | 6.75 | 13.50 | 20.25 | 27.00 |
| **Total** | | | | 55.71 | 111.42 | 167.13 | 222.84 |
| Fixed Operating Costs (million USD/yr) | | | | | | | |
| Labor (SF from V-R) | DFC x MF | 0.0165 | 0.25 | 0.07 | 0.08 | 0.09 | 0.10 |
| Maintenance (3% of A) | | | 0.03 | 0.12 | 0.20 | 0.27 | 0.33 |
| General Overhead (65% of labor + maint) | | | 0.65 | 0.13 | 0.18 | 0.23 | 0.28 |
| Direct Overhead (45% of Labor) | | | 0.45 | 0.03 | 0.04 | 0.04 | 0.04 |
| Insurance (0.5% of TIC) | | | 0.005 | 0.04 | 0.06 | 0.08 | 0.09 |
| **Total** | | | | 0.39 | 0.56 | 0.71 | 0.84 |

## Table 103. Argentina renewable diesel, capital and operating costs

| **Country** | Argentina | **Code =** | 5 | | | | |
|---|---|---|---|---|---|---|---|
| **Cost component** | **Units** | **Cost Factor** | **Scale Factor** | **Argentina Renewable Diesel** | **Argentina Renewable Diesel** | **Argentina Renewable Diesel** | **Argentina Renewable Diesel** |
| Year $ | $ | | | 2005 | 2005 | 2005 | 2005 |
| Cost Index (2 = M&S, 3 = CE) | | | 3 | 130.93 | 130.93 | 130.93 | 130.93 |
| Plant Size | dry short ton/day | | | 282 | 565 | 847 | 1,130 |
| | dry tonne/day | | | 256 | 512 | 769 | 1,025 |
| Stream Factor | % | | | 95% | 95% | 95% | 95% |
| Feed | Dry short ton/yr | | | 9.792E+04 | 1.958E+05 | 2.938E+05 | 3.917E+05 |
| | GJ/short ton | | | 3.592E+01 | 3.592E+01 | 3.592E+01 | 3.592E+01 |
| | GJ/yr | | | 3.518E+06 | 7.035E+06 | 1.055E+07 | 1.407E+07 |
| Feed Cost - Soy oil | USD/dt | | | **$ 157.24** | **$ 157.24** | **$ 157.24** | **$ 157.24** |
| | Bbl/SD | | | 1,753 | 3,505 | 5,258 | 7,010 |
| Renewable Diesel | gal/short ton soybean oil | | | 255.3 | 255.3 | 255.3 | 255.3 |
| | gal/ton soybeans | | | 47.0 | 47.0 | 47.0 | 47.0 |
| Process Efficiency - to diesel | % HHV | | | 92.3% | 92.3% | 92.3% | 92.3% |
| Process Effciency - with hydrogen | | | | | | | |
| Diesel | gal/yr | | | 2.500E+07 | 5.000E+07 | 7.500E+07 | 1.000E+08 |
| | Mil Gal/YR | | | **25.00** | **50.00** | **75.00** | **100.00** |
| | tpy | | | 8.127E+04 | 1.625E+05 | 2.438E+05 | 3.251E+05 |
| | GJ/yr | | | 3.248E+06 | 6.495E+06 | 9.743E+06 | 1.299E+07 |
| | gal/Stream day | | | 7.210E+04 | 1.442E+05 | 2.163E+05 | 2.884E+05 |
| | bbl/s stream day | | | 1,717 | 3,433 | 5,150 | 6,866 |
| Electricity requirement | kWh/yr | | | 6.077E+05 | 1.215E+06 | 1.823E+06 | 2.431E+06 |
| **Capital Cost (million USD)** | | | | | | | |
| Process | | 0.9 | 0.70 | 4.01 | 6.51 | 8.65 | 10.58 |
| | | | 0.80 | | | | |
| TIC from Hydrocarbon Processing | 2367 | $/BPSD | 0.70 | | | | |
| | | | 0.85 | | | | |
| | | | 0.65 | | | | |
| | | | 0.65 | | | | |
| | | | 0.65 | | | | |
| Utiliies - 10% assumed | | 0.1 | 0.70 | 0.45 | 0.72 | 0.96 | 1.18 |
| | | | 0.70 | | | | |
| **Direct Fixed Capital (DFC), also called TIC** | | | | 4.45 | 7.24 | 9.61 | 11.75 |
| Engineering | DFC x MF | 0.12 | | 0.53 | 0.87 | 1.15 | 1.41 |
| Construction | DFC x MF | 0.13 | | 0.58 | 0.94 | 1.25 | 1.53 |
| Contractor & Legal | DFC x MF | 0.08 | | 0.36 | 0.58 | 0.77 | 0.94 |
| Process/Project Contingency | DFC x MF | 0.0412 | | 0.18 | 0.30 | 0.40 | 0.48 |
| **Total Plant Cost (TPC)** | | | | 6.11 | 9.92 | 13.18 | 16.12 |
| AFUDC | | | | | | | |
| **Total Plant Investment (TPI)** | | | | 6.11 | 9.92 | 13.18 | 16.12 |
| Land | | | 0.60 | 2.21 | 2.21 | 2.82 | 3.76 |
| Startup | | | | 0.00 | 0.00 | 0.00 | 0.00 |
| **Total Capital Cost (TCC)** | | | | 8.32 | 12.13 | 16.00 | 19.88 |
| **Contingency/TPI** | | | | | | | |
| Working Capital | DFC x MF | 0.05 | | 0.31 | 0.50 | 0.66 | 0.81 |
| **Variable Operating Costs (million USD/yr)** | | | | | | | |
| Feed | | | | 15.40 | 30.79 | 46.19 | 61.59 |
| Utilities (electricity) | | 0.04 | | 0.02 | 0.05 | 0.07 | 0.097 |
| Other | | | | 1.30 | 2.60 | 3.90 | 5.20 |
| Hydrogen | | | | 6.75 | 13.50 | 20.25 | 27.00 |
| **Total** | | | | 23.47 | 46.94 | 70.41 | 93.88 |
| Fixed Operating Costs (million USD/yr) | | | | | | | |
| Labor (SF from V-R) | DFC x MF | 0.0165 | 0.25 | 0.06 | 0.07 | 0.08 | 0.08 |
| Maintenance (3% of A) | | | 0.03 | 0.13 | 0.22 | 0.29 | 0.35 |
| General Overhead (65% of labor + maint) | | | 0.65 | 0.12 | 0.19 | 0.24 | 0.28 |
| Direct Overhead (45% of Labor) | | | 0.45 | 0.03 | 0.03 | 0.03 | 0.04 |
| Insurance (0.5% of TIC) | | | 0.005 | 0.04 | 0.06 | 0.08 | 0.10 |
| **Total** | | | | 0.38 | 0.56 | 0.72 | 0.85 |

## Table 104. Brazil renewable diesel capital and operating costs

| Country | Brazil | Code = | 2 | | | | |
|---|---|---|---|---|---|---|---|
| | | | | Brazil | Brazil | Brazil | Brazil |
| **Cost component** | **Units** | **Cost Factor** | **Scale Factor** | **Renewable Diesel** | **Renewable Diesel** | **Renewable Diesel** | **Renewable Diesel** |
| Year $ | $ | | | 2005 | 2005 | 2005 | 2005 |
| Cost Index (2 = M&S, 3 = CE) | | | 3 | 130.93 | 130.93 | 130.93 | 130.93 |
| Plant Size | dry short ton/day | | | 282 | 565 | 847 | 1,130 |
| | dry tonne/day | | | 256 | 512 | 769 | 1,025 |
| Stream Factor | % | | | 95% | 95% | 95% | 95% |
| Feed | Dry short ton/yr | | | 9.792E+04 | 1.958E+05 | 2.938E+05 | 3.917E+05 |
| | GJ/short ton | | | 3.592E+01 | 3.592E+01 | 3.592E+01 | 3.592E+01 |
| | GJ/yr | | | 3.518E+06 | 7.035E+06 | 1.055E+07 | 1.407E+07 |
| Feed Cost - Soy oil | USD/dt | | | **$ 453.47** | **$ 453.47** | **$ 453.47** | **$ 453.47** |
| | Bbl/SD | | | 1,753 | 3,505 | 5,258 | 7,010 |
| Renewable Diesel | gal/short ton soybean oil | | | 255.3 | 255.3 | 255.3 | 255.3 |
| | gal/ton soybeans | | | 47.0 | 47.0 | 47.0 | 47.0 |
| Process Efficiency - to diesel | % HHV | | | 92.3% | 92.3% | 92.3% | 92.3% |
| Process Effciency - with hydrogen | | | | | | | |
| Diesel | gal/yr | | | 2.500E+07 | 5.000E+07 | 7.500E+07 | 1.000E+08 |
| | Mil Gal/YR | | | **25.00** | **50.00** | **75.00** | **100.00** |
| | tpy | | | 8.127E+04 | 1.625E+05 | 2.438E+05 | 3.251E+05 |
| | GJ/yr | | | 3.248E+06 | 6.495E+06 | 9.743E+06 | 1.299E+07 |
| | gal/Stream day | | | 7.210E+04 | 1.442E+05 | 2.163E+05 | 2.884E+05 |
| | bbl/s stream day | | | 1,717 | 3,433 | 5,150 | 6,866 |
| Electricity requirement | kWh/yr | | | 6.077E+05 | 1.215E+06 | 1.823E+06 | 2.431E+06 |
| **Capital Cost (million USD)** | | | | | | | |
| Process | | 0.9 | 0.70 | 3.82 | 6.20 | 8.23 | 10.07 |
| | | | 0.80 | | | | |
| TIC from Hydrocarbon Processing | 2367 | $/BPSD | 0.70 | | | | |
| | | | 0.85 | | | | |
| | | | 0.65 | | | | |
| | | | 0.65 | | | | |
| | | | 0.65 | | | | |
| Utiliies - 10% assumed | | 0.1 | 0.70 | 0.42 | 0.69 | 0.91 | 1.12 |
| | | | 0.70 | | | | |
| **Direct Fixed Capital (DFC), also called TIC** | | | | 4.24 | 6.89 | 9.15 | 11.19 |
| Engineering | DFC x MF | 0.12 | | 0.51 | 0.83 | 1.10 | 1.34 |
| Construction | DFC x MF | 0.13 | | 0.55 | 0.90 | 1.19 | 1.45 |
| Contractor & Legal | DFC x MF | 0.08 | | 0.34 | 0.55 | 0.73 | 0.89 |
| Process/Project Contingency | DFC x MF | 0.0412 | | 0.17 | 0.28 | 0.38 | 0.46 |
| **Total Plant Cost (TPC)** | | | | 5.81 | 9.44 | 12.54 | 15.34 |
| AFUDC | | | | | | | |
| **Total Plant Investment (TPI)** | | | | 5.81 | 9.44 | 12.54 | 15.34 |
| Land | | | 0.60 | 2.21 | 2.21 | 2.82 | 3.76 |
| Startup | | | | 0.00 | 0.00 | 0.00 | 0.00 |
| **Total Capital Cost (TCC)** | | | | 8.02 | 11.65 | 15.36 | 19.10 |
| **Contingency/TPI** | | | | | | | |
| Working Capital | DFC x MF | 0.05 | | 0.29 | 0.47 | 0.63 | 0.77 |
| **Variable Operating Costs (million USD/yr)** | | | | | | | |
| Feed | | | | 44.40 | 88.81 | 133.21 | 177.61 |
| Utilities (electricity) | | 0.04 | | 0.02 | 0.05 | 0.07 | 0.097 |
| Other | | | | 1.30 | 2.60 | 3.90 | 5.20 |
| Hydrogen | | | | 6.75 | 13.50 | 20.25 | 27.00 |
| **Total** | | | | 52.48 | 104.95 | 157.43 | 209.91 |
| Fixed Operating Costs (million USD/yr) | | | | | | | |
| Labor (SF from V-R) | DFC x MF | 0.0165 | 0.25 | 0.03 | 0.03 | 0.04 | 0.04 |
| Maintenance (3% of A) | | | 0.03 | 0.13 | 0.21 | 0.27 | 0.34 |
| General Overhead (65% of labor + maint) | | | 0.65 | 0.10 | 0.16 | 0.20 | 0.24 |
| Direct Overhead (45% of Labor) | | | 0.45 | 0.01 | 0.02 | 0.02 | 0.02 |
| Insurance (0.5% of TIC) | | | 0.005 | 0.04 | 0.06 | 0.08 | 0.10 |
| **Total** | | | | 0.31 | 0.47 | 0.61 | 0.74 |

## Table 105. China renewable diesel capital and operating costs

| Cost component | Units | Cost Factor | Scale Factor | China Renewable Diesel | China Renewable Diesel | China Renewable Diesel | China Renewable Diesel |
|---|---|---|---|---|---|---|---|
| Country | China | Code = | 3 | | | | |
| Year $ | $ | | | 2005 | 2005 | 2005 | 2005 |
| Cost Index (2 = M&S, 3 = CE) | | | 3 | 130.93 | 130.93 | 130.93 | 130.93 |
| Plant Size | dry short ton/day | | | 282 | 565 | 847 | 1,130 |
| | dry tonne/day | | | 256 | 512 | 769 | 1,025 |
| Stream Factor | % | | | 95% | 95% | 95% | 95% |
| Feed | Dry short ton/yr | | | 9.792E+04 | 1.958E+05 | 2.938E+05 | 3.917E+05 |
| | GJ/short ton | | | 3.592E+01 | 3.592E+01 | 3.592E+01 | 3.592E+01 |
| | GJ/yr | | | 3.518E+06 | 7.035E+06 | 1.055E+07 | 1.407E+07 |
| Feed Cost - Soy oil | USD/dt | | | **$ 829.62** | **$ 829.62** | **$ 829.62** | **$ 829.62** |
| | Bbl/SD | | | 1,753 | 3,505 | 5,258 | 7,010 |
| Renewable Diesel | gal/short ton soybean oil | | | 255.3 | 255.3 | 255.3 | 255.3 |
| | gal/ton soybeans | | | 47.0 | 47.0 | 47.0 | 47.0 |
| Process Efficiency - to diesel | % HHV | | | 92.3% | 92.3% | 92.3% | 92.3% |
| Process Effciency - with hydrogen | | | | | | | |
| Diesel | gal/yr | | | 2.500E+07 | 5.000E+07 | 7.500E+07 | 1.000E+08 |
| | Mil Gal/YR | | | **25.00** | **50.00** | **75.00** | **100.00** |
| | tpy | | | 8.127E+04 | 1.625E+05 | 2.438E+05 | 3.251E+05 |
| | GJ/yr | | | 3.248E+06 | 6.495E+06 | 9.743E+06 | 1.299E+07 |
| | gal/Stream day | | | 7.210E+04 | 1.442E+05 | 2.163E+05 | 2.884E+05 |
| | bbl/s stream day | | | 1,717 | 3,433 | 5,150 | 6,866 |
| Electricity requirement | kWh/yr | | | 6.077E+05 | 1.215E+06 | 1.823E+06 | 2.431E+06 |
| **Capital Cost (million USD)** | | | | | | | |
| Process | | 0.9 | 0.70 | 4.11 | 6.67 | 8.86 | 10.84 |
| | | | 0.80 | | | | |
| TIC from Hydrocarbon Processing | 2367 | $/BPSD | 0.70 | | | | |
| | | | 0.85 | | | | |
| | | | 0.65 | | | | |
| | | | 0.65 | | | | |
| | | | 0.65 | | | | |
| Utiliies - 10% assumed | | 0.1 | 0.70 | 0.46 | 0.74 | 0.98 | 1.20 |
| | | | 0.70 | | | | |
| **Direct Fixed Capital (DFC), also called TIC** | | | | 4.57 | 7.42 | 9.85 | 12.05 |
| Engineering | DFC x MF | 0.12 | | 0.55 | 0.89 | 1.18 | 1.45 |
| Construction | DFC x MF | 0.13 | | 0.59 | 0.96 | 1.28 | 1.57 |
| Contractor & Legal | DFC x MF | 0.08 | | 0.37 | 0.59 | 0.79 | 0.96 |
| Process/Project Contingency | DFC x MF | 0.0412 | | 0.19 | 0.31 | 0.41 | 0.50 |
| **Total Plant Cost (TPC)** | | | | 6.26 | 10.17 | 13.51 | 16.52 |
| AFUDC | | | | | | | |
| **Total Plant Investment (TPI)** | | | | 6.26 | 10.17 | 13.51 | 16.52 |
| Land | | | 0.60 | 2.21 | 2.21 | 2.82 | 3.76 |
| Startup | | | | 0.00 | 0.00 | 0.00 | 0.00 |
| **Total Capital Cost (TCC)** | | | | 8.47 | 12.38 | 16.32 | 20.28 |
| **Contingency/TPI** | | | | | | | |
| Working Capital | DFC x MF | 0.05 | | 0.31 | 0.51 | 0.68 | 0.83 |
| **Variable Operating Costs (million USD/yr)** | | | | | | | |
| Feed | | | | 81.24 | 162.47 | 243.71 | 324.94 |
| Utilities (electricity) | | 0.04 | | 0.02 | 0.05 | 0.07 | 0.097 |
| Other | | | | 1.30 | 2.60 | 3.90 | 5.20 |
| Hydrogen | | | | 6.75 | 13.50 | 20.25 | 27.00 |
| **Total** | | | | 89.31 | 178.62 | 267.93 | 357.24 |
| Fixed Operating Costs (million USD/yr) | | | | | | | |
| Labor (SF from V-R) | DFC x MF | 0.0165 | 0.25 | 0.08 | 0.09 | 0.10 | 0.11 |
| Maintenance (3% of A) | | | 0.03 | 0.14 | 0.22 | 0.30 | 0.36 |
| General Overhead (65% of labor + maint) | | | 0.65 | 0.14 | 0.20 | 0.26 | 0.30 |
| Direct Overhead (45% of Labor) | | | 0.45 | 0.03 | 0.04 | 0.05 | 0.05 |
| Insurance (0.5% of TIC) | | | 0.005 | 0.04 | 0.06 | 0.08 | 0.10 |
| **Total** | | | | 0.43 | 0.62 | 0.78 | 0.92 |

## Table 106. Renewable diesel summary yields, capital and operating costs, 25 MM GPY

| | Units | USA | Argentina | Brazil | Canada | Caribbean Basin | China | Colombia | India | Mexico |
|---|---|---|---|---|---|---|---|---|---|---|
| **Plant Size** | MM gal/yr | 25 | 25 | 25 | | | 25 | | | |
| **Yields** | | | | | | | | | | |
| Primary Product Yield | gal/ton soy oil | 255.3 | 255.3 | 255.3 | | | 255.3 | | | |
| | gal/ton soybeans | 47.0 | 47.0 | 47.0 | | | 47.0 | | | |
| | gal/bu soybeans | 1.4 | 1.4 | 1.4 | | | 1.4 | | | |
| By Product A | | | | | | | | | | |
| Unit | | meal | meal | meal | | | meal | | | |
| Yield unit | ton/ton beans | 0.7370 | 0.7370 | 0.7370 | | | 0.7370 | | | |
| Annual Yield | ton/yr | 3.92E+05 | 3.92E+05 | 3.92E+05 | | | 3.92E+05 | | | |
| Price unit | USD/ton | 171.55 | 131.42 | 171.55 | | | 171.55 | | | |
| MM USD/yr | | 67.19 | 51.47 | 67.19 | | | 67.19 | | | |
| By Product B | | | | | | | | | | |
| Unit | | | | | | | | | | |
| Yield unit | | | | | | | | | | |
| Annual Yield | | | | | | | | | | |
| Price unit | | | | | | | | | | |
| Annual Cost | | | | | | | | | | |
| **Energy Inputs** | | | | | | | | | | |
| Electricity | | | | | | | | | | |
| Cost | $/kWh | 0.04 | 0.04 | 0.04 | | | 0.04 | | | |
| Amount | kWh/yr | 2.43E+06 | 2.43E+06 | 2.43E+06 | | | 2.43E+06 | | | |
| Annual Cost | MM USD/yr | 0.024 | 0.024 | 0.024 | | | 0.024 | | | |
| Fuel | | | | | | | | | | |
| Type | | Hydrogen | Hydrogen | Hydrogen | | | Hydrogen | | | |
| Unit | | ton/ton feed | ton/ton feed | ton/ton feed | | | ton/ton feed | | | |
| Amount | lb/ton soy oil | 76 | 76 | 76 | | | 76 | | | |
| Cost | USD/MMBtu | 14.92 | 14.92 | 14.92 | | | 14.92 | | | |
| Annual Cost | MM USD/yr | 6.75 | 6.75 | 6.75 | | | 6.75 | | | |
| **Capital & Operating Costs** | | | | | | | | | | |
| Capital | MMUSD | 7.90 | 8.32 | 8.02 | | | 8.47 | | | |
| Operating w/o feed | MM USD/yr | 8.46 | 8.46 | 8.38 | | | 8.50 | | | |
| Feed | | | | | | | | | | |
| Amt soybeans/yr | ton/yr | 5.31E+05 | 5.31E+05 | 5.31E+05 | | | 5.31E+05 | | | |
| Cost soybeans | USD/bu | 5.60 | 4.29 | 5.60 | | | 5.60 | | | |
| T Beans/t soyoil | ton/ton | 5.427 | 5.427 | 5.427 | | | 5.427 | | | |
| $ ton oil/$ ton beans | USD/USD | 2.345 | 2.345 | 2.345 | | | 2.345 | | | |
| Unit | ton soy oil/yr | 9.79E+04 | 9.79E+04 | 9.79E+04 | | | 9.79E+04 | | | |
| Cost/unit | USD/lb | 0.24 | 0.18 | 0.24 | | | 0.24 | | | |
| Rate | unit/day | 256 | 256 | 256 | | | 256 | | | |
| Beans yearly cost | MM USD/yr | 99.19 | 75.99 | 99.19 | | | 99.19 | | | |
| **Product Plant Gate Cost** | | | | | | | | | | |
| Basis | | USTL | 10 yr SL | 10 yr SL | | | 9 yr SL | | | |
| Value | USD/gal | 2.29 | 1.85 | 2.16 | | | 3.64 | | | |

## Table 107. Renewable diesel summary yields, capital and operating costs, 100 MM GPY

| | Units | USA | Argentina | Brazil | Canada | Caribbean Basin | China | Colombia | India | Mexico |
|---|---|---|---|---|---|---|---|---|---|---|
| **Plant Size** | MM gal/yr | 100 | 100 | 100 | | | 100 | | | |
| **Yields** | | | | | | | | | | |
| Primary Product Yield | gal/ton | 255.3 | 255.3 | 255.3 | | | 255.3 | | | |
| | gal/ton soybeans | 47.0 | | | | | | | | |
| | gal/bu soybeans | 1.4 | | | | | | | | |
| By Product A | | | | | | | | | | |
| Unit | | meal | meal | meal | | | meal | | | |
| Yield unit | ton/ton beans | 0.7370 | 0.7370 | 0.7370 | | | 0.7370 | | | |
| Annual Yield | tons | 1.57E+06 | 1.57E+06 | 1.57E+06 | | | 1.57E+06 | | | |
| Price unit | USD/ton | 171.55 | 131.42 | 159.91 | | | 292.55 | | | |
| MM USD/yr | | 268.80 | 205.92 | 250.56 | | | 458.39 | | | |
| By Product B | | | | | | | | | | |
| Unit | | N/A | N/A | N/A | | | N/A | | | |
| Yield unit | | | | | | | | | | |
| Annual Yield | | | | | | | | | | |
| Price unit | | | | | | | | | | |
| Annual Cost | MM USD/yr | | | | | | | | | |
| **Energy Inputs** | | | | | | | | | | |
| Electricity | | | | | | | | | | |
| Cost | $/kWh | 0.04 | 0.04 | 0.04 | | | 0.04 | | | |
| Amount | kWh/yr | 2.43E+06 | 2.43E+06 | 2.43E+06 | | | 2.43E+06 | | | |
| Annual Cost | MM USD/yr | 0.097 | 0.097 | 0.097 | | | 0.097 | | | |
| Fuel | | | | | | | | | | |
| Type | | Hydrogen | Hydrogen | Hydrogen | | | Hydrogen | | | |
| Unit | | ton/ton feed | ton/ton feed | ton/ton feed | | | ton/ton feed | | | |
| Amount | lb/ton soy oil | 76 | 76 | 76 | | | 76 | | | |
| Cost | USD/unit | 14.92 | 14.92 | 14.92 | | | 14.92 | | | |
| Annual Cost | MM USD/yr | 27.00 | 27.00 | 27.00 | | | 27.00 | | | |
| **Capital & Operating Costs** | | | | | | | | | | |
| Capital | MMUSD | 18.77 | 19.88 | 19.1 | | | 20.28 | | | |
| Operating w/o feed | MM USD/yr | 33.13 | 33.15 | 33.03 | | | 33.22 | | | |
| Feed | | | | | | | | | | |
| Amt soybeans/yr | ton/yr | 2.13E+06 | 2.13E+06 | 2.13E+06 | | | 2.13E+06 | | | |
| Cost soybeans | USD/bu | 5.60 | 4.29 | 5.22 | | | 9.55 | | | |
| T Beans/t soyoil | ton/ton | 5.427 | 5.427 | 5.427 | | | 5.427 | | | |
| $ ton oil/$ ton beans | $/ | 2.345 | 2.345 | 2.345 | | | 2.345 | | | |
| Unit | ton soybean oil/yr | 3.92E+05 | 3.92E+05 | 3.92E+05 | | | 3.92E+05 | | | |
| Cost/unit | USD/lb | 0.24 | 0.18 | 0.22 | | | 0.40928571 | | | |
| Rate | ton/day | 1025 | 1025 | 1025 | | | 1025 | | | |
| Beans yearly cost | MM USD/yr | 396.85 | 304.02 | 369.92 | | | 676.78 | | | |
| **Product Plant Gate Cost** | | | | | | | | | | |
| Basis | | USTL | 10 yr SL | 10 yr SL | | | 9 yr SL | | | |
| Value | USD/gal | 2.27 | 1.82 | 2.14 | | | 3.61 | | | |

**Table 108. USA base case wheat ethanol capital and operating costs**

| Country | USA | Code = | 1 | | | | |
|---|---|---|---|---|---|---|---|
| | | | | USA | USA | USA | USA |
| Cost component | Units | Cost Factor | Scale Factor | Wheat Ethanol | Wheat Ethanol | Wheat Ethanol | Wheat Ethanol |
| Year $ | $ | | | 2005 | 2005 | 2005 | 2005 |
| Cost Index (2 = M&S, 3 = CE) | | | 3 | 130.93 | 130.93 | 130.93 | 130.93 |
| Plant Size | dry short ton/day | | | 772 | 1,545 | 3,091 | 6,179 |
| | dry tonne/day | | | 700 | 1,402 | 2,804 | 5,606 |
| Stream Factor | % | | | 95% | 95% | 95% | 95% |
| Feed | Dry short ton/yr | | | 2.676E+05 | 5.359E+05 | 1.072E+06 | 2.143E+06 |
| | GJ/short ton | | | 1.698E+01 | 1.698E+01 | 1.698E+01 | 1.698E+01 |
| | GJ/yr | | | 4.542E+06 | 9.098E+06 | 1.820E+07 | 3.638E+07 |
| Feed Cost | USD/dt | | | $ 113.33 | $ 113.33 | $ 113.33 | $ 113.33 |
| Yield (gal/Dry US Ton) | | | | | | | |
| Ethanol | gal/short ton | | | 93.3 | 93.3 | 93.3 | 93.3 |
| DDGs | dry ton/ton | | | 0.37 | 0.37 | 0.37 | 0.37 |
| Process Efficiency - to ethanol | % HHV | | | 48.8% | 48.8% | 48.8% | 48.8% |
| Process efficiency - overall | % HHV | | | | | | |
| Ethanol | gal/yr | | | 2.497E+07 | 5.002E+07 | 1.000E+08 | 2.000E+08 |
| | Mil Gal/YR | | | **25.0** | **50.0** | **100.0** | **200.0** |
| | tpy | | | 8.278E+04 | 1.658E+05 | 3.316E+05 | 6.630E+05 |
| | GJ/yr | | | 2.216E+06 | 4.438E+06 | 8.875E+06 | 1.774E+07 |
| | gal/Stream day | | | 7.202E+04 | 1.442E+05 | 2.885E+05 | 5.768E+05 |
| | bbl/s stream day | | | 1.715E+03 | 3.434E+03 | 6.869E+03 | 1.373E+04 |
| DDGs | dry ton/yr | | | 9.900E+04 | 1.983E+05 | 3.965E+05 | 7.928E+05 |
| Wheat Density | lb/bu | | | 60.0 | 60.0 | 60.0 | 60.0 |
| Wheat Density | $/bu | | | **3.40** | **3.40** | **3.40** | **3.40** |
| Wheat HHV | Btu/lb | | | 8.05E+03 | 8.05E+03 | 8.05E+03 | 8.05E+03 |
| | MMBtu/ton | | | 1.61E+01 | 1.61E+01 | 1.61E+01 | 1.61E+01 |
| **Capital Cost (million USD)** | | | | | | | |
| Feed Handling | | | 0.70 | 1.95 | 3.17 | 5.15 | 8.36 |
| Saccharification | | | 0.70 | 1.61 | 2.63 | 4.27 | 6.93 |
| Fermentation | | | 0.70 | 3.53 | 5.75 | 9.34 | 15.16 |
| Distillation | | | 0.70 | 3.91 | 6.37 | 10.34 | 16.80 |
| Solid/Syrup Separation | | | 0.70 | 8.63 | 14.04 | 22.80 | 37.03 |
| Storage/Load out | | | 0.70 | 1.09 | 1.77 | 2.88 | 4.67 |
| Wastewater Treatment | | | 0.70 | 0.52 | 0.85 | 1.37 | 2.23 |
| Air compressor | | | 0.70 | 0.10 | 0.16 | 0.26 | 0.43 |
| Steam Gen & Cooling Water | | | 0.70 | 1.37 | 2.23 | 3.61 | 5.87 |
| **Direct Fixed Capital (DFC), also called TIC** | | | | 22.72 | 36.95 | 60.02 | 97.48 |
| Engineering | DFC x MF | 0.12 | | 2.73 | 4.43 | 7.20 | 11.70 |
| Construction | DFC x MF | 0.13 | | 2.95 | 4.80 | 7.80 | 12.67 |
| Contractor & Legal | DFC x MF | 0.08 | | 1.82 | 2.96 | 4.80 | 7.80 |
| Process/Project Contingency | DFC x MF | 0.0412 | | 0.94 | 1.52 | 2.47 | 4.02 |
| **Total Plant Cost (TPC)** | | | | 31.15 | 50.66 | 82.30 | 133.66 |
| AFUDC | | | | | | | |
| **Total Plant Investment (TPI)** | | | | 31.15 | 50.66 | 82.30 | 133.66 |
| Land | | | 0.60 | 2.21 | 3.35 | 5.08 | 10.16 |
| **Total Capital Cost (TCC)** | | | | 33.36 | 54.01 | 87.38 | 143.82 |
| **Contingency/TPI** | | | | | | | |
| Working Capital | DFC x MF | 0.05 | | 1.56 | 2.53 | 4.11 | 6.68 |
| **Variable Operating Costs (million USD/yr)** | | | | | | | |
| Feed | | | | 30.32 | 60.73 | 121.47 | 242.84 |
| Nat Gas, GJ/y | 8.51E+05 | 6.33 | USD/GJ | 5.39 | 10.79 | 21.59 | 43.16 |
| Electricity, kWh/y | 2.01E+07 | 0.05 | $/kWh | 1.01 | 2.01 | 4.03 | 8.05 |
| Catalysts and Chemicals, Misc | | | | 1.79 | 3.59 | 7.17 | 14.34 |
| **Total** | | | | 38.51 | 77.12 | 154.25 | 308.38 |
| Fixed Operating Costs (million USD/yr) | | | | | | | |
| Labor (SF from V-R) | DFC x MF | 0.0255 | 0.25 | 0.58 | 0.69 | 0.82 | 0.97 |
| Maintenance (3% of A) | | | 0.03 | 0.68 | 1.11 | 1.80 | 2.92 |
| General Overhead (65% of labor + maint) | | | 0.65 | 0.82 | 1.17 | 1.70 | 2.53 |
| Direct Overhead (45% of Labor) | | | 0.45 | 0.26 | 0.31 | 0.37 | 0.44 |
| Insurance (0.5% of TIC) | | | 0.005 | 0.17 | 0.27 | 0.44 | 0.72 |
| **Total** | | | | 2.51 | 3.55 | 5.13 | 7.59 |

## 10. WHEAT ETHANOL

The corn dry mill ethanol model was used to estimate capital costs, operating costs, and plant gate prices for ethanol from a wheat dry mill. The ethanol yield was estimated from Shapouri 2006, and DDG yield (0.37 ton DDG/ton wheat) from OECD 2006. The USA wheat price was taken from the USDA Wheat Yearbook (2007).

Table 108 gives the USA base case capital and operating cost summary. Figure 30 shows the breakdown of plant gate price for the USA as a function of plant size. Tables 109-111 give capital and operating costs for other countries. Table 112 and Table 113 give cost summaries for 25 and 100 MM GPY plants.

**Table 109 . Argentina wheat ethanol capital and operating costs**

| Country | Argentina | Code = | 5 | | | | |
|---|---|---|---|---|---|---|---|
| | | | | Argentina | Argentina | Argentina | Argentina |
| Cost component | Units | Cost Factor | Scale Factor | Wheat Ethanol | Wheat Ethanol | Wheat Ethanol | Wheat Ethanol |
| Year $ | $ | | | 2005 | 2005 | 2005 | 2005 |
| Cost Index (2 = M&S, 3 = CE) | | | 3 | 130.93 | 130.93 | 130.93 | 130.93 |
| Plant Size | dry short ton/day | | | 772 | 1,545 | 3,091 | 6,179 |
| | dry tonne/day | | | 700 | 1,402 | 2,804 | 5,606 |
| Stream Factor | % | | | 95% | 95% | 95% | 95% |
| Feed | Dry short ton/yr | | | 2.676E+05 | 5.359E+05 | 1.072E+06 | 2.143E+06 |
| | GJ/short ton | | | 1.698E+01 | 1.698E+01 | 1.698E+01 | 1.698E+01 |
| | GJ/yr | | | 4.542E+06 | 9.098E+06 | 1.820E+07 | 3.638E+07 |
| Feed Cost | USD/dt | | | $ 92.67 | $ 92.67 | $ 92.67 | $ 92.67 |
| Yield (gal/Dry US Ton) | | | | | | | |
| Ethanol | gal/short ton | | | 93.3 | 93.3 | 93.3 | 93.3 |
| DDGs | dry ton/ton | | | 0.37 | 0.37 | 0.37 | 0.37 |
| Process Efficiency - to ethanol | % HHV | | | 48.8% | 48.8% | 48.8% | 48.8% |
| Process efficiency - overall | % HHV | | | | | | |
| Ethanol | gal/yr | | | 2.496E+07 | 5.000E+07 | 9.999E+07 | 1.999E+08 |
| | Mil Gal/YR | | | **25.0** | **50.0** | **100.0** | **199.9** |
| | tpy | | | 8.275E+04 | 1.657E+05 | 3.315E+05 | 6.627E+05 |
| | GJ/yr | | | 2.215E+06 | 4.436E+06 | 8.872E+06 | 1.774E+07 |
| | gal/Stream day | | | 7.199E+04 | 1.442E+05 | 2.884E+05 | 5.765E+05 |
| | bbl/s stream day | | | 1.714E+03 | 3.433E+03 | 6.866E+03 | 1.373E+04 |
| DDGs | dry ton/yr | | | 9.900E+04 | 1.983E+05 | 3.965E+05 | 7.928E+05 |
| Wheat density | lb/bu | | | 60.0 | 60.0 | 60.0 | 60.0 |
| Wheat density | $/bu | | | **2.78** | **2.78** | **2.78** | **2.78** |
| Wheat HHV | Btu/lb | | | 8.05E+03 | 8.05E+03 | 8.05E+03 | 8.05E+03 |
| | MMBtu/ton | | | 1.61E+01 | 1.61E+01 | 1.61E+01 | 1.61E+01 |
| **Capital Cost (million USD)** | | | | | | | |
| Feed Handling | | | 0.70 | 2.09 | 3.40 | 5.53 | 8.97 |
| Saccharification | | | 0.70 | 1.73 | 2.82 | 4.58 | 7.44 |
| Fermentation | | | 0.70 | 3.80 | 6.17 | 10.03 | 16.28 |
| Distillation | | | 0.70 | 4.20 | 6.84 | 11.10 | 18.03 |
| Solid/Syrup Separation | | | 0.70 | 9.27 | 15.07 | 24.48 | 39.77 |
| Storage/Load out | | | 0.70 | 1.17 | 1.90 | 3.09 | 5.02 |
| Wastewater Treatment | | | 0.70 | 0.56 | 0.91 | 1.48 | 2.40 |
| Air compressor | | | 0.70 | 0.11 | 0.17 | 0.28 | 0.46 |
| Steam Gen & Cooling Water | | | 0.70 | 1.47 | 2.39 | 3.88 | 6.30 |
| **Direct Fixed Capital (DFC), also called TIC** | | | | 24.40 | 39.67 | 64.45 | 104.67 |
| Engineering | DFC x MF | 0.12 | | 2.93 | 4.76 | 7.73 | 12.56 |
| Construction | DFC x MF | 0.13 | | 3.17 | 5.16 | 8.38 | 13.61 |
| Contractor & Legal | DFC x MF | 0.08 | | 1.95 | 3.17 | 5.16 | 8.37 |
| Process/Project Contingency | DFC x MF | 0.0412 | | 1.01 | 1.63 | 2.66 | 4.31 |
| **Total Plant Cost (TPC)** | | | | 33.45 | 54.40 | 88.37 | 143.52 |
| AFUDC | | | | | | | |
| **Total Plant Investment (TPI)** | | | | 33.45 | 54.40 | 88.37 | 143.52 |
| Land | | | 0.60 | 2.21 | 3.35 | 5.08 | 10.16 |
| **Total Capital Cost (TCC)** | | | | 35.66 | 57.75 | 93.45 | 153.68 |
| **Contingency/TPI** | | | | | | | |
| Working Capital | DFC x MF | 0.05 | | 1.67 | 2.72 | 4.42 | 7.18 |
| **Variable Operating Costs (million USD/yr)** | | | | | | | |
| Feed | | | | 24.79 | 49.66 | 99.32 | 198.56 |
| Nat Gas, GJ/y | 8.51E+05 | 6.33 | USD/GJ | 5.39 | 10.79 | 21.59 | 43.16 |
| Electricity, kWh/y | 2.01E+07 | 0.05 | $/kWh | 1.01 | 2.01 | 4.03 | 8.05 |
| Catalysts and Chemicals, Misc | | | | 1.79 | 3.59 | 7.17 | 14.34 |
| **Total** | | | | 32.98 | 66.05 | 132.10 | 264.10 |
| Fixed Operating Costs (million USD/yr) | | | | | | | |
| Labor (SF from V-R) | DFC x MF | 0.0255 | 0.25 | 0.49 | 0.58 | 0.70 | 0.83 |
| Maintenance (3% of A) | | | 0.03 | 0.73 | 1.19 | 1.93 | 3.14 |
| General Overhead (65% of labor + maint) | | | 0.65 | 0.80 | 1.15 | 1.71 | 2.58 |
| Direct Overhead (45% of Labor) | | | 0.45 | 0.22 | 0.26 | 0.31 | 0.37 |
| Insurance (0.5% of TIC) | | | 0.005 | 0.18 | 0.29 | 0.47 | 0.77 |
| **Total** | | | | 2.42 | 3.48 | 5.12 | 7.69 |

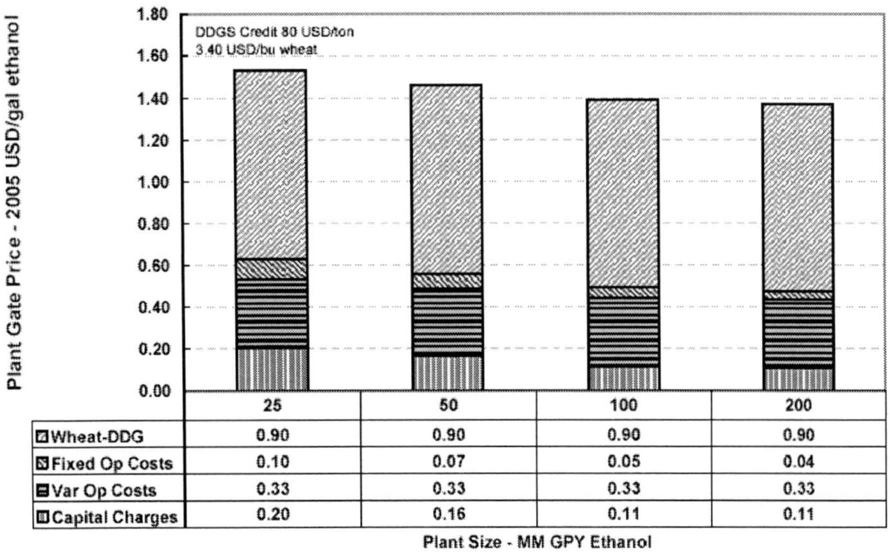


Figure 30. Plant gate price components, USA base case wheat ethanol

## Table 110. Canada wheat ethanol capital and operating costs

| Country | Canada | Code = | 7 | | | | |
|---|---|---|---|---|---|---|---|
| Cost component | Units | Cost Factor | Scale Factor | Canada Wheat Ethanol | Canada Wheat Ethanol | Canada Wheat Ethanol | Canada Wheat Ethanol |
| Year $ | $ | | | 2005 | 2005 | 2005 | 2005 |
| Cost Index (2 = M&S, 3 = CE) | | | 3 | 130.93 | 130.93 | 130.93 | 130.93 |
| Plant Size | dry short ton/day | | | 772 | 1,545 | 3,091 | 6,179 |
| | dry tonne/day | | | 700 | 1,402 | 2,804 | 5,606 |
| Stream Factor | % | | | 95% | 95% | 95% | 95% |
| Feed | Dry short ton/yr | | | 2.676E+05 | 5.359E+05 | 1.072E+06 | 2.143E+06 |
| | GJ/short ton | | | 1.688E+01 | 1.698E+01 | 1.698E+01 | 1.698E+01 |
| | GJ/yr | | | 4.516E+06 | 9.098E+06 | 1.820E+07 | 3.638E+07 |
| Feed Cost | USD/dt | | | $ 102.00 | $ 102.00 | $ 102.00 | $ 102.00 |
| Yield (gal/Dry US Ton) | | | | | | | |
| Ethanol | gal/short ton | | | 93.3 | 93.3 | 93.3 | 93.3 |
| DDGs | dry ton/ton | | | 0.37 | 0.37 | 0.37 | 0.37 |
| Process Efficiency - to ethanol | % HHV | | | 49.0% | 48.8% | 48.8% | 48.8% |
| Process efficiency - overall | % HHV | | | | | | |
| Ethanol | gal/yr | | | 2.496E+07 | 5.000E+07 | 9.999E+07 | 1.999E+08 |
| | Mil Gal/YR | | | **25.0** | **50.0** | **100.0** | **199.9** |
| | tpy | | | 8.275E+04 | 1.657E+05 | 3.315E+05 | 6.627E+05 |
| | GJ/yr | | | 2.215E+06 | 4.436E+06 | 8.872E+06 | 1.774E+07 |
| | gal/Stream day | | | 7.199E+04 | 1.442E+05 | 2.884E+05 | 5.765E+05 |
| | bbl/s stream day | | | 1.714E+03 | 3.433E+03 | 6.866E+03 | 1.373E+04 |
| DDGs | dry ton/yr | | | 9.900E+04 | 1.983E+05 | 3.965E+05 | 7.928E+05 |
| Wheat density | lb/bu | | | 60.0 | 60.0 | 60.0 | 60.0 |
| Wheat density | $/bu | | | **3.06** | **3.06** | **3.06** | **3.06** |
| Wheat HHV | Btu/lb | | | 8.00E+03 | 8.05E+03 | 8.05E+03 | 8.05E+03 |
| | MMBtu/ton | | | 1.60E+01 | 1.61E+01 | 1.61E+01 | 1.61E+01 |
| **Capital Cost (million USD)** | | | | | | | |
| Feed Handling | | | 0.70 | 2.54 | 4.13 | 6.70 | 10.89 |
| Saccharification | | | 0.70 | 2.10 | 3.42 | 5.56 | 9.03 |
| Fermentation | | | 0.70 | 4.61 | 7.49 | 12.17 | 19.76 |
| Distillation | | | 0.70 | 5.10 | 8.29 | 13.47 | 21.88 |
| Solid/Syrup Separation | | | 0.70 | 11.25 | 18.29 | 29.71 | 48.25 |
| Storage/Load out | | | 0.70 | 1.42 | 2.31 | 3.75 | 6.09 |
| Wastewater Treatment | | | 0.70 | 0.68 | 1.10 | 1.79 | 2.91 |
| Air compressor | | | 0.70 | 0.13 | 0.21 | 0.34 | 0.55 |
| Steam Gen & Cooling Water | | | 0.70 | 1.78 | 2.90 | 4.71 | 7.65 |
| **Direct Fixed Capital (DFC), also called TIC** | | | | 29.60 | 48.14 | 78.20 | 127.01 |
| Engineering | DFC x MF | 0.12 | | 3.55 | 5.78 | 9.38 | 15.24 |
| Construction | DFC x MF | 0.13 | | 3.85 | 6.26 | 10.17 | 16.51 |
| Contractor & Legal | DFC x MF | 0.08 | | 2.37 | 3.85 | 6.26 | 10.16 |
| Process/Project Contingency | DFC x MF | 0.0412 | | 1.22 | 1.98 | 3.22 | 5.23 |
| **Total Plant Cost (TPC)** | | | | 40.59 | 66.01 | 107.23 | 174.15 |
| AFUDC | | | | | | | |
| **Total Plant Investment (TPI)** | | | | 40.59 | 66.01 | 107.23 | 174.15 |
| Land | | | 0.60 | 2.21 | 3.35 | 5.08 | 10.16 |
| **Total Capital Cost (TCC)** | | | | 42.80 | 69.36 | 112.31 | 184.31 |
| **Contingency/TPI** | | | | | | | |
| Working Capital | DFC x MF | 0.05 | | 2.03 | 3.30 | 5.36 | 8.71 |
| **Variable Operating Costs (million USD/yr)** | | | | | | | |
| Feed | | | | 27.29 | 54.66 | 109.32 | 218.56 |
| Nat Gas, GJ/y | 8.51E+05 | 6.33 | USD/GJ | 5.39 | 10.79 | 21.59 | 43.16 |
| Electricity, kWh/y | 2.01E+07 | 0.05 | $/kWh | 1.01 | 2.01 | 4.03 | 8.05 |
| Catalysts and Chemicals, Misc | | | | 1.79 | 3.59 | 7.17 | 14.34 |
| **Total** | | | | 35.47 | 71.05 | 142.10 | 284.10 |
| Fixed Operating Costs (million USD/yr) | | | | | | | |
| Labor (SF from V-R) | DFC x MF | 0.0255 | 0.25 | 0.97 | 1.16 | 1.38 | 1.64 |
| Maintenance (3% of A) | | | 0.03 | 0.89 | 1.44 | 2.35 | 3.81 |
| General Overhead (65% of labor + maint) | | | 0.65 | 1.21 | 1.69 | 2.42 | 3.54 |
| Direct Overhead (45% of Labor) | | | 0.45 | 0.44 | 0.52 | 0.62 | 0.74 |
| Insurance (0.5% of TIC) | | | 0.005 | 0.21 | 0.35 | 0.56 | 0.92 |
| **Total** | | | | 3.72 | 5.16 | 7.32 | 10.64 |

## Table 111. China wheat ethanol capital and operating costs

| Country | China | Code = | 3 | | | | |
|---|---|---|---|---|---|---|---|
| **Cost component** | **Units** | **Cost Factor** | **Scale Factor** | **China Dry Mill Ethanol** | **China Dry Mill Ethanol** | **China Dry Mill Ethanol** | **China Dry Mill Ethanol** |
| Year $ | $ | | | 2005 | 2005 | 2005 | 2005 |
| Cost Index (2 = M&S, 3 = CE) | | | 3 | 130.93 | 130.93 | 130.93 | 130.93 |
| Plant Size | dry short ton/day | | | 772 | 1,545 | 3,091 | 6,179 |
| | dry tonne/day | | | 700 | 1,402 | 2,804 | 5,606 |
| Stream Factor | % | | | 95% | 95% | 95% | 95% |
| Feed | Dry short ton/yr | | | 2.676E+05 | 5.359E+05 | 1.072E+06 | 2.143E+06 |
| | GJ/short ton | | | 1.698E+01 | 1.698E+01 | 1.698E+01 | 1.698E+01 |
| | GJ/yr | | | 4.542E+06 | 9.098E+06 | 1.820E+07 | 3.638E+07 |
| Feed Cost | USD/dt | | | $ 159.00 | $ 159.00 | $ 159.00 | $ 159.00 |
| Yield (gal/Dry US Ton) | | | | | | | |
| Ethanol | gal/short ton | | | 93.3 | 93.3 | 93.3 | 93.3 |
| DDGs | dry ton/ton | | | 0.37 | 0.37 | 0.37 | 0.37 |
| Process Efficiency - to ethanol | % HHV | | | 48.8% | 48.8% | 48.8% | 48.8% |
| Process efficiency - overall | % HHV | | | | | | |
| Ethanol | gal/yr | | | 2.496E+07 | 5.000E+07 | 9.999E+07 | 1.999E+08 |
| | Mil Gal/YR | | | **25.0** | **50.0** | **100.0** | **199.9** |
| | tpy | | | 8.275E+04 | 1.657E+05 | 3.315E+05 | 6.627E+05 |
| | GJ/yr | | | 2.215E+06 | 4.436E+06 | 8.872E+06 | 1.774E+07 |
| | gal/Stream day | | | 7.199E+04 | 1.442E+05 | 2.884E+05 | 5.765E+05 |
| | bbl/s stream day | | | 1.714E+03 | 3.433E+03 | 6.866E+03 | 1.373E+04 |
| DDGs | dry ton/yr | | | 9.900E+04 | 1.983E+05 | 3.965E+05 | 7.928E+05 |
| Wheat density | lb/bu | | | 60.0 | 60.0 | 60.0 | 60.0 |
| Wheat density | $/bu | | | **4.77** | **4.77** | **4.77** | **4.77** |
| Wheat HHV | Btu/lb | | | 8.05E+03 | 8.05E+03 | 8.05E+03 | 8.05E+03 |
| | MMBtu/ton | | | 1.61E+01 | 1.61E+01 | 1.61E+01 | 1.61E+01 |
| **Capital Cost (million USD)** | | | | | | | |
| Feed Handling | | | 0.70 | 2.14 | 3.49 | 5.66 | 9.20 |
| Saccharification | | | 0.70 | 1.78 | 2.89 | 4.69 | 7.62 |
| Fermentation | | | 0.70 | 3.89 | 6.33 | 10.28 | 16.69 |
| Distillation | | | 0.70 | 4.31 | 7.01 | 11.38 | 18.48 |
| Solid/Syrup Separation | | | 0.70 | 9.50 | 15.45 | 25.09 | 40.75 |
| Storage/Load out | | | 0.70 | 1.20 | 1.95 | 3.17 | 5.14 |
| Wastewater Treatment | | | 0.70 | 0.57 | 0.93 | 1.51 | 2.46 |
| Air compressor | | | 0.70 | 0.11 | 0.18 | 0.29 | 0.47 |
| Steam Gen & Cooling Water | | | 0.70 | 1.51 | 2.45 | 3.98 | 6.46 |
| **Direct Fixed Capital (DFC), also called TIC** | | | | 25.00 | 40.66 | 66.05 | 107.27 |
| Engineering | DFC x MF | 0.12 | | 3.00 | 4.88 | 7.93 | 12.87 |
| Construction | DFC x MF | 0.13 | | 3.25 | 5.29 | 8.59 | 13.95 |
| Contractor & Legal | DFC x MF | 0.08 | | 2.00 | 3.25 | 5.28 | 8.58 |
| Process/Project Contingency | DFC x MF | 0.0412 | | 1.03 | 1.68 | 2.72 | 4.42 |
| **Total Plant Cost (TPC)** | | | | 34.28 | 55.75 | 90.57 | 147.09 |
| AFUDC | | | | | | | |
| **Total Plant Investment (TPI)** | | | | 34.28 | 55.75 | 90.57 | 147.09 |
| Land | | | 0.60 | 2.21 | 3.35 | 5.08 | 10.16 |
| **Total Capital Cost (TCC)** | | | | 36.49 | 59.10 | 95.65 | 157.25 |
| **Contingency/TPI** | | | | | | | |
| Working Capital | DFC x MF | 0.05 | | 1.71 | 2.79 | 4.53 | 7.35 |
| **Variable Operating Costs (million USD/yr)** | | | | | | | |
| Feed | | | | 42.54 | 85.20 | 170.41 | 340.70 |
| Nat Gas, GJ/y | 8.51E+05 | 6.33 | USD/GJ | 5.39 | 10.79 | 21.59 | 43.16 |
| Electricity, kWh/y | 2.01E+07 | 0.05 | $/kWh | 1.01 | 2.01 | 4.03 | 8.05 |
| Catalysts and Chemicals, Misc | | | | 1.79 | 3.59 | 7.17 | 14.34 |
| **Total** | | | | 50.73 | 101.60 | 203.19 | 406.24 |
| Fixed Operating Costs (million USD/yr) | | | | | | | |
| Labor (SF from V-R) | DFC x MF | 0.0255 | 0.25 | 0.64 | 0.77 | 0.91 | 1.08 |
| Maintenance (3% of A) | | | 0.03 | 0.75 | 1.22 | 1.98 | 3.22 |
| General Overhead (65% of labor + maint) | | | 0.65 | 0.91 | 1.29 | 1.88 | 2.80 |
| Direct Overhead (45% of Labor) | | | 0.45 | 0.29 | 0.34 | 0.41 | 0.49 |
| Insurance (0.5% of TIC) | | | 0.005 | 0.18 | 0.30 | 0.48 | 0.79 |
| **Total** | | | | 2.77 | 3.92 | 5.66 | 8.37 |

## Table 112. Wheat ethanol summary yields, capital and operating costs, 25 MM GPY

| Ethanol-Wheat Dry Mill | Units | USA | Argentina | Brazil | Canada | Caribbean Basin | China | Columbia | India | Mexico |
|---|---|---|---|---|---|---|---|---|---|---|
| **Plant Size** | MM gal/yr | 25 | 25 | | 25 | | 25 | | | |
| **Yields** | | | | | | | | | | |
| Primary Product Yield | gal/ton | 93.3 | 93.3 | | 93.3 | | 93.3 | | | |
| By Product A | | DDG | DDG | | DDG | | DDG | | | |
| Unit | | ton | ton | | ton | | ton | | | |
| Annual Yield | | 9.90E+04 | 9.90E+04 | | 9.90E+04 | | 9.90E+04 | | | |
| Price unit | | 80 | 80 | | 80 | | 80 | | | |
| MM USD/yr | | 7.92 | 7.92 | | 7.92 | | 7.92 | | | |
| By Product B | | | | | | | | | | |
| Unit | | | | | | | | | | |
| Annual Yield | | | | | | | | | | |
| Price unit | | | | | | | | | | |
| Annual Cost | MM USD/yr | | | | | | | | | |
| **Energy Inputs** | | | | | | | | | | |
| Electricity | | | | | | | | | | |
| Cost | $/kWh | 0.05 | 0.05 | | 0.05 | | 0.05 | | | |
| Amount | kWh/yr | 2.01E+07 | 2.01E+07 | | 2.01E+07 | | 2.01E+07 | | | |
| Annual Cost | MM USD/yr | 5.39E+00 | 5.39E+00 | | 5.39E+00 | | 5.39E+00 | | | |
| Fuel | | | | | | | | | | |
| Type | | Nat Gas | Nat Gas | | Nat Gas | | Nat Gas | | | |
| Unit | | GJ | GJ | | GJ | | GJ | | | |
| Amount | | 8.51E+05 | 8.51E+05 | | 8.51E+05 | | 8.51E+05 | | | |
| Cost | USD/unit | 6.33 | 6.33 | | 6.33 | | 6.33 | | | |
| Annual Cost | MM USD/yr | 5.39 | 5.39 | | 5.39 | | 5.39 | | | |
| **Capital & Operating Costs** | | | | | | | | | | |
| Capital | MMUSD | 33.36 | 35.66 | | 42.80 | | 36.49 | | | |
| Operating w/o feed | MM USD/yr | 10.69 | 10.6 | | 11.9 | | 10.96 | | | |
| Feed | | | | | | | | | | |
| Unit | | ton | | | | | | | | |
| Cost/unit | USD/unit | 113.33 | 92.67 | | 102.00 | | 159.00 | | | |
| | USD/bu | 3.40 | 2.78 | | 3.06 | | 4.77 | | | |
| Rate | tonne/day | 700 | 700 | | 700 | | 700 | | | |
| Yearly Cost | MM USD/yr | 30.32 | 24.79 | | 27.29 | | 42.54 | | | |
| **Product Plant Gate Cost** | | | | | | | | | | |
| Basis | | USTL | 10 SL | | 10 SL | | 10 SL | | | |
| Value | USD/gal | 1.53 | 1.31 | | 1.49 | | 2.03 | | | |

## Table 113. Wheat ethanol summary yields, capital and operating costs, 100 MM GPY

| Ethanol-Wheat Dry Mill | Units | USA | Argentina | Brazil | Canada | Caribbean Basin | China | Columbia | India | Mexico |
|---|---|---|---|---|---|---|---|---|---|---|
| **Plant Size** | MM gal/yr | 100 | 100 | | 100 | | 100 | | | |
| **Yields** | | | | | | | | | | |
| Primary Product Yield | gal/ton | 101.8 | 101.8 | | 101.8 | | 101.8 | | | |
| By Product A | | | | | | | | | | |
| Unit | | DDG | DDG | | DDG | | DDG | | | |
| Yield unit | | ton | ton | | ton | | ton | | | |
| Annual Yield | | 7.93E+05 | 7.93E+05 | | 7.93E+05 | | 7.93E+05 | | | |
| Price unit | | 80.00 | 80.00 | | 80.00 | | 80.00 | | | |
| MM USD/yr | | 63.42 | 63.42 | | 63.42 | | 63.42 | | | |
| By Product B | | | | | | | | | | |
| Yield unit | | | | | | | | | | |
| Annual Yield | | | | | | | | | | |
| Price unit | | | | | | | | | | |
| Annual Cost | MM USD/yr | | | | | | | | | |
| **Energy Inputs** | | | | | | | | | | |
| Electricity | | | | | | | | | | |
| Cost | $/kWh | 0.05 | 0.05 | | 0.05 | | 0.05 | | | |
| Amount | kWh/yr | 8.04E+07 | 8.04E+07 | | 8.04E+07 | | 8.04E+07 | | | |
| Annual Cost | MM USD/yr | 4.02 | 4.02 | | 4.02 | | 4.02 | | | |
| Fuel | | | | | | | | | | |
| Type | | Nat Gas | Nat Gas | | Nat Gas | | Nat Gas | | | |
| Unit | | GJ/yr | GJ/yr | | GJ/yr | | GJ/yr | | | |
| Amount | | 3.40E+06 | 3.40E+06 | | 3.40E+06 | | 3.40E+06 | | | |
| Cost | USD/unit | 6.33 | 6.33 | | 6.33 | | 6.33 | | | |
| Annual Cost | MM USD/yr | 21.67 | 21.67 | | 21.67 | | 21.67 | | | |
| **Capital & Operating Costs** | | | | | | | | | | |
| Capital | MMUSD | 87.38 | 93.45 | | 112.31 | | 85.65 | | | |
| Operating w/o feed | MM USD/yr | 73.13 | 37.9 | | | | 38.44 | | | |
| Feed | | | | | | | | | | |
| Unit | tonne/day | | | | | | | | | |
| Cost/unit | USD/unit | 113.33 | 92.67 | | 102.00 | | 159.00 | | | |
| | USD/bu | 3.40 | 2.78 | | 3.06 | | 4.77 | | | |
| Rate | unit/day | 2804 | 2804 | | 2804 | | 2804 | | | |
| Yearly Cost | MM USD/yr | 121.47 | 99.32 | | 109.32 | | 170.41 | | | |
| **Product Plant Gate Cost** | | | | | | | | | | |
| Basis | | USTL | 10 SL | | 10 SL | | 10 SL | | | |
| Value | USD/gal | 1.39 | 1.20 | | 1.31 | | 1.89 | | | |

# 11. Palm Oil Biodiesel

The palm oil biodiesel estimates were made assuming that yields are equal to yields from soybean oil. The soybean spreadsheet was modified to an oil basis instead of a soybean basis. The capital and operating cost estimate for the Caribbean Basin is given in Table 114; and Colombia is given in Table 115. Cost summaries for 25 and 100 MM GPY plants are given in Table 116 and Table 117.

**Table 114. Caribbean basin palm oil biodiesel capital and operating costs**

**Country** **Caribbean** **Code =** 9

| Cost component | Units | Cost Factor | Scale Factor | Caribbean Palm Oil Biodiesel | Caribbean Palm Oil Biodiesel | Caribbean Palm Oil Biodiesel | Caribbean Palm Oil Biodiesel |
|---|---|---|---|---|---|---|---|
| Year $ | $ | | | 2005 | 2005 | 2005 | 2005 |
| Cost Index (2 = M&S, 3 = CE) | | | 3 | 130.93 | 130.93 | 130.93 | 130.93 |
| Plant Size | dry short ton/day | | | 118 | 295 | 589 | 1,178 |
| | dry tonne/day | | | 107 | 267 | 534 | 1,068 |
| Stream Factor | % | | | 95% | 95% | 95% | 95% |
| Feed | Dry short ton/yr | | | 3.704E+04 | 9.259E+04 | 1.852E+05 | 3.704E+05 |
| | GJ/short ton | | | 3.592E+01 | 3.592E+01 | 3.592E+01 | 3.592E+01 |
| | GJ/yr | | | 1.331E+06 | 3.326E+06 | 6.653E+06 | 1.331E+07 |
| Palm Oil Cost | USD/dt | | | **$ 275.00** | **$ 275.00** | **$ 275.00** | **$ 275.00** |
| | USD/gal | | | $ 1.06 | $ 1.06 | $ 1.06 | $ 1.06 |
| Biodiesel | gal/short ton | | | 270.0 | 270.0 | 270.0 | 270.0 |
| | GJ/yr | | | 1.366E+06 | 3.416E+06 | 6.831E+06 | 1.366E+07 |
| | gal/Stream day | | | 3.181E+04 | 7.952E+04 | 1.590E+05 | 3.181E+05 |
| | bbl/s stream day | | | 7.574E+02 | 1.893E+03 | 3.787E+03 | 7.574E+03 |
| | gal/yr | | | 1.000E+07 | 2.500E+07 | 5.000E+07 | 1.000E+08 |
| | Mil Gal/YR | | | **10** | **25** | **50** | **100** |
| Glycerin | ton/day | | | 12.61 | 31.53 | 63.06 | 126.12 |
| | ton/ton feed | | | 0.107 | 0.107 | 0.107 | 0.107 |
| | gal/yr | | | 7.55E+05 | 1.89E+06 | 3.77E+06 | 7.55E+06 |
| Process Efficiency - to biodiesel | % HHV | | | 102.7% | 102.7% | 102.7% | 102.7% |
| Process efficiency - overall | % HHV | | | | | | |
| **Capital Cost (million USD)** | | | | | | | |
| Caustic Refining | | | 0.70 | 3.39 | 7.62 | 12.38 | 20.11 |
| Feed Handling | | | 0.70 | 0.17 | 0.32 | 0.53 | 0.86 |
| Esterification | | | 0.70 | 0.48 | 0.91 | 1.48 | 2.40 |
| Separation | | | 0.70 | 2.39 | 4.54 | 7.37 | 11.98 |
| Biodiesel Post-Treatment | | | 0.70 | 1.82 | 3.45 | 5.60 | 9.10 |
| Glycerin Purification & MeOH Recovery | | | 0.70 | 1.13 | 2.14 | 3.48 | 5.65 |
| Storage Facilities | | | 0.70 | 1.62 | 3.08 | 5.00 | 8.12 |
| **Direct Fixed Capital (DFC), also called TIC** | | | | 10.99 | 22.06 | 35.84 | 58.21 |
| Engineering | DFC x MF | 0.12 | | 1.32 | 2.65 | 4.30 | 6.99 |
| Construction | DFC x MF | 0.13 | | 1.43 | 2.87 | 4.66 | 7.57 |
| Contractor & Legal | DFC x MF | 0.08 | | 0.88 | 1.76 | 2.87 | 4.66 |
| Process/Project Contingency | DFC x MF | 0.0412 | | 0.45 | 0.91 | 1.48 | 2.40 |
| **Total Plant Cost (TPC)** | | | | 15.07 | 30.25 | 49.14 | 79.82 |
| AFUDC | | | | | | | |
| **Total Plant Investment (TPI)** | | | | 15.07 | 30.25 | 49.14 | 79.82 |
| Land | | | 0.60 | 2.21 | 3.83 | 5.80 | 11.61 |
| Startup | | | | | | | |
| **Total Capital Cost (TCC)** | | | | 17.28 | 34.08 | 54.94 | 91.43 |
| **Contingency/TPI** | | | | | | | |
| Working Capital | DFC x MF | 0.05 | | 0.75 | 1.51 | 2.46 | 3.99 |
| **Variable Operating Costs (million USD/yr)** | | | | | | | |
| Feed | | | | 10.19 | 25.46 | 50.93 | 101.85 |
| Utilities | | | | 0.83 | 2.07 | 4.14 | 8.28 |
| Methanol | | | | 1.024 | 2.56 | 5.12 | 10.24 |
| Catalysts and Chemicals | | | | 1.57 | 3.94 | 7.87 | 15.74 |
| **Total** | | | | 13.61 | 34.03 | 68.06 | 136.11 |
| Fixed Operating Costs (million USD/yr) | | | | | | | |
| Labor (SF from V-R) | DFC x MF | 0.072 | 0.25 | 0.68 | 0.85 | 1.01 | 1.20 |
| Maintenance (3% of A) | | | 0.03 | 0.33 | 0.66 | 1.08 | 1.75 |
| General Overhead (65% of labor + maint) | | | 0.65 | 0.65 | 0.98 | 1.36 | 1.92 |
| Direct Overhead (45% of Labor) | | | 0.45 | 0.30 | 0.38 | 0.46 | 0.54 |
| Insurance (0.5% of TIC) | | | 0.005 | 0.09 | 0.17 | 0.27 | 0.46 |
| **Total** | | | | 2.05 | 3.05 | 4.18 | 5.87 |

## Table 115. Colombia palm oil biodiesel capital and operating costs

**Country** **Colombia** **Code =** 6

| Cost component | Units | Cost Factor | Scale Factor | Colombia Palm Oil Biodiesel | Colombia Palm Oil Biodiesel | Colombia Palm Oil Biodiesel | Colombia Palm Oil Biodiesel |
|---|---|---|---|---|---|---|---|
| Year $ | $ | | | 2005 | 2005 | 2005 | 2005 |
| Cost Index (2 = M&S, 3 = CE) | | | 3 | 130.93 | 130.93 | 130.93 | 130.93 |
| Plant Size | dry short ton/day | | | 118 | 295 | 589 | 1,178 |
| | dry tonne/day | | | 107 | 267 | 534 | 1,068 |
| Stream Factor | % | | | 95% | 95% | 95% | 95% |
| Feed | Dry short ton/yr | | | 3.704E+04 | 9.259E+04 | 1.852E+05 | 3.704E+05 |
| | GJ/short ton | | | 3.592E+01 | 3.592E+01 | 3.592E+01 | 3.592E+01 |
| | GJ/yr | | | 1.331E+06 | 3.326E+06 | 6.653E+06 | 1.331E+07 |
| Palm Oil Cost | USD/dt | | | **$ 406.00** | **$ 406.00** | **$ 406.00** | **$ 406.00** |
| | USD/gal | | | $ 1.56 | $ 1.56 | $ 1.56 | $ 1.56 |
| Biodiesel | gal/short ton | | | 270.0 | 270.0 | 270.0 | 270.0 |
| | GJ/yr | | | 1.366E+06 | 3.416E+06 | 6.831E+06 | 1.366E+07 |
| | gal/Stream day | | | 3.181E+04 | 7.952E+04 | 1.590E+05 | 3.181E+05 |
| | bbl/s stream day | | | 7.574E+02 | 1.893E+03 | 3.787E+03 | 7.574E+03 |
| | gal/yr | | | 1.000E+07 | 2.500E+07 | 5.000E+07 | 1.000E+08 |
| | Mil Gal/YR | | | **10** | **25** | **50** | **100** |
| Glycerin | ton/day | | | 12.61 | 31.53 | 63.06 | 126.12 |
| | ton/ton feed | | | 0.107 | 0.107 | 0.107 | 0.107 |
| | gal/yr | | | 7.55E+05 | 1.89E+06 | 3.77E+06 | 7.55E+06 |
| Process Efficiency - to biodiesel | % HHV | | | 102.7% | 102.7% | 102.7% | 102.7% |
| Process efficiency - overall | % HHV | | | | | | |
| **Capital Cost (million USD)** | | | | | | | |
| Caustic Refining | | | 0.70 | 2.99 | 6.72 | 10.91 | 17.72 |
| Feed Handling | | | 0.70 | 0.15 | 0.29 | 0.46 | 0.75 |
| Esterification | | | 0.70 | 0.42 | 0.80 | 1.30 | 2.11 |
| Separation | | | 0.70 | 2.11 | 4.00 | 6.50 | 10.56 |
| Biodiesel Post-Treatment | | | 0.70 | 1.60 | 3.04 | 4.94 | 8.02 |
| Glycerin Purification & MeOH Recovery | | | 0.70 | 0.99 | 1.89 | 3.07 | 4.98 |
| Storage Facilities | | | 0.70 | 1.43 | 2.71 | 4.41 | 7.16 |
| **Direct Fixed Capital (DFC), also called TIC** | | | | 9.69 | 19.44 | 31.58 | 51.31 |
| Engineering | DFC x MF | 0.12 | | 1.16 | 2.33 | 3.79 | 6.16 |
| Construction | DFC x MF | 0.13 | | 1.26 | 2.53 | 4.11 | 6.67 |
| Contractor & Legal | DFC x MF | 0.08 | | 0.78 | 1.56 | 2.53 | 4.10 |
| Process/Project Contingency | DFC x MF | 0.0412 | | 0.40 | 0.80 | 1.30 | 2.11 |
| **Total Plant Cost (TPC)** | | | | 13.29 | 26.66 | 43.31 | 70.35 |
| AFUDC | | | | | | | |
| **Total Plant Investment (TPI)** | | | | 13.29 | 26.66 | 43.31 | 70.35 |
| Land | | | 0.60 | 2.21 | 3.83 | 5.80 | 11.61 |
| Startup | | | | | | | |
| **Total Capital Cost (TCC)** | | | | 15.50 | 30.49 | 49.11 | 81.96 |
| **Contingency/TPI** | | | | | | | |
| Working Capital | DFC x MF | 0.05 | | 0.66 | 1.33 | 2.17 | 3.52 |
| **Variable Operating Costs (million USD/yr)** | | | | | | | |
| Feed | | | | 15.04 | 37.59 | 75.19 | 150.37 |
| Utilities | | | | 0.83 | 2.07 | 4.14 | 8.28 |
| Methanol | | | | 1.024 | 2.56 | 5.12 | 10.24 |
| Catalysts and Chemicals | | | | 1.57 | 3.94 | 7.87 | 15.74 |
| **Total** | | | | 18.46 | 46.16 | 92.32 | 184.63 |
| Fixed Operating Costs (million USD/yr) | | | | | | | |
| Labor (SF from V-R) | DFC x MF | 0.072 | 0.25 | 0.45 | 0.56 | 0.67 | 0.80 |
| Maintenance (3% of A) | | | 0.03 | 0.29 | 0.58 | 0.95 | 1.54 |
| General Overhead (65% of labor + maint) | | | 0.65 | 0.48 | 0.75 | 1.05 | 1.52 |
| Direct Overhead (45% of Labor) | | | 0.45 | 0.20 | 0.25 | 0.30 | 0.36 |
| Insurance (0.5% of TIC) | | | 0.005 | 0.08 | 0.15 | 0.25 | 0.41 |
| **Total** | | | | 1.50 | 2.30 | 3.22 | 4.63 |

**Table 116. Palm oil biodiesel summary yields, capital and operating costs, 25 MM GPY**

| | Units | USA | Argentina | Brazil | Canada | Caribbean Basin | China | Colombia | India | Mexico |
|---|---|---|---|---|---|---|---|---|---|---|
| **Plant Size** | MM gal/yr | | | | | 25 | | 25 | | |
| **Yields** | | | | | | | | | | |
| Primary Product Yield | gal/ton palm oil | | | | | 270 | | 270 | | |
| By Product A | | | | | | glycerin | | glycerin | | |
| Unit | | | | | | | | | | |
| Yield unit | | | | | | | | | | |
| Annual Yield | MM gpy | | | | | 0.755 | | 0.755 | | |
| Price unit | USD/gal | | | | | 1.27 | | 1.27 | | |
| MM USD/yr | | | | | | 2.4 | | 2.4 | | |
| By Product B | | | | | | | | | | |
| Unit | | | | | | | | | | |
| Yield unit | | | | | | | | | | |
| Annual Yield | | | | | | | | | | |
| Price unit | | | | | | | | | | |
| Annual Cost | | | | | | | | | | |
| **Energy Inputs** | | | | | | | | | | |
| Electricity | | | | | | | | | | |
| Cost | $/kWh | | | | | 0.04 | | 0.04 | | |
| Amount | kWh/yr | | | | | 2.50E+06 | | 2.50E+06 | | |
| Annual Cost | MM USD/yr | | | | | 0.10 | | 0.10 | | |
| Fuel | | | | | | | | | | |
| Type | | | | | | Methanol | | Methanol | | |
| Unit | | | | | | gal | | gal | | |
| Amount | Mmgpy | | | | | 1.14 | | 1.14 | | |
| Cost | USD/unit | | | | | 0.90 | | 0.90 | | |
| Annual Cost | MM USD/yr | | | | | 1.024 | | 1.024 | | |
| **Capital & Operating Costs** | | | | | | | | | | |
| Capital | MMUSD | | | | | 34.08 | | 30.49 | | |
| Operating w/o feed | MM USD/yr | | | | | 11.82 | | 10.87 | | |
| Feed | | | | | | | | | | |
| | USD/ton | | | | | 275.00 | | 4.06E+02 | | |
| | ton/yr | | | | | 92590 | | 92590 | | |
| | MM USD/yr | | | | | 25.46 | | 37.59 | | |
| **Product Plant Gate Cost** | | | | | | | | | | |
| Basis | | | | | | 10% SL | | 10% SL | | |
| Value | USD/gal | | | | | 1.58 | | 2.02 | | |

**Table 117. Palm oil biodiesel summary yields, capital and operating costs, 100 MM GPY**

| | Units | USA | Argentina | Brazil | Canada | Caribbean Basin | China | Colombia | India | Mexico |
|---|---|---|---|---|---|---|---|---|---|---|
| **Plant Size** | MM gal/yr | | | | | 100 | | 100 | | |
| **Yields** | | | | | | | | | | |
| Primary Product Yield | gal/ton soy oil | | | | | 270 | | 270 | | |
| By Product A | | | | | | glycerin | | glycerin | | |
| Unit | | | | | | | | | | |
| Yield unit | | | | | | 7.55 | | 7.55 | | |
| Annual Yield | MM gpy | | | | | 1.27 | | 1.27 | | |
| Price unit | USD/gal | | | | | 9.59 | | 9.59 | | |
| MM USD/yr | | | | | | | | | | |
| By Product B | | | | | | | | | | |
| Unit | | | | | | | | | | |
| Yield unit | ton/ton beans | | | | | | | | | |
| Annual Yield | tons | | | | | | | | | |
| Price unit | USD/ton | | | | | | | | | |
| Annual Cost | MM USD/yr | | | | | | | | | |
| **Energy Inputs** | | | | | | | | | | |
| Electricity | | | | | | | | | | |
| Cost | $/kWh | | | | | 0.04 | | 0.04 | | |
| Amount | kWh/yr | | | | | 1.00E+07 | | 1.00E+07 | | |
| Annual Cost | MM USD/yr | | | | | 0.40 | | 0.40 | | |
| Fuel | | | | | | | | | | |
| Type | | | | | | Methanol | | Methanol | | |
| Unit | | | | | | gal | | gal | | |
| Amount | MM gpy | | | | | 4.56 | | 4.56 | | |
| Cost | USD/unit | | | | | 0.90 | | 0.90 | | |
| Annual Cost | MM USD/yr | | | | | 4.10 | | 4.10 | | |
| **Capital & Operating Costs** | | | | | | | | | | |
| Capital | MMUSD | | | | | | | | | |
| Operating w/o feed | MM USD/yr | | | | | 91.43 | | 81.96 | | |
| Feed | | | | | | 40.13 | | 38.89 | | |
| Feed | | | | | | | | | | |
| | USD/ton | | | | | | | | | |
| | ton/yr | | | | | 275 | | 408 | | |
| | MM USD/yr | | | | | 3.70E+05 | | 3.70E+05 | | |
| | | | | | | 101.85 | | 150.37 | | |
| **Product Plant Gate Cost** | | | | | | | | | | |
| Basis | | | | | | 10% SL | | 10% SL | | |
| Value | USD/gal | | | | | 1.44 | | 1.91 | | |

## 12. SUMMARY COMPARISONS

Figure 31 and Figure 32 present comparisons of the plant gate price of ethanol from three biomass sources–sugar cane, corn, and bagasse. For sugar cane, Figure 31, plant gate prices are primarily impacted by feedstock costs; the fraction of PGP representing feedstock ranges from 61% in the Caribbean to 78% in Mexico. There are small variations in capital and non-feedstock operating costs between countries. Comparable trends are seen for ethanol from corn dry mills, Figure 32. Again, capital and non-feedstock PGP contribution variations are small between countries. Feedstock has the largest contribution to PGP, and shows the most variation between countries. Feedstock cost percentages range from 46% in Canada to 66% in Brazil.

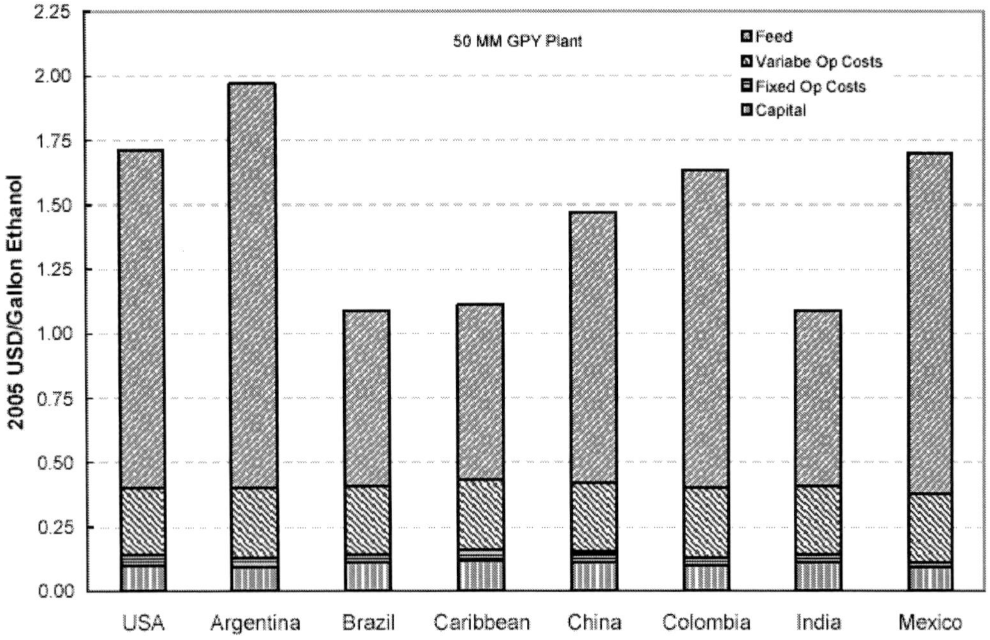


Figure 31. Comparison of sugar cane ethanol plant gate price

Figure 33 shows a comparison for cellulosic ethanol from bagasse. The USA PGP assumes a high cost for bagasse, $35/ton, which is a value equal to that assumed in modeling for woody feedstocks. At this time there is no USA supply curve for bagasse. For the USA case feedstock represents 35% and capital 40%. For other countries, the bagasse cost estimates were estimated from the ORNL report, and represent between 11% (Caribbean Basin), and 23% (Colombia).

Figure 34 shows a comparison of biodiesel PGPs for both biodiesel from soybeans and from palm oil. PGPs are primarily a function of feedstock cost.

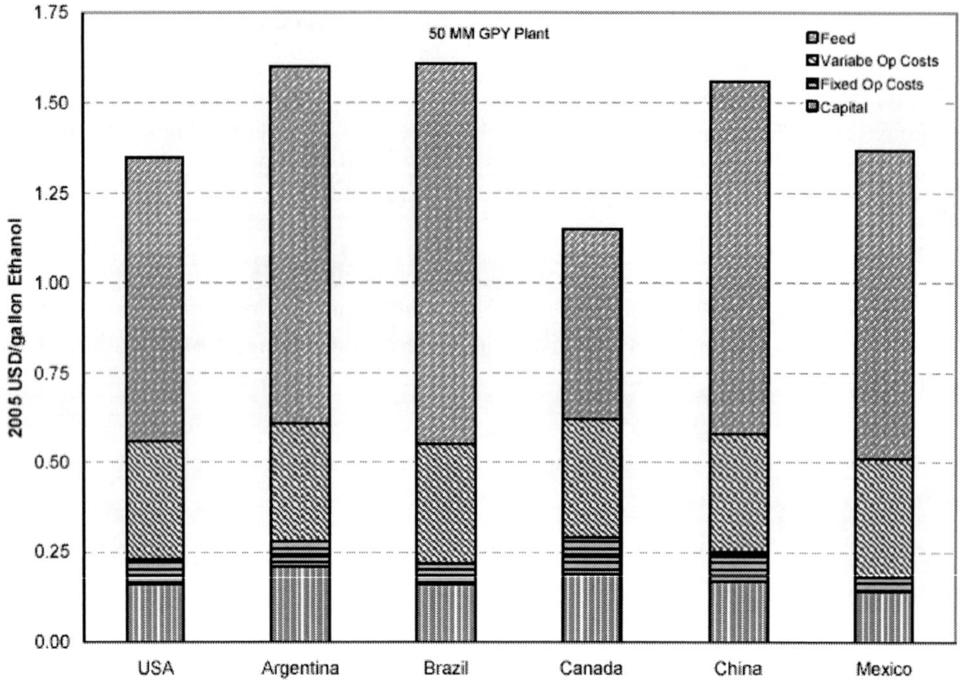


Figure 32. Comparison of corn dry mill ethanol plant gate prices

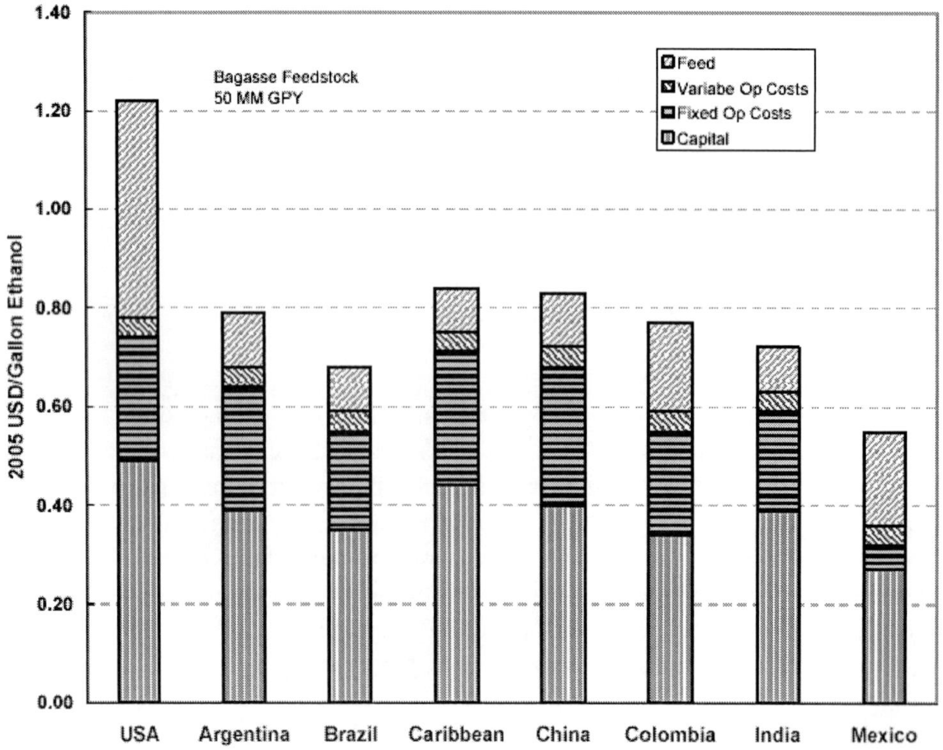


Figure 33. Comparison of cellulosic ethanol (TC) plant gate prices

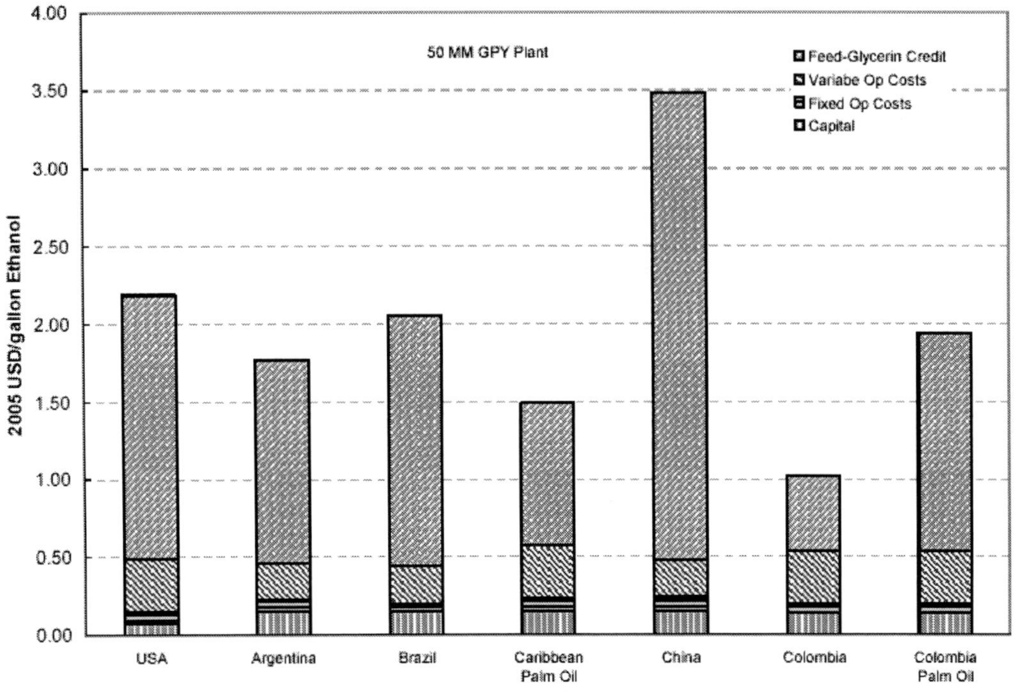


Figure 34. Comparison of biodiesel plant gate prices

# 13 References


Aden, A; Ruth, M; Ibsen, K; Jechura, J; Neeves, K; Sheehan, J; Wallace, B; Montague, L; Slayton, A; & Lukas, J. (2002). "*Lignocellulosic Biomass to Ethanol Process Design and Economics Utilizing Co-Current Dilute Acid Prehydrolysis and Enzymatic Hydrolysis for Corn Stover," 154 pp*, NREL Report No. TP-510-32438.

Arena, B; Holmgren, J; Marinangeli, R; Marker, T; McCall, M; Petri, J; Czernik, S; Elliott, D; & Shonnard, D; (2006). "Opportunities for Biorenewables in Petroleum Refineries," *Rio Oil and Gas Expo and Conference*, Instituto Brasileiro de Petróleo e Gás, Rio de Janeiro, Brazil, Sep.

ASPEN Tech (2007). *Icarus Process Estimator*

Bain, R. L; Craig, KR; & Overend, RP; (2000). "*Gasification for Heat and Power, Methanol, and Hydrogen*," Chapter 9.2 of Industrial Uses of Biomass Energy, ed. F. Rosillo-Calle, et.al., Taylor and France, London, UK, ISBN-0-7484-0884-3.

Bohlmann, GM; & Cesar, MA; (2006). "*The Brazilian Opportunity for Biorefineries," Industrial Biotechnology, 2(2), p* 127132.

Chemical Engineering (2007). http://www.che.com/onc/CEeconomic Indicators/html.

Chem Systems "*Assessment of Costs and Benefits of Flexible and Alternative Fuel Use in the U.S. Transportation Sector", by Chem Systems, Tarrytown, NY, for the U.S. Department of Energy, Washington, D.C., Report No. DOE/PE--0093, 46 pp*, November (1989).

Chicago Board of Trade (2006). "CBOT® Soybean Crush Reference Guide," Board of Trade of the City of Chicago.

Delta-T Corporation (1997). proprietary information, Williamsburg, *VA, 23185.*

Feldmann, HF; Paisley, MA; Appelbaum, HR; & Taylor, DR; "Conversion of Forest Residues to a Methane-Rich Gas in a High- Throughput Gasifier", *prepared by Battelle Columbus Division, Columbus, Ohio, for the Pacific Northwest Laboratory, Richland,* WA, PNL Report no. PNL-6570, May 1988.

GoCurrancy.COM (2006). "Historic Exchange Rates," http://www.gocurrancy.com/v2/historic-exchange-rates.php?ccode2=..., 7/31/2007.

Haas, MJ; McAloon, AJ; Yee, WC; & Foglia TA; (2006). " A Process Model to Estimate Biodiesel Production Costs," *Bioresource Technology, 97*, 671-678.

Hamelinck, CN; & Faaij, APC; (2001). "*Future Prospects for Production of Methanol and Hydrogen from Biomass*," Utrecht University, Utrecht, the Netherlands, Report No. NWS-E-2001-49, ISBN-90-73958-84-9.

Humphries, KK; (2005a). "*Sources of International Cost Data," NORDET '97 Conference*, with partial updates through September 2005, International Cost Engineering Council, http://www.icoste.org.intldata.htm.

Humphries, KK; (2005b). *Project and Cost Engineer's Handbook, Marcel Decker, NY, NY*, ISBN: 0-8247-5746-7.

Ibsen, KR; Wallace, S; Jones, & Werpy, T;(2005). "*Evaluating Progressive Technology Scenarios in the Development of the Advanced Dry Mill Biorefinery*, NREL Milestone Completion Report, ID: FY05-630, National Renewable Energy Laboratory, Golden, CO.

Jechura, J. (2005). ASPEN Case Summary: 25 MMgal_USDA_DCFROR.xls, National Renewable Energy Laboratory, Golden, CO.

KMPG (2006). "KMPG's Corporate Tax Rate Survey," KMPG, http://www.kpmg.com/NR/rdonlyres/D8CBA9FF-C953-45FA-940A-FAAC86729554/0/KPMGCorporateTaxRateSurvey.pdf

Larson, ED; Jin, H; & Celika, FE. (2006)"Large-Scale Gasification-Based Co-Production of Fuels and Electricity from Switchgrass." Draft manuscript for *Biomass and Bioenergy*. March.

McAloon, A; Taylor, F; Yee, W; Ibsen, K; & Wooley, R; "Determining the Cost of Producing Ethanol from Corn Starch and Lignocellulosic Feedstocks. *A Joint Study Sponsored by U.S. Department of Agriculture and U.S. Department of Energy,"* 44 pp, NREL Report No. TP-580-28893.

Mohan, DCU; Pittman, Jr; & Steele, PH; (2006). "Pyrolysis of Wood/Biomass for Bio-oil: A Critical Review," *Energy and Fuels,* web preprint, March 10.

National Corn Growers Association (2007). "World of Corn," http:/www.ncga.com/WorldOfCorn/main/production1.asp, 8/28/2007.

Peters, SM; & Timmerhaus, KD; (2003). *Plant Design and Economics for Chemical Engineers*, McGraw-Hill, NY, NY, ISBN 0-07-049613-7.

Phillips, S; Aden, A; Jechura, J; Dayton, D; & Eggeman, T. (2007). "*Thermochemical Ethanol via Indirect Gasification and Mixed Alcohol Synthesis of Lignocellulosic Biomass," 132 pp.*, NREL Report No. TP-510-41168.

OECD (2002) Canada Biofuel Report.

Oliverio, JL; (2006). "Technological Evolution of the Brazilian Sugar and Alcohol Sector: Dedini's Contribution," *International Sugar Journal, 108(1287),* p 120-129.

Oliverio, JL; & Ribeiro, JE; (2006). "Cogeneration in Brazilian Sugar and Bioethanol Mills: Past, Present and Challenges," *International Sugar Journal, 108(191),* p391-401.

ORNL Draft (2007) "*Study of the Worldwide Potential to Produce Ethanol and Other Biofuels in this Decade and Beyond: Biofuel Feedstock Assessment for Selected Countries.*"

Phillips, S; Aden, A; Jechura, J; Dayton, D; & Eggeman, T. (2007). *Thermochemical Ethanol via Indirect Gasification and Mixed Alcohol Synthesis of Lignocellulosic Biomass. 132 pp.*; NREL Report No. TP-510-41168.

Richardson Engineering (1997). "Richardson's International Construction Factors"

Rosillo-Calle, F; Bajay, SV; & Rothman, H; (2000). *Industrial Uses of Biomass Energy: The Example of Brazil*, Taylor and Francis, London, UK, ISBN 0-7484-0884-3.

Shapouri, H; & Salassi, M; (2006). "*The Economic Feasibility of Ethanol Production in the United States, USDA, 69 p*, Jul.

Spath, P; Aden, A; Eggeman, T; Ringer, M; Wallace, B; & Jechura, J; (2005). "*Biomass to Hydrogen Production Detailed Design and Economics Utilizing the Battelle Columbus Laboratory Indirectly-Heated Gasifier." 161 pp.*; NREL Report No. TP-510-37408.

Turn, S. (2007). *University of Hawaii, personal communication*, July.

USDA (2006). " OilCrops Yearbook/OCS-2006/Mar21, Economic Research Service.

USDA (2007). "*Wheat Data: Yearbook Tables*," Economic Research Service, Aug.

Valle-Riestra, JF; (1983). Project Evaluation in the Chemical Process Industries, McGraw-Hill, NY, NY, ISBN 0-07-066840.

Van den Wall Bake (2006). "*Cane as Key in Brazilian Ethanol Industry,*" Master Thesis, University of Utrecht, Utrecht, the Netherlands, NWS-1-2006-14.

Walter, A., (2007). UNICAMP, personal communication, July.

Wan, EI; & Fraser, MD; (1989)"*Economic Assessment of Advanced Biomass Gasification Systems*", by Science Applications International Corporation, McLean, VA, IGT Biomass Conference, New Orleans, LA, Feb.

Williams, RH; Larson, ED; Katofsky, RE; & Chen, J; (1995). "*Methanol and Hydrogen from Biomass for Transportation, with Comparisons to Methanol and Hydrogen from Natural Gas and Coal,*" PU/CEES Report No. 292, Princeton University, Princeton, NJ, July.

Wyman, CE; Bain, RL; Hinman, ND; & Stevens, DJ; (1993) "*Ethanol and Methanol from Cellulosic Materials*, "Chapter 21 of Renewable Energy: Sources for Fuels and Electricity, ed T.B. Johansson, et al, Island Press.

# CHAPTER SOURCE

The following chapters have been previously published:

These chapters is an edited, reformatted and augmented versión of World Biofuels Assessment publication, Milestone Report NREL/MP-510-42467, dated December 2007.

# Index

## D



## E



## F



## G

## H



## I



## L



## M



## N



## O

## P



## R



## S



## T



## U

## V



## W



## Y